CAMBRIDGE LIBRARY COLLECTION

Books of enduring scholarly value

Physical Sciences

From ancient times, humans have tried to understand the workings of the world around them. The roots of modern physical science go back to the very earliest mechanical devices such as levers and rollers, the mixing of paints and dyes, and the importance of the heavenly bodies in early religious observance and navigation. The physical sciences as we know them today began to emerge as independent academic subjects during the early modern period, in the work of Newton and other 'natural philosophers', and numerous sub-disciplines developed during the centuries that followed. This part of the Cambridge Library Collection is devoted to landmark publications in this area which will be of interest to historians of science concerned with individual scientists, particular discoveries, and advances in scientific method, or with the establishment and development of scientific institutions around the world.

The Story of the Heavens

An Irish astronomer and talented mathematician, Sir Robert Stawell Ball (1840–1913) was also a prolific writer of popular astronomy. As a young man, Ball conducted observations of nebulae using Lord Rosse's telescope – at the time the largest in the world. His *Story of the Heavens* displays the same fascination with the beauties and mysteries of the sky, providing a detailed survey of the history and contemporary situation of the solar system, and speculating about the possibility of life on other planets. Originally published in 1885, when Ball was Andrews Professor of Astronomy in the University of Dublin and Royal Astronomer of Ireland, this beautifully illustrated volume covers all eight planets, the Sun, as well as double stars, distant suns, comets, and the Milky Way. Extremely popular in its time, this book remains relevant today for its historical account of astronomy as a science.

The Story of the Heavens

Robert S. Ball

CAMBRIDGE UNIVERSITY PRESS

Cambridge, New York, Melbourne, Madrid, Cape Town, Singapore,
São Paolo, Delhi, Dubai, Tokyo

Published in the United States of America by Cambridge University Press, New York

www.cambridge.org
Information on this title: www.cambridge.org/9781108014144

This edition first published 1885
This digitally printed version 2010

ISBN 978-1-108-01414-4 Paperback

THE

STORY OF THE HEAVENS

FRONTISPIECE.

THE PLANET SATURN,

IN 1872.

THE

STORY OF THE HEAVENS.

BY

ROBERT STAWELL BALL, LL.D.,

FELLOW OF THE ROYAL SOCIETY OF LONDON, FELLOW OF THE ROYAL ASTRONOMICAL SOCIETY,
HONORARY MEMBER OF THE CAMBRIDGE PHILOSOPHICAL SOCIETY, VICE-PRESIDENT
OF THE ROYAL IRISH ACADEMY, ANDREWS PROFESSOR OF ASTRONOMY IN THE
UNIVERSITY OF DUBLIN, AND ROYAL ASTRONOMER OF IRELAND.

With Coloured Plates and numerous Illustrations.

CASSELL & COMPANY, LIMITED:

LONDON, PARIS, NEW YORK & MELBOURNE.

1885.

PREFACE.

I HAVE to acknowledge the kind aid which I have received in the preparation of this book.

Mr. Nasmyth has permitted me to use some of the beautiful drawings of the Moon, which have appeared in the well-known work published by him in conjunction with Mr. Carpenter. To this source I am indebted for Plates VII., VIII., IX., X., and Figs. 26, 27, 28.

Professor Pickering has allowed me to copy some of the drawings made at Harvard College Observatory by Mr. Trouvelot, and I have availed myself of his kindness for Plates I., IV., XI., XII., XV.

I am indebted to Professor Langley for Plate II., to Mr. De la Rue for Plates III., and XIV., to Dr. Huggins for Fig. 16, to Professor C. Piazzi Smyth for Fig. 89, to Mr. Chambers for Fig. 6, which has been borrowed from his "Handbook of Descriptive Astronomy," to Dr. G. J. Stoney for Fig. 65, to Mr. Grubb for Fig. 4, and to Dr. Copeland and Dr. Dreyer for Fig. 59.

I have also to thank Dr. Copeland and Mr. L. H. Steele for their kindness in reading through the entire proofs; while I have also occasionally availed myself of the help of Mr. G. Cathcart.

ROBERT S. BALL.

OBSERVATORY,
DUNSINK,
CO. DUBLIN.

17th April, 1885.

TABLE OF CONTENTS.

CHAPTER IV.

THE SOLAR SYSTEM.

CHAPTER V.

THE LAW OF GRAVITATION.

CHAPTER IX.

THE EARTH.

CHAPTER X.

MARS.

CHAPTER XI.

THE MINOR PLANETS.

CHAPTER XII.

JUPITER.

CHAPTER XVI.

COMETS.

CHAPTER XVII.

SHOOTING STARS.

CHAPTER XVIII.

THE STARRY HEAVENS.

CHAPTER XIX.

THE DISTANT SUNS.

CHAPTER XX.

DOUBLE STARS.

CHAPTER XXI.

THE DISTANCES OF THE STARS.

CHAPTER XXII.

THE SPECTROSCOPE.

CHAPTER XXIII.

STAR CLUSTERS AND NEBULÆ.

CHAPTER XXIV.

THE PRECESSION AND NUTATION OF THE EARTH'S AXIS.

CHAPTER XXV.

THE ABERRATION OF LIGHT.

LIST OF ILLUSTRATIONS.

PLATES.

ENGRAVINGS.

THE

Story of the Heavens.

"The Story of the Heavens" is the title of our book. We have indeed a wondrous story to narrate; and could we tell it adequately, it would prove of boundless interest and of exquisite beauty. It leads to the contemplation of the mightiest efforts of nature and the greatest achievements of human genius.

Let us enumerate a few of the questions which will be naturally asked by one who seeks to learn something of those glorious bodies which adorn our skies: What is the Sun—how hot, how big, and how distant? whence comes its heat? What is the Moon? What scenery do its landscapes show? how does the moon move? how is it related to the earth? What of the planets—are they globes like the earth? how large are they, and how far off? What do we know of the satellites of Jupiter and of the rings of Saturn? What was the memorable discovery of Uranus? and what was the supreme intellectual triumph which brought the planet Neptune to light? Then, as to the other bodies of our system, what are we to say of those mysterious objects, the comets? can we perceive order to reign in their seemingly capricious movements? do we know anything of their nature and of the marvellous tails with which they are often decorated? What can be told about the familiar shooting-star which so often dashes into our atmosphere to perish in a streak of splendour? What do we know of those constellations which have been from all antiquity, and of the myriad hosts of smaller stars which our telescopes disclose?

Can it be true that these countless orbs are really majestic suns, sunk to an appalling depth in the abyss of unfathomable space? What have we to tell of all the different varieties of stars—of coloured stars, of variable stars, of double stars, of multiple stars, of stars that move, and of stars that seem at rest? What of those most supremely glorious objects, the great star clusters? What of the milky way? And lastly, what can we tell of those marvellous nebulæ which our telescopes disclose, poised at an immeasurable distance on the very confines of the universe? Such are a few of the questions which occur, when we ponder on the mysteries of the heavens.

The history of Astronomy is, in one respect, only too like many other histories. The earliest part of it is completely and hopelessly unknown. The stars had been studied, and some great astronomical discoveries had been made, untold ages before those to which our earliest historical records extend. For example, the perception of the apparent movements, of the sun and of the moon, and the recognition of the planets by their movements, are both to be classed among these discoveries of the pre-historic ages. Nor is it to be said that these achievements were all of a very obvious or elementary character. To us of the present day who have been familiar with such truths from childhood, they may now seem simple and rudimentary; but in the infancy of science, the first man who arose to demonstrate one of these great doctrines was indeed a most sagacious philosopher.

Of all the phenomena of Astronomy, the first and the most obvious is that of the rising and the setting of the sun. We may fairly conjecture, that in the dawn of the growth of the human intellect this was probably one of the very first problems to engage the attention of those whose thoughts rose above the animal anxieties of everyday existence. A sun sets and disappears in the west; that sun is obviously a very brilliant body, and the simplest reflection suggests that it is a body of very considerable importance. The following morning a sun arises in the east, moves across the heavens, and it too disappears in the west; the same process happens every day. To us it is obvious that the sun, which appears each day, is the same sun; but this would not be an obvious truth to one who thought his senses showed him that the earth was a flat plane of indefinite extent, and that around the inhabited regions on all sides extended, to vast distances, either desert wastes or trackless oceans. How could the sun, which plunged into the ocean at a fabulous distance in the west, reappear the next morning at an equally great distance to the east? The old mythological account asserted that after the sun had dipped in the western ocean at sunset (the Iberians, and other ancient nations, actually imagined that they could hear the hissing of the waters when the glowing globe was plunged therein), he was seized by Vulcan and placed in a golden goblet, and thus navigated the ocean round by the north, so as to reach the east again in time for sunrise the following morning. Even the more sober physicists of old, as we are told by Aristotle, believed that in some manner the sun was conveyed round over the earth's surface by the north, and that the darkness of night arose from the elevation of the northern lands, which cut off the sun's light during his midnight voyage.

Even in very early times it was found more rational to suppose that the sun actually pursued his course down below the solid earth during the darkness of night. The earliest astronomers had, moreover, learned to recognise the fixed stars. It was seen that, like the sun, many of these stars rose and set in the course of the diurnal movement, while the moon obviously followed the same law. It thus became plain that the various heavenly bodies possessed the power of actually going below the solid earth. Once it was realised that the whole contents of the heavens performed these movements, it became possible to take a very important step in the knowledge of the constitution of the universe. It was clear that the earth could not be a plane extending to an indefinitely great distance. It was also obvious that there must be a finite depth to the earth below our feet. Nay, more, it became certain that whatever be the shape of the earth, it was at all events something detached from all other bodies, and poised without visible support in space. When first presented to the mind of man, this must have appeared a very startling truth. It was surely difficult to realise that the solid earth on which we stand reposed on nothing! What is to keep it from falling? How can it be poised, like the legendary coffin of Mahomet, without tangible support? But difficult as it may have been to receive this doctrine, yet its necessary truth commanded assent, and the first great step in Astronomy had been made.

The changes of the seasons and the recurrence of seed-time and of harvest must, from the earliest times, have been associated with certain changes in the position of the sun. In the summer at midday the sun rises high in the heavens, in the winter the sun is always low. The sun, therefore, had an annual movement up and down in the heavens, combined with the diurnal movement of rising and setting. But besides these movements of the sun there was another of no less importance, which was not quite so obvious, though still capable of being detected by the simplest observations, when combined with a philosophical habit of reflection. The very earliest observers of the stars can hardly fail to have noticed, that the constellations visible at night varied with the season of the year. For instance, the constellation of Orion, which is so well seen during the winter nights, becomes invisible in the summer, and the place it occupied is then taken by quite different stars. So it is with other constellations; and, indeed, in ancient days, the time for commencing the cycle of agricultural occupations was sometimes indicated by the position of the constellations in the evening.

Reflection on this subject must have demonstrated in very early times the apparent annual movement of the sun. It was seen that the places of the stars, relatively to each other, did not alter appreciably, and there could be no explanation of the changes in the constellations with the seasons, except by supposing that the place of the sun was altering, so as to make the complete circuit of the heavens in the course of the year. The same conclusion is easily confirmed by looking from time to time at the west after sunset, and watching the stars. As the season progresses, it will be noticed that each evening the western constellations sink lower and lower towards the sun, until at length they come so near the sun that they set at the same time as he does. This is simply explained by the supposition that the sun is gradually but continually rising up from the west to meet the stars. This motion is of course not to be confounded with the ordinary diurnal motion, in which all the heavenly bodies alike participate; inasmuch as besides this motion of the whole heavens, the sun has a slow motion in the opposite direction; so that while the sun and a star may set to-day at the same time, by to-morrow the sun will have moved a little towards the east, relatively to the star, and thus the star will set a few minutes before the sun.*

The patient watchings of the early astronomers enabled the sun's track through the heavens to be ascertained, and it was found that in its annual circuit the sun invariably pursued the same path and traversed the same constellations. The belt of constellations thus specially distinguished is known by the name of the *zodiac*, while the circle traversed by the sun is called the *ecliptic*. The zodiac was divided into twelve equal portions or "signs," and thus the stages on the sun's great journey were conveniently indicated. In the very earliest ages, also, it seems that the duration of the year, or the period required by the sun to run its course around the heavens, became accurately known. The skill of the ancient geometers was also demonstrated by the accurate measures they succeeded in making of the position of the ecliptic with regard to the equator, and in measuring the angle between these two most important circles on the heavens.

The principal phenomena presented by the motion of the moon have also been understood from an antiquity beyond all historical record. The slightest attention reveals the important truth that the moon does not occupy a fixed position in the starry heavens. Indeed, the motion of the moon among the stars is a phenomenon much more easy to recognise than that of the sun among the stars, as during the course of a single night the movement of the moon from west to east across the heavens can be perceived with but very moderate attention. It is most probable that the motion of

* It may, however, be remarked that a star is never *seen* to set, as owing to our atmosphere it ceases to be visible before it reaches the horizon.

the moon among the stars was a discovery prior to that of the annual motion of the sun, inasmuch as it depends upon simple observation, and involves but little exercise of any intellectual power. The time of revolution of the moon had also been discovered, and the phases of the moon had been correctly attributed to the varying aspect under which the sun-illuminated side of the moon is turned towards the earth.

But even this does not exhaust the list of great discoveries which have come down to us from prehistoric times. The striking phenomenon of a lunar eclipse, in which the brilliant surface is plunged temporarily into darkness, and the still more imposing spectacle of a solar eclipse, in which the sun himself undergoes a partial, or even a total obscuration, had also been correctly explained. Then, too, the acuteness of the early astronomers had detected the five wandering stars or planets: they had traced the movements of Mercury and Venus, Mars, Jupiter, and Saturn. They had observed with awe the various configurations of these planets; and just as the sun, and in a lesser degree the moon, were intimately associated with the affairs of daily life, so in the imagination of these early investigators the movements of the planets were thought to be pregnant with human weal or human woe. At length a certain degree of order was perceived to govern the capricious movements of the planets. It was found that they obeyed certain laws. The cultivation of the science of geometry went hand in hand with the study of astronomy; and as we emerge from the dim pre-historic ages into the historical period, we find that a theory possessing some degree of coherence had been established, to explain the phenomena of the heavens.

Although the Ptolemaic doctrine is now known to be framed on an utterly extravagant estimate of the true place of the earth in the scheme of the heavens, yet the apparent movements of the celestial bodies are accounted for by the theory with considerable accuracy. This theory is described in the great work of Ptolemy, known as the "Almagest," which was written in the second century of our era, and for fourteen centuries was regarded as the final authority on all questions of astronomy.

Ptolemy saw that the shape of the earth was globular, and he demonstrated this by the arguments which we employ at the present day. He also saw how this mighty globe was poised, in what he believed to be the centre of the universe. He admitted that the diurnal movement of the whole heavens could be accounted for by the revolution of the earth upon its axis, but he assigned reasons for the deliberate rejection of this view. The earth according to him was a fixed body; it possessed neither rotation nor translation, but remained constantly at rest at the centre of the universe. The sun and the moon he supposed to move in circular orbits around the earth in the centre. The movements of the planets were more complicated, as it was necessary to account for the occasional retrograde motions as well as for the direct motions. The ancient geometry refused to admit that any movement, except circular, could be perfect, and accordingly a contrivance was devised by which each planet revolved in a circle, while the centre of that circle described another circle around the earth. It must be admitted that this scheme, though so widely divergent from what is now known to be the truth, did really present a fairly accurate account of the movements of the planets.

Such was the system of Astronomy which prevailed during the Middle Ages, and which was only finally overturned by the great work to which Copernicus devoted his lifetime. The discovery of the true system of the universe was nearly simultaneous with the discovery of the New World by Columbus. The first principles which were established by the labours of Copernicus, stated that the diurnal movement of the heavens was really due to the rotation of the earth on its axis. He showed the difference between real motions and apparent motions; he proved that all the appearances of the daily rising and setting of the sun and the stars could be just as well accounted for by the supposition that the earth rotated, as by the more cumbrous supposition of Ptolemy. He showed that the latter supposition would attribute an almost infinite velocity to the stars, and that the rotation of the entire universe around the earth was really a preposterous supposition. The second great point, which it is the immortal glory of Copernicus to have demonstrated, assigned to the earth its true position in the fabric of the universe. He transferred the centre, about which all the planets revolve, from the earth to the sun; and he established the somewhat humiliating truth, that our earth is after all merely one of the system of planets revolving around the sun, and pursuing a track between the paths of Venus and of Mars.

Such was, in brief outline, the great revolution which swept from astronomy those distorted views of the earth's importance, arising from the fact that we are domiciled on that particular planet. The achievement of Copernicus was soon to be followed by the invention of the telescope, that wondrous instrument by which the modern science of astronomy has been created. To the consideration of this most important subject we may well devote the first chapter of our book.

PLATE II.

A TYPICAL SUN-SPOT.

(AFTER LANGLEY.)

CHAPTER I.

THE ASTRONOMICAL OBSERVATORY.

Early Astronomical Observations—The Observatory of Tycho Brahe—The Pupil of the Eye—Vision of Faint Objects—The Telescope—The Object-Glass—Advantages of Large Telescopes—The Equatorial—The Observatory—The Power of a Telescope—Reflecting Telescopes—Lord Rosse's Great Reflector at Parsonstown—How the mighty Telescope is used—The Instruments of Precision—The Meridian Circle—The Spider Lines—Delicacy of pointing a Telescope—The Precautions necessary in making Observations—The Ideal Instrument and the Practical one—The Elimination of Error—The ordinary Opera-Glass as an Astronomical Instrument—The Great Bear—Counting the Stars in the Constellation—How to become an Observer.

THE earliest traces of the Astronomical Observatory are as little known, as the earliest discoveries in astronomy itself. Probably the first application of instruments to the observations of the heavenly bodies, consisted in the extremely simple operation of measuring the length of the shadow cast by the sun at noonday. The variations in the length of this shadow from day to day, and its periodical maxima and minima, furnished valuable information in the early attempts to investigate the movements of the sun. But even in very early times there were astronomical instruments employed which possessed considerable complexity, and showed no small amount of astronomical knowledge.

The first great advance in this subject was made by the celebrated Tycho Brahe, who was born in 1546, three years after the death of Copernicus. His attention seems first to have been directed to astronomy by the eclipse of the sun which occurred on the 21st August, 1560. It amazed his reflective spirit to find that so surprising a phenomenon admitted of actual prediction, and he determined to devote his life to the study of a science possessed of such wonderful precision. In the year 1576 the King of

Denmark had established Tycho Brahe on the island of Huen, and had furnished him with the splendid observatory of Uraniberg. It was here that Tycho assiduously observed the places of the heavenly bodies for some twenty years, and accumulated the observations which were destined, in the hands of Kepler, to lead to the great discovery of the planetary movements. Compared with our modern astronomical equipment the great instruments of Tycho are but quaint and primitive apparatus. In his days the telescope had not yet been invented, and he could only determine the places of the heavenly bodies in a comparatively crude manner; but his skill and patience in a great degree compensated for the imperfection of his instruments, and with him it may be said that the epoch of accurate astronomical observation commences.

The application of the telescope by Galileo gave a most wonderful impulse to the study of the heavenly bodies. This extraordinary man stands out prominently in the history of astronomy, not alone for his connection with this supreme invention, but for his achievements in the more abstract parts of astronomy. It was Galileo who first laid with any solidity the foundation of the science of Dynamics, of which astronomy is the most splendid illustration; and it was he who expounded and upheld the great doctrine of Copernicus, and thereby drew down upon himself the penalties of the Inquisition.

The structure of the eye itself, and more particularly the exquisite adaptation of the pupil, presents us with an apt illustration of the principle of the telescope. To see an object, it is necessary that the light from that object should enter the eye. The portal through which the light enters the eye is the pupil. In daytime, when the light is abundant, the iris gradually decreases the size of the pupil, and as the portal is thus contracted, less light can enter. At night, on the other hand, when the light is scarce, the eye requires to grasp all it can. The pupil then expands, more and more light is admitted according as the pupil grows larger, until at length the pupil is dilated to its utmost extent. The admission of light is thus controlled in the most perfect manner.

The stars send us their feeble rays of light, and those rays form an image on the retina; but, even with the most widely-opened pupil, it may happen that the image is still not bright enough to excite the sensation of vision. Here the telescope comes to our aid: it catches all the rays in a beam of dimensions far too large to enter the pupil, and concentrates those rays into a small beam which can enter the pupil. We thus have the image on the retina intensified in brilliancy; in fact, it is illuminated with nearly as much light as would be obtained through a pupil as large as the object-glass of the telescope.

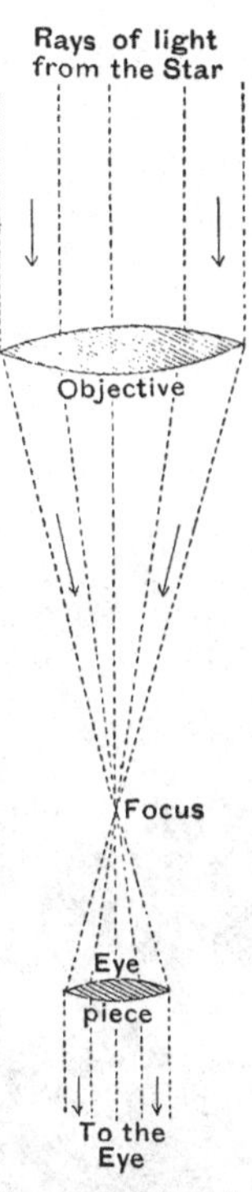

Fig. 1.—Principle of the Refracting Telescope.

In our astronomical observatories we find two entirely different classes of telescopes. The more familiar forms are those known as refractors, in which the operation of condensing the rays of light is effected by refraction. The same object can, however, be attained in a wholly different manner by the aid of the laws of reflection, and accordingly many telescopes, including the most gigantic instruments yet erected, are known as reflectors. The character of the refractor is shown in Fig. 1. The rays from the star fall upon the object-glass which is at the end of the telescope, and after passing through it they are refracted into a converging beam, so that all intersect at the focus. Diverging from thence, they encounter the eye-piece, which has the effect of again reducing them to parallelism. The large cylindrical beam which poured down on the object-glass is thus concentrated into a small one, which can enter the pupil. The composite nature of light requires a more complex form of object-glass than the simple lens here shown. In modern telescopes we employ what is known as the achromatic object-glass, which consists of one lens of flint glass and one of crown glass, combined together.

It will thus be apparent, that the larger the object-glass, the greater the quantity of light grasped, and the greater will be the success of the telescope in revealing very faint objects. Hence it is that in the efforts to increase the powers of their telescopes,

Fig. 2.—The Dome of the South Equatorial at Dunsink Observatory, Co. Dublin.

each succeeding race of astronomers has sought to obtain larger object-glasses than those which were used by their predecessors.

The appearance of an astronomical observatory, built to hold an instrument of moderate dimensions, is shown in the adjoining figures. The first (Fig. 2) represents the dome erected at Dunsink

Observatory for the equatorial telescope, the object-glass of which was presented to the Board of Trinity College, Dublin, by the late Sir James South. The main part of the building is a circular wall, on the top of which reposes a hemispherical roof. In this roof is a shutter, which can be opened so as to allow the telescope in

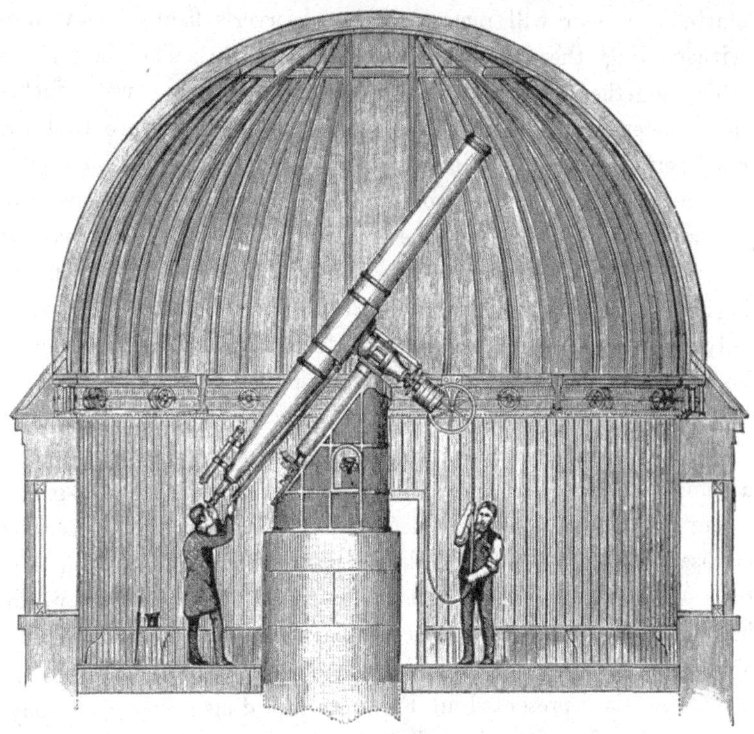

Fig. 3.—Section of the Dome of Dunsink Observatory.

the interior to be directed towards the heavens. The whole structure revolves, so that the opening may be pointed to any part of the sky which it is desired to examine. The next view (Fig. 3) exhibits a section of the roof, showing the machinery by which the attendant causes it to revolve, as well as the telescope itself. The eye of the observer is at the eye-piece, and he is in the act of turning a handle, which has the power of slowly moving the telescope, in order to direct the instrument towards any point that may be desired. A telescope mounted in the manner here shown, is called an equatorial. The convenience of the equatorial form of mounting lies in the ease with which the telescope can be moved so as to follow any celestial object in its journey around the sky. The necessary movements are given by clockwork, so that, once the instrument has been correctly pointed, and the clockwork started, the star will remain in the observer's field of view notwithstanding the apparent diurnal movement. The two lenses which together form the object-glass are in this case twelve inches in diameter, and it is on the correctness of the objective that the good performance of the telescope mainly depends. The eye-piece consists merely of one or two small lenses; various eye-pieces can be employed, according to the magnifying power which may be desired. It is to be observed that for many purposes of astronomy highly magnifying powers are not desirable. The object-glass can only grasp a certain quantity of light, and if the magnifying power be too great, the light will be thinly dispersed over a large surface, and the result will be unsatisfactory.

The power of a refracting telescope—so far as the expression has a definite meaning—is measured by the diameter of its object-glass. There has, indeed, been some degree of rivalry between the various civilised nations as to which should possess the greatest refracting telescope. Among the largest telescopes of this type the world has yet seen, is that recently constructed by Mr. Howard Grubb, of Dublin, for the splendid observatory at Vienna. This great instrument is represented in Fig. 4. The dimensions of it may be estimated from the fact that the object-glass is two feet and three inches in diameter. Many ingenious contrivances obviate the inconveniences incident to the use of an instrument of such vast proportions. We may here only notice the method by which the graduated circles attached to the telescope are brought within easy view of the observer. These circles are situated at parts of the instrument very remote from the eye-piece at which the observer is stationed. They can, however, be readily seen by small auxiliary telescope tubes (shown in the figure, close to the eye-piece), which, by suitable reflectors, conduct the rays of

Fig. 4.—The Great Vienna Telescope.

light from the illuminated circles to the eye of the observer. The clock movement of this great instrument is also noteworthy, as it is controlled by electricity, so that the mighty tube follows the star with almost mathematical precision.

Numerous refracting telescopes of exquisite perfection have been produced by Messrs. Alvan Clark, of Cambridgeport, Boston, Mass. The size of their instruments has been gradually increasing, and they have recently completed a gigantic telescope with an object-glass of no less than thirty inches in diameter for the Russian astronomers.

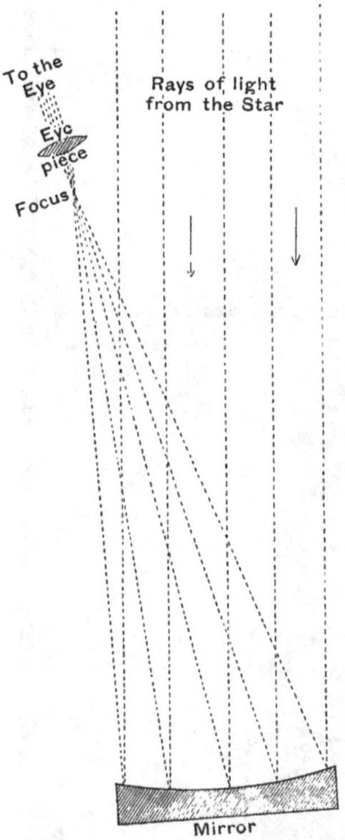

Fig. 5.—Principle of Herschel's Reflecting Telescope.

Can refracting telescopes be constructed of still greater dimensions? The present limit to the size of the refractor chiefly lies in the material of the object-glass. Glass manufacturers experience great difficulties in any attempts to form large discs of optical glass pure enough and uniform enough to be suitable for telescopes. These difficulties increase with every increase in the size of the instrument, and at the present moment this is the chief impediment to the construction of refracting telescopes of the largest dimensions.

There is, however, the alternative method of constructing a telescope, in which this difficulty does not arise. The simplest form of reflector is that shown in Fig. 5, which represents the Herschelian instrument. The rays from the star fall on a beautifully polished and carefully shaped mirror, so that, after the reflection, they proceed to a focus, and diverging thence, fall on the eye-piece, from which they emerge, reduced to parallelism and fitted for reception by the eye. It is essentially on

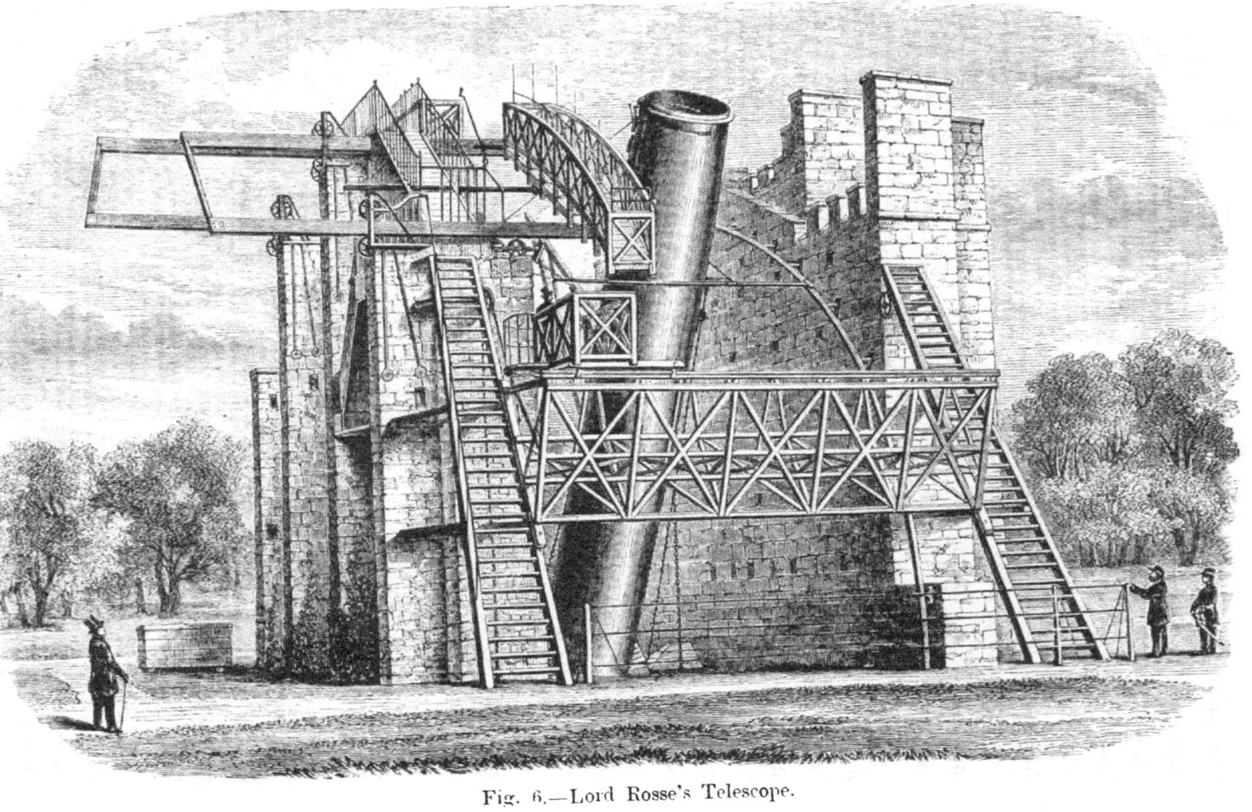

Fig. 6.—Lord Rosse's Telescope.

this principle, though with an additional reflection, that the mightiest telescope in existence has been constructed. This renowned instrument, known wherever science is known, was built, forty years ago, by the late Earl of Rosse at Parsonstown. The colossal dimensions of this instrument have never been surpassed; they have, indeed, never been rivalled. The reflector in this case is a thick metallic disc, consisting of an alloy of two parts of copper to one of tin, forming a hard and brittle metal intractable for mechanical operations, but admitting of a brilliant polish, and of receiving and retaining an accurate figure. The great reflector—six feet in diameter—reposes at the end of a tube sixty feet long. This tube is mounted between two castellated walls of masonry, which form an imposing feature on the lawn at Birr Castle, as represented in Fig. 6. This instrument does not admit of being directed towards any part of the sky like the equatorials we have recently been considering. The great reflector is only capable of an up and down movement along the meridian, and of a small lateral movement east and west of the meridian. A little consideration will, however, show that, though the telescope cannot at any moment be directed to any particular star, yet that each star visible in the latitude of Parsonstown can be observed when looked for at the right time.

As the object is approaching the meridian, be it planet or comet, star or nebula, the telescope is raised to the right height. This is accomplished by a chain passing from the mouth of the instrument to a windlass at the northern end of the walls. By this windlass the telescope can be raised or lowered, and an ingenious system of counterpoises renders the movement equally easy at all altitudes. The observer then takes his station in the lofty gallery which gives access to the eye-piece; and when the right moment has arrived, the object enters the field of view. A vast clockwork mechanism at the lower end of the tube gives movement to the great instrument, so that the object can be followed by the observer until he has made his measurements, or finished his drawing.

It will thus be seen that, notwithstanding the stupendous size of this telescope (the tube is large enough for a tall man to walk through without stooping), it is comparatively easy to observe with. It must not, however, be assumed that for all the purposes of

Fig. 7.—Meridian Circle.

astronomy an instrument so colossal is the most suitable. The mighty reflector is chiefly of use where very faint objects are to be sought for; but where accurate measurements are required of objects not unusually faint, telescopes of smaller dimensions and of different construction are more suitable. Among the other great reflectors, we may mention that constructed by

Mr. Common, of Ealing, three feet in aperture, which possesses great optical perfection and has done excellent astronomical work.

The fundamental truths of the movements of the heavenly bodies have been chiefly learned from the work of instruments of comparatively moderate telescopic power, specially arranged to enable precise measures of position to be secured. Indeed, in the early stages of astronomy, important observations of position were obtained by contrivances which showed the direction of the object without any telescopic aid.

In our modern observatories the most important measurements are those obtained by that most accurate of all instruments of precision, known as the meridian circle. It would be out of place to attempt to give here any minute description of this instrument, even in any of its multitudinous forms. It is, however, equally impossible, in any adequate account of the Story of the Heavens, to avoid some reference to this fundamental instrument; and therefore we shall give a very brief account of one of the simpler forms, choosing for this purpose a great instrument in the Paris Observatory, which is represented in the illustration (Fig. 7).

The telescope is attached at its centre to an axis at right angles to its length. The pivots at the extremities of this axis rotate in fixed bearings, so that the movements of the telescope are completely restricted to the plane of the meridian. Inside the eye-piece of the telescope extremely fine vertical lines are stretched. The observer watches the moon, or star, or planet, or whatever may be the object, enter the field of view; and he notes the second, or fraction of a second, by the clock, as the star passes over each of the lines. The circle attached to the telescope is divided into degrees and subdivisions of a degree, and this circle, which moves with the telescope, will indicate the elevation at which the telescope is pointed. For the accurate reading of the circle, microscopes are used. These microscopes are shown in the sketch, each one being fixed into an aperture in the wall which supports one of the pivots. At the opposite side is a lamp, the light from which passes through the perforated pivot and is thence deflected to illuminate the lines at the focus.

The lines, which the observer sees stretched over the field of view of the telescope, demand a few words of explanation. We require for this purpose a line which shall be very fine and durable, elastic, and of little or no weight. These conditions cannot be completely fulfilled by any metallic wire, but they are most exquisitely fulfilled in the beautiful thread which is spun by the spider. These gossamer threads are stretched with nice skill across the field of view of the telescope, and secured in their proper places. With instruments so furnished, it is easy to understand the precision of modern observations. The telescope is directed towards a star, and the image of the star is a minute point of light. When that point is made to coincide with the intersection of the two central spider lines, the telescope is properly sighted.

We use the word sighted designedly, because we wish to suggest a comparison between the sighting of a rifle at the target and the sighting of a telescope at a star. Instead of the large bull's-eye of a rifle-target, suppose that the target only contained an ordinary watch-dial; the rifleman would not be able to sight the dial. But with the telescope of the meridian circle we could easily see the watch-dial at the distance of a mile. The meridian circle has, indeed, such delicacy as a sighting instrument, that it could be pointed separately to each of two stars, which subtend at the eye an angle no greater than that subtended by an adjoining pair of the sixty minute dots around the circumference of a watch dial a mile away.

This delicacy of sighting would be of little use were it not combined with arrangements by which, when the telescope has been pointed correctly, its position can be ascertained and recorded. One element is secured by the astronomical clock, which gives the moment when the object crosses the central vertical wire; the other element is given by the graduated circle which reads the zenith distance.

Superb meridian instruments adorn our great observatories, and are nightly devoted to those measurements upon which the great truths of astronomy are mainly based. These instruments are made with every refinement of skill; but it is the duty of the painstaking astronomer to distrust the accuracy of his instrument in every conceivable way. The great tube may be as rigid a structure as mechanical engineers can produce; the divisions on the circle may have been engraved by the most perfect mechanical contrivance; but the conscientious astronomer will not rely upon mechanical precision. That meridian circle which, to the uninitiated, seems a marvellous piece of workmanship possessing almost illimitable accuracy, is presented in a different light to the astronomer who makes use of it. No one can appreciate, indeed, so fully as he, the skill of the artist who has made it, and the numerous beautiful contrivances for illumination and reading off, which give to the instrument its perfection; but while he recognises the beauty of the actual machine he is using, the astronomer has always before his mind's eye an ideal instrument of absolute perfection, to which the actual meridian circle only makes an approximation. Contrasted with this ideal instrument the best meridian circle is little more than a mass of imperfections. The ideal tube is perfectly rigid, the actual tube is flexible; the ideal divisions of the circle are all perfectly uniform; the actual divisions are not uniform. The ideal instrument is a geometrical embodiment of perfect circles, perfect straight lines, and perfect right angles; the actual instrument can only give us approximate circles, approximate straight lines, and approximate right angles. Perhaps the spider's part of the work is on the whole the best; he gives us the nearest mechanical approach to a perfectly straight line; but we mar his work by not being able to put in his beautiful threads with perfect uniformity, while our attempts to stretch two of them across the field of view at right angles, do not succeed in producing an angle of exactly ninety degrees. Nor are the difficulties encountered by the meridian observer solely due to his instrument. He has to contend with his own want of skill; he has often to allow for personal peculiarities of an unexpected nature; the troubles that the atmosphere can give him are notorious; while the levelling of his instrument tells him that he cannot even rely on the solid earth itself. The meridian circle shows that the earthquakes, which sometimes startle us, are merely the more conspicuous instances of incessant and universal movements in the earth, which every night in the year derange the delicacy of the instrument.

When the existence of these errors has been recognised, the first great step has been taken. By an alliance between the astronomer and the mathematician it is possible to measure the differences and the irregularities which separate the ideal meridian circle from the actual meridian circle. Once this has been done, it is possible to estimate the effect which all the irregularities can produce on the observations, and finally, to purge the observations from the grosser errors with which they are contaminated. We thus have observations, not indeed mathematically accurate, but still close approximations to those which would be obtained by a perfect observer, using an ideal instrument of geometrical accuracy, standing on an earth of absolute rigidity, and viewing the heavens without the intervention of the atmosphere.

It is not, however, necessary to use such great instruments as those just described in order to obtain some idea of the aid the telescope will afford in showing the celestial glories. The most suitable instrument for commencing astronomical studies is within ordinary reach. It is the ordinary binocular that a captain uses on board ship; or if that cannot be had, then the common opera-glass will answer nearly as well. This is, no doubt, not nearly so powerful as a large telescope, but it has some compensating advantages which the telescope does not possess. The opera-glass will survey a large region of the sky at once, while a telescope only looks at a small part of the sky. Let us suppose that the observer is provided with an opera-glass and is about to commence his astronomical studies.

The first step is to become acquainted with the very renowned group of seven stars which is represented in Fig. 8. It is often called the Plough, but astronomers prefer to regard it as a portion of the constellation of the Great Bear (Ursa Major).

There are many features of interest in this constellation, and the beginner should learn as soon as possible to identify the seven remarkable stars. Of these the two stars, a and β, at the head of the bear are generally called the "pointers." They are of special use in astronomy, because they enable us to find out the most important star in the whole sky, which is known as the "pole star." We shall return in a later chapter to the study of the different constellations. Our present object is a simpler one; it is merely to employ the Great Bear as a means of teaching us how vast is

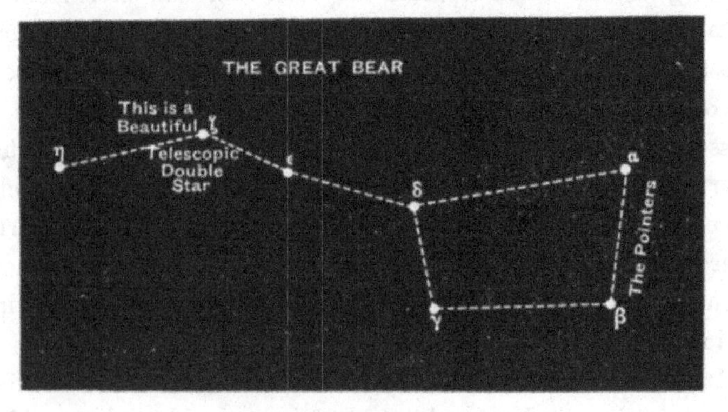

Fig. 8.—The Great Bear.

the richness of the heavens in stars. Each student of astronomy is recommended to make one very simple observation on the Great Bear, which will give a wondrous conception of what the telescope can do, and will also reveal in a very impressive manner the glories of the starry heavens.

Fix the attention on that region in the Great Bear bounded by the four stars $a\ \beta\ \gamma\ \delta$. They form a sort of rectangle, of which the stars named are the corners. The next fine night try and count how many stars are visible within that rectangle. There are no really bright stars, but there are two or three sufficiently bright to be easily seen. On a very fine night, when there is no moon, perhaps a dozen might be perceived, or even more, according to the keenness of the eyesight. But when the opera-glass is directed to the same region, a most interesting, and indeed astonishing sight will be witnessed. Instead of the few stars which were seen before with difficulty, a hundred stars or more can now be seen with the greatest ease. The opera-glass will, indeed, easily disclose ten times as many stars as could be seen with the unaided eye.

But even the opera-glass will not show nearly all the stars in this region. Any good telescope will reveal hundreds of stars too faint for the opera-glass. The greater the telescope, the more numerous the stars; so that in one of the colossal instruments this region would be found studded with thousands of stars.

We have chosen the Great Bear for the purpose of this illustration, because it is more generally known than any other constellation. But the Great Bear is not exceptionally rich in stars; any other part of the sky would equally well have demonstrated the grand truth, that the stars which our unaided eyes disclose, are only an exceedingly small fraction of the entire number with which the whole heaven is teeming. To tell the number of the stars is a task which no man has accomplished; but various estimates have been made. Our great telescopes can probably show at least 50,000,000 stars. There would be a star apiece for every man, woman, and child in the United Kingdom, and there would still remain a liberal margin for distribution elsewhere.

The student of the heavens who uses a good refracting telescope, having an object-glass about three inches in diameter, will find ample and delightful occupation for many a fine evening. He should also be provided with an atlas of the stars, while a copy of the "Nautical Almanac," and of Webb's "Celestial Objects for Common Telescopes," will form a sufficiently complete astronomical equipment for much interesting occupation.

CHAPTER II.

THE SUN.

The vast Size of the Sun—Hotter than Fusing Platinum—Is the Sun the Source of Heat for the Earth ?—The Sun is 92,700,000 miles distant—How to realise the magnitude of this distance—Day and Night Luminous and Non-Luminous Bodies—Contrast between the Sun and the Stars—The Sun a Star—The Spots on the Sun—Changes in the Form of a Spot—They are Depressions on the Surface—The Rotation of the Sun on its Axis—The Size and Weight of the Sun—Is the Sun a Solid Body ?—View of a Typical Sun-Spot—Periodicity of the Sun-Spots—Connection between the Sun-Spots—Terrestrial Magnetism—The Faculæ—The Granulated Appearance of the Sun—The Prominences surrounding the Sun—Total Eclipse of the Sun—Size and Movement of the Prominences—Drawings of the Objects, coloured—The Corona Surrounding the Sun—The Heat of the Sun.

In commencing our examination of the orbs which surround us, we naturally begin with our peerless sun. His splendid brilliance gives him the pre-eminence over all other celestial bodies. The proportions of the sun are commensurate with his importance.

Astronomers are actually able to measure the sun; and they find that his dimensions are so great as to tax our imagination to realise them. The diameter of the sun, or the length of the axis, passing through the centre from one side to the other, is 865,000 miles. Yet this bare statement of the dimensions of the great globe fails to awaken an adequate idea of its vastness. If a railway were laid round the sun, and if we were to start in an express train moving sixty miles an hour, we should have to travel night and day for five years without intermission before we had accomplished our journey.

If the sun be compared with the size of the earth, its stupendous bulk becomes still more apparent. Suppose his globe were cut up into one million parts: each of these parts would appreciably exceed the bulk of our earth. Were the sun placed in one pan of a mighty weighing balance, and were 300,000 bodies as heavy as our earth placed in the other, the sun would still turn the scale. Fig. 9 exhibits a large white circle and a very small one. These circles are drawn to exhibit the comparative size of the earth and the sun, the small circle being the earth and the large one the sun.

The temperature of the sun has an intensity far surpassing the greatest temperature we can artificially produce. In our laboratories we send a galvanic current through a piece of platinum

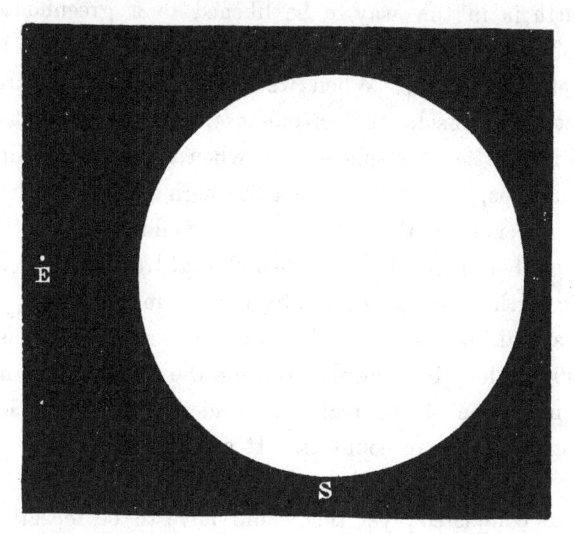

Fig. 9.—Comparative Sizes of the Earth and the Sun.

wire. The wire first becomes red-hot, then white-hot; then it is almost of dazzling brilliancy, until it fuses and breaks. The temperature of the melting platinum wire could hardly be surpassed in the most elaborate furnaces, but it falls far short of the temperature of the sun.

It must, however, be admitted that there is one seeming discrepancy between the fact of the sun's high temperature and a well-known physical fact. "If the sun were hot," it has been said, "then the nearer we get to the sun the hotter we should be; yet this is not the case. On the top of a high mountain we are nearer to the sun, and yet everybody knows that it is much colder on the top of the mountain than in the valley at its foot. If the mountain be as high as Mont Blanc, then we are two or three miles nearer the sun; yet up there, instead of additional warmth, we find eternal snow." A simple illustration will dispose of this difficulty. Go into a greenhouse on a sunshiny day, and we find the temperature much hotter there than outside. The glass will allow the hot sunbeams to enter, but it refuses to allow them out again with equal freedom, and consequently the temperature rises. Our whole earth is in this way to be likened to a greenhouse, only, instead of the panes of glass, we are enveloped by an enormous coating of atmosphere. When we are on the earth's surface, we are, as it were, inside the greenhouse, and we benefit by the interposition of the atmosphere; but when we begin to climb very high mountains, we gradually get through the atmosphere, and then we suffer from the cold. If we could imagine the earth to be stripped of its coat of air, then eternal frost would reign over the whole earth as well as on the tops of the mountains.

The actual distance of the sun from the earth is about 92,700,000 miles; but merely reciting the figures does not give a vivid impression of the real magnitude. 92,700,000 is a very large quantity. Try to count it. It would be necessary to count as quickly as possible for three days and three nights before one million was completed; yet this would have to be repeated nearly ninety-three times before we had even counted all the miles between the earth and the sun.

Every clear night we see a vast host of stars scattered over the sky. Some are bright, some are faint, some are grouped into remarkable forms. With regard to this vast host we can now ask an important question. Are they bodies which shine by their own light like the sun, or do they only shine with light borrowed from the sun? The answer is easily stated. Most of these bodies shine by their own light, and they are properly called *stars*. If, then, our sun and the multitude of stars, properly so called, are each and all self-luminous brilliant bodies, what is the great distinction between the sun and the stars? There is, of course, a vast and obvious difference between the unrivalled splendour of the sun and the feeble twinkle of the stars. Yet this distinction does not necessarily indicate that the sun has an intrinsic splendour superior to that of the stars. The fact is that we are nestled up comparatively close to the sun for the benefit of his warmth and light, while we are separated from even the nearest of the stars by a mighty abyss. If the sun were gradually to retreat from the earth his light would decrease, so that when he had penetrated the depths of space to a distance comparable with that by which we are separated from the stars, his glory would have utterly departed. No longer would the sun be the majestic orb with which we are familiar. No longer would he be a source of genial heat, or a luminary to dispel the darkness of night. Our great sun would have shrunk to the unimportance of a star, not so bright as many of those which we see every night.

Momentous indeed is the conclusion to which we are now led. That myriad host of stars which stud our sky every night, has sprung into vast importance. Each one of those stars is itself a mighty sun, actually rivalling, and in many cases surpassing, the splendour of our own luminary. We thus open up a majestic conception of the vast dimensions of space, and of the dignity and splendour of the myriad globes by which that space is tenanted.

There is another aspect of the picture not without its utility. We must from henceforth remember that our sun is only a star, and not by any means an important star. If the sun and the earth, and all which it contains, were to vanish, the effect in the universe would merely be, that a tiny star had ceased its twinkling. Viewed merely as a star, the sun thus assumes a place of insignificance in the mighty fabric of the universe. But it is not as a star that we have to deal with the sun. To us his proximity gives him an importance incalculably transcending that of all the other stars. We receded from the sun to obtain his true perspective in the universe; let us now draw near, and give to him that attention which his importance requires.

To the unaided eye the sun appears to be a flat circle. If, however, it be examined with the telescope, taking care of course to employ a piece of deep-coloured glass, or some similar precaution to screen the eye from injury, it will then be seen that the sun is really not a flat surface but a glowing globe, of which one hemisphere is presented to us. The first question which we must attempt to answer is with reference to the constitution of that globe; and the first branch of that question is, whether the glowing

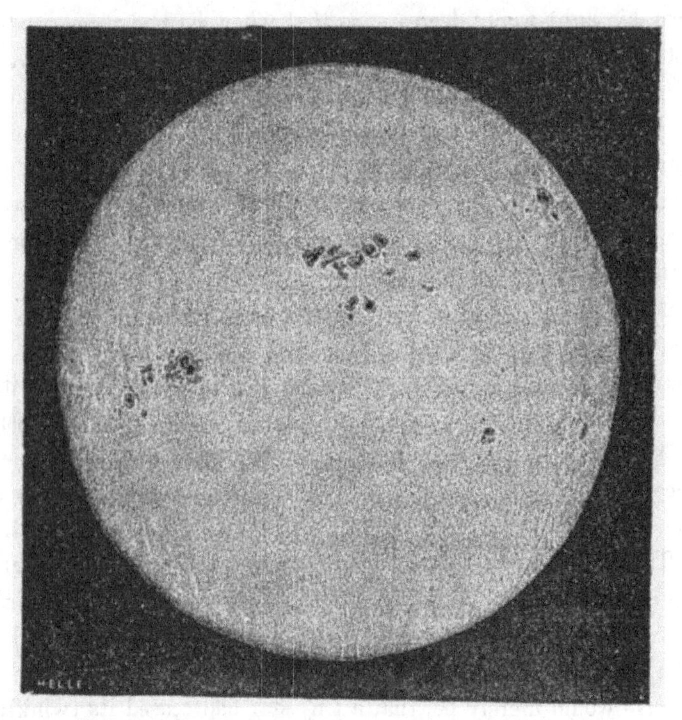

Fig. 10.—The Sun, photographed on September 22, 1870.

matter which forms the globe is a solid mass, and, if it be not solid, whether it is liquid or gaseous? At the first glance we might think that the sun must certainly be solid. We have all seen white-hot iron, and we might naturally think that the sun was a stupendous ball of solid white-hot substance, or something analogous thereto. But this view could not be correct; and our first task will be to show that the sun is certainly not a solid body so far as we can see it.

Look at the general view of the sun shown by a telescope of moderate dimensions. It is represented in Fig. 10.* We observe the circular outline and the general bright surface, but the brilliancy of the surface is not quite uninterrupted. There are, here and there on the surface, small, dark objects called *spots*, which can be made to render a great deal of information with respect to the sun. These spots vary both as to size and as to number—indeed, the sun seems sometimes almost devoid of spots. In the early days of telescopes the discoverers of the sun's spots were laughed at. They were told that the sun was far too perfect to have any blemishes, and that it was absurd to suppose that "the eye of the universe could suffer from ophthalmia."

The general character of a sun spot, as seen in a moderate telescope under ordinary circumstances, is illustrated in Fig. 11,

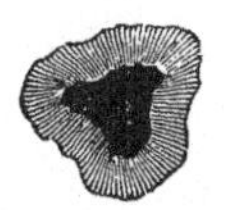

Fig. 11.—An Ordinary Sun-Spot.

in which the dark central part is seen well contrasted with the lighter margin. Various theories have been propounded as to the nature of these curious features, and one of the early suppositions was that they were merely objects situated between the earth and the sun, and thus projected on the sun as a background. It is easy to prove that this cannot be the case, by carefully watching the same spot for a few days. Look first at the spot marked A in Fig. 12, which exhibits a small portion of the sun and a small part of its edge as seen in a good telescope; A represents a spot of an ordinary type on the sun. The central portion of the spot seems black by contrast with the brilliant background in which the spot is shown. Around the black centre we have a shaded region of a somewhat lighter hue. We carefully observe the spot and note how far off it appears

* This picture is a copy of a very beautiful photograph of the sun, taken by Mr. Rutherfurd at New York on the 22nd September, 1870.

to be from the edge of the sun. The next day we repeat the observation, and find that the spot is no longer in its original position. It has travelled nearer the edge of the sun, to the place marked B. Repeat the same observation the third day, and it will be found that the spot has attained the position C, again nearer to the edge of the sun. It will also be noticed that the appearance of the spot has changed. The shaded portion at one side has diminished and, indeed, disappeared. Day after day the spot gradually approaches the edge of the sun until, finally, it is on

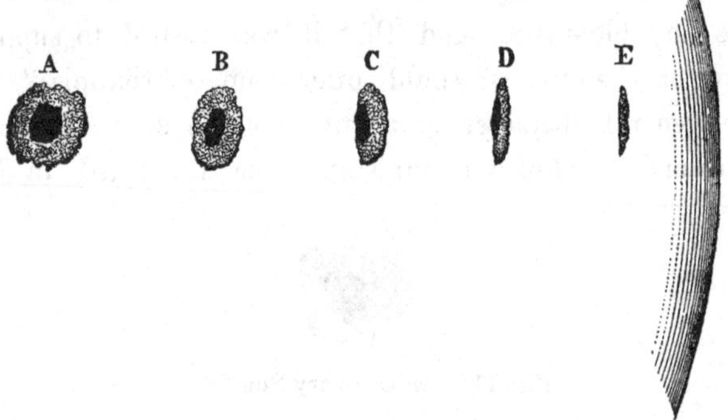

Fig. 12.—Successive Appearances of a Sun-Spot.

rare occasions actually seen on the sun's edge, though it is not so represented in this drawing.

It is by such observations we learn that the spot cannot be any body floating aloft above the surface of the sun, for then an interval between the spot and the sun would be seen after the spot attained the edge. Yet, though we must admit that the spots are really on the surface of the sun, we cannot agree with those old philosophers who said that these spots are blemishes which detract from the perfection of the eye of the heavens. We ought rather, in these days of scientific activity, to feel grateful to the spots; for they teach us much about the glories of the sun of which we otherwise must have remained in ignorance.

If an artist were to view the changes in the appearance of the spot as it approached the sun's edge, he might think that the apparent alterations of the form of the spot were chiefly due to what he would call the effect of fore-shortening, and he would draw the conclusion that the spot was really a cavity in the surface of the sun. If the interior parts of the sun were much darker than the exterior, and if the spots were really basin-shaped apertures through the outer portions of the sun, then some of the changes in

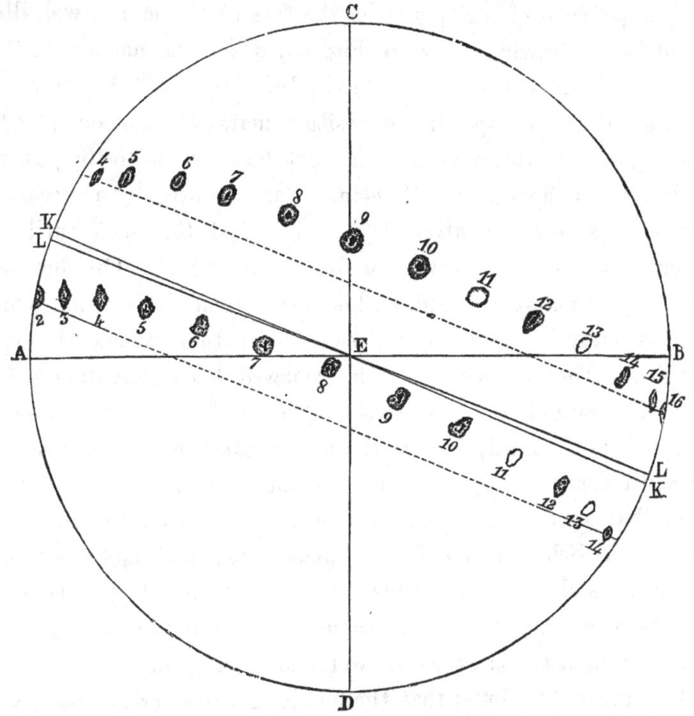

Fig. 13.—Scheiner's Observations on Sun-Spots.

the appearances of the spots could be readily explained. When the region is turned directly towards the observer he sees the bottom of the basin which exposes the dark interior of the sun. The view of the spot is then represented by A. As the spot is carried nearer to the limb of the sun, one side of the basin becomes much fore-shortened, and the appearance of the spot is represented at B, in which we observe that the shaded edge of the spot on the left is narrower than that on the other side. Finally, when the spot is extremely close to the limb, then one side of the basin would be entirely hid from view, and only a glimpse be had of the dark interior; while the opposite side of the basin is distorted into undue prominence. It cannot, however, be regarded as proved that the sun-spots are really depressions in the surface; indeed many astronomers hold different views on the subject.

The progress of the spots over the face of the sun is well illustrated in the drawing shown in Fig. 13, which was made more than 250 years ago. On the 2nd March, 1627, the skilful astronomer Scheiner observed a spot in the position marked 2, just on the edge of the sun. On the next day, the spot had moved to the position marked 3. It appears that Scheiner was favoured by a succession of fine days, for day after day he followed the spot until the eleventh, when his observations were interrupted. On this day, therefore, he marked a blank on his drawing, in the position which he reasonably conjectured to have been that which the spot occupied. The following day he renewed his observation. The 13th was again cloudy, but on the 14th another and final view of the spot was obtained, just as it was approaching the edge of the sun and before its disappearance. In the same month it so happened that another conspicuous spot was visible on the sun, and the faithful Scheiner recorded its place also; and again we find interruptions due to the clouds on the 11th and 13th. In each case the spot travelled in the same direction and crossed the face of the sun in a period of about twelve or thirteen days.

It is invariably found that these objects move across the sun in the same direction. It is also noticed that when the spots disappear at the edge of the sun, and remain invisible for twelve or thirteen days, the same spots often reappear at the other edge. It is therefore obvious that the spots must move round the back of the sun in about the same time that they occupy in crossing his face. Further inquiries on this subject have enabled the movements of the spots to be measured with accuracy, and it has been shown that each spot accomplishes a complete revolution around the sun in about twenty-five days and five hours.

So remarkable a characteristic of the movements of the spots demands satisfactory explanation. How does it come to pass that the spots—both large and small, regular or irregular—all accomplish their revolutions in nearly the same time? A very simple explanation of the phenomenon conducts at once to a very important discovery.

We know that the sun is a globe, and that our earth is also a globe. We also know that the earth performs a daily rotation on its axis, and it is natural to ask, whether it may not be likewise possible for the sun to perform a rotation. If the sun slowly rotated in a period of about twenty-five days and five hours, then the hemisphere of the sun directed towards the earth would be completely turned to the other side in a fortnight, and we should at once be able to account for the apparent movement of the spots. This explanation is so simple, and so satisfactory—it is confirmed by so many other lines of reasoning—that no doubt can any longer be attached to it; and hence we have, as the first-fruits of the study of the spots, the very interesting and remarkable discovery of the rotation of the sun on its axis. It has, however, been shown that the time of rotation of the sun-spot varies slightly with the position of the spot. Observations made with spots at the Equator give for the period of rotation a value of $25\frac{1}{3}$ days, while, judging from spots at the latitude of 30°, the period of rotation of the sun is a day longer. It thus follows that we cannot state the period of the sun's rotation with the same accuracy that we can that of the earth. It can only be said to lie somewhere between the two extremes of about 25 and $26\frac{1}{2}$ days.

It must not, however, be imagined that the only changes in the spots are those variations of perspective which arise from the rotation of the sun. In this movement all spots alike participate; but there are other movements and changes constantly going on in the individual spots. Some of these objects may last for days, for weeks, or for months, but they are in no sense permanent ; and after an existence of greater or less duration, the spots on one part of the sun will disappear, while as frequently fresh spots will become visible in other places. The inference from these various facts is irresistible. It tells us that the visible surface of the sun is not a solid mass—is not even a liquid mass—but that the sun, as far as we can see it, consists of matter in the gaseous or vaporous condition.

The sun-spots are confined to certain limited regions of the surface. They are nearly always found in two zones on each side of the Equator, and between the latitudes of 10 and 30 degrees. Spots are comparatively rare on the Equator, E E′ (Fig. 14) and very few are found beyond the latitude of 35 degrees, while we have the authority of Professor Young for the statement that there is only a single recorded instance of a spot more than 45 degrees from the Solar

Fig. 14.—Zones on the Sun's surface in which Spots appear.

Equator—one observed in 1846 by Dr. Peters at Naples. The average duration of a sun-spot is about two or three months, and the longest life of a spot that has been recorded is one which in 1840 and 1841 lasted for eighteen months. There are, however, some spots which last only for a day or two, and some only for a few hours.

It should also be observed that the sun-spots usually appear in groups, and very often a large spot is attended or followed by a number of smaller ones, more or less imperfect. It often happens that a large spot divides into two or more smaller spots, and these parts have been sometimes seen to fly apart, with a velocity in some cases not less than a thousand miles an hour. On rare occasions a phenomenon of the most surprising character has been witnessed in connection with the sun-spots, where patches of intense brightness suddenly break out, remain visible for a few minutes, and travel with a velocity over 100 miles a second. One of these events has become celebrated for the extraordinary character of the phenomena, as well as for the fortunate circumstance that it has been authenticated by the independent testimony of two skilled witnesses. On the forenoon of the 1st September, 1859, two well-known observers of the sun, Mr. Carrington and Mr. Hodgson, were both engaged in observation. Mr. Carrington was employed at his self-imposed daily task of observing the positions, the configuration, and the size of the spots by means of an image of the sun upon a screen. Mr. Hodgson, many miles away, was at the same moment sketching some details of sun-spot structure. They saw simultaneously two luminous objects, shaped something like two new moons, each about eight thousand miles long and two thousand miles wide, at a distance of about twelve thousand miles apart: these suddenly burst into view near the edge of a great sun-spot, with a brightness at least five or six times that of the neighbouring parts of the sun, and travelled eastward over the spot in parallel lines, growing smaller and fainter, until in about five minutes they disappeared, after a journey of about thirty-six thousand miles.

We have still to note one very extraordinary feature, which points to an intimate connection between the phenomena of sun-spots, and the purely terrestrial phenomena of magnetism. It has been noticed that the occurrence of the maximum of sun-spots occurs simultaneously with an unusual amount of disturbance of the magnetic needle. The latter are well known to be connected with the phenomena of the aurora borealis, inasmuch as an unusual aurora seems to be invariably accompanied by a great magnetic disturbance. It has also been shown that there is an almost perfect parallelism between the intensity of auroral phenomena and the abundance of sun-spots. Besides these general coincidences, there have been also special cases in which a peculiar outbreak on the sun has been associated with remarkable auroral or magnetic phenomena. Thus, the occurrence cited above as witnessed by Mr. Carrington and Mr. Hodgson in 1859, was immediately followed by a magnetic storm of unusual intensity, as well as by splendid auroras, not only in Europe and America, but even in the Southern Hemisphere. A very interesting instance of a similar kind is recorded by Professor Young, who, when observing at Sherman on the 3rd August, 1872, perceived a very violent disturbance of the sun's surface. He was told the same day by the photographer of the party, who was engaged in magnetic observations, and who was quite in ignorance of what Professor Young had seen, that he had been obliged to desist from the magnetic observations, in con-

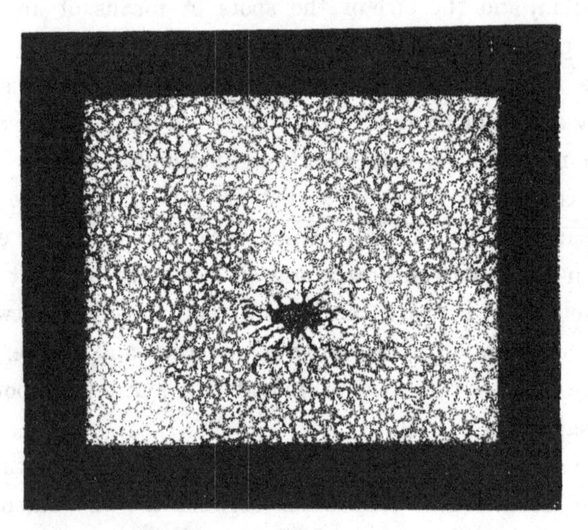

Fig. 15.—The Texture of the Sun and a small Spot.

sequence of the violent fluctuations of the needle. Subsequent inquiry showed that in England on the same day a magnetic storm was also witnessed.

These observations demonstrate that there is *some* connection between solar phenomena and terrestrial magnetism, but what the nature of that connection may be is quite unknown, and will form a problem of deep interest for the future labours of astronomers and physicists.

Another mysterious law governs the sun-spots. Their number fluctuates from year to year, but it would seem that the epochs of maximum sun-spots succeed each other with a certain degree of regularity. The observations of sun-spots for nearly three centuries show that the recurrence of a maximum takes place, on an average, every eleven years. The course of one of these cycles is somewhat as follows :—For two or three years the sun-spots are both larger and more numerous than on the average ; then they begin to diminish, until in about five or six years from the maximum

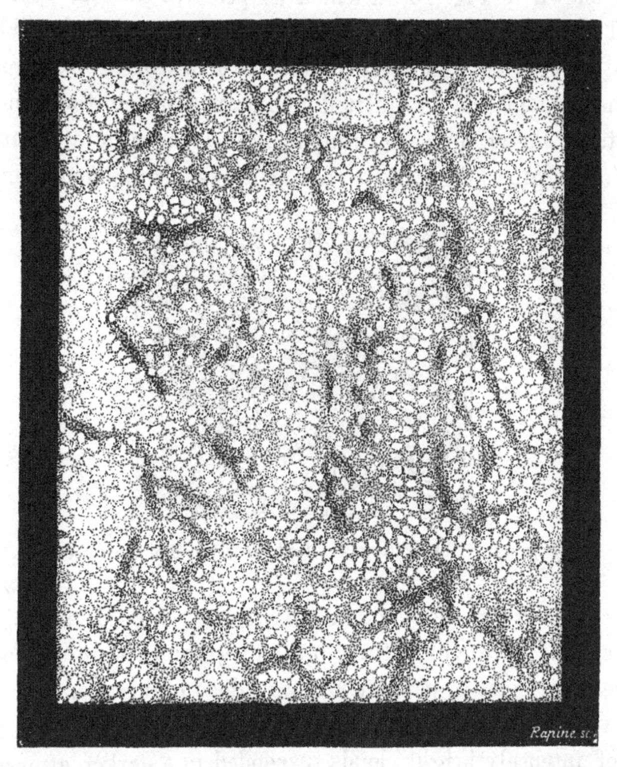

Fig. 16.—Dr. Huggins' drawing of a remarkable arrangement of Solar Granules.

they reach a minimum ; then the spots begin to increase, and in another five or six years the maximum is once more attained. The cause of this periodicity is a question of the most profound interest, but at present the answer must be regarded as unknown. It has, indeed, to be admitted that the real nature of sun-spots is still a matter of uncertainty. No theory yet proposed will account in a thoroughly satisfactory manner for *all* the phenomena which they present, when viewed with the telescope and the spectroscope, as well as for their peculiar distribution over the sun, and the marvellous phenomena of periodicity.

When the atmosphere will allow of very good vision, we can see that the sun's surface is mottled in a remarkable manner. This is well shown in Fig. 15, in which we perceive that the spot in the central part of the picture is merely an enlargement of one of the minute pores with which the surface is marked. A very remarkable instance of the granulated appearance which the sun often presents is shown in a drawing made by the accurate pencil of Dr. W. Huggins (Fig. 16). This curious arrangement has also

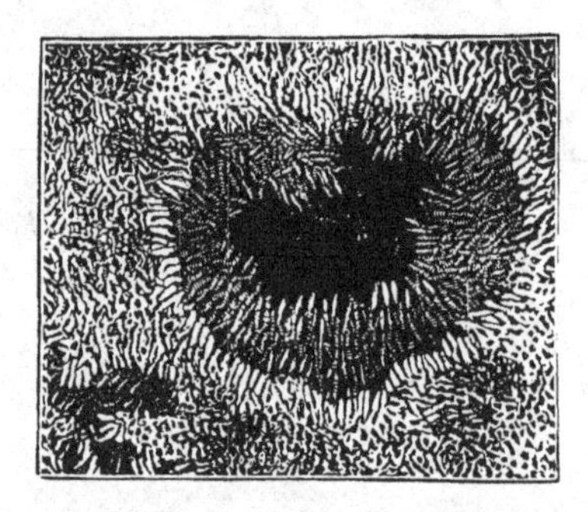

Fig. 17.—The Willow-leaf texture of the Sun's surface.

been witnessed by many other observers. Indeed, photographs have been taken in which these brilliant granules seem disposed to arrange themselves in patterns of marvellous regularity.

It would thus appear as if the luminous surface of the sun was composed of intensely bright clouds suspended in a darker atmosphere. Some observers have thought that these floating objects are, occasionally at all events, of a characteristic size and shape, variously known as "willow leaves" or "rice grains." In Fig. 17, the curious willow-leaf texture is shown surrounding a sun-spot. But the spot itself seldom fails to give the impression of violent disturbance, as is well shown in Professor Langley's fine drawing (Plate II.) of a spot which he observed on December 23—24, 1873.

Near the edge of the sun, as represented in Plate III., will be seen some of those brighter streaks or patches which are called faculæ

PLATE III.

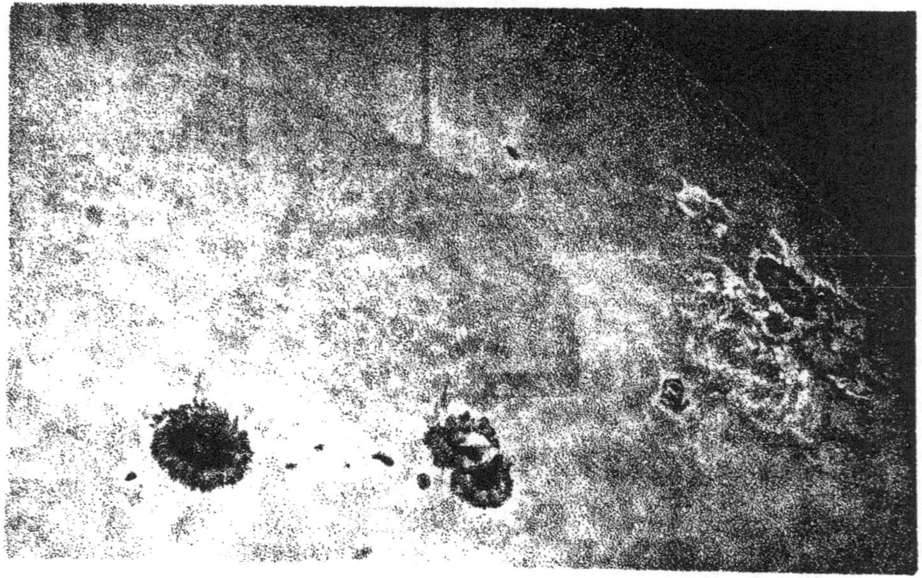

SPOTS AND FACULÆ ON THE SUN.

(FROM A PHOTOGRAPH BY MR. WARREN DE LA RUE, 20TH SEPT., 1861.)

(little torches). They are often of enormous dimensions, covering areas vastly larger than any of our continents.

The margin of the sun is fringed with objects of very great interest. They are so faint that in the full blaze of sunlight they cannot be seen. They are invisible for the same reason that stars are invisible in daylight. We see the stars at night, when the sun is gone, and so we can see the fringe surrounding the sun when the brilliant central portion is obscured by the rare occurrence of a total eclipse.

For an eclipse of the sun to occur, the moon must actually come between the earth and the sun. The occurrence of an eclipse will

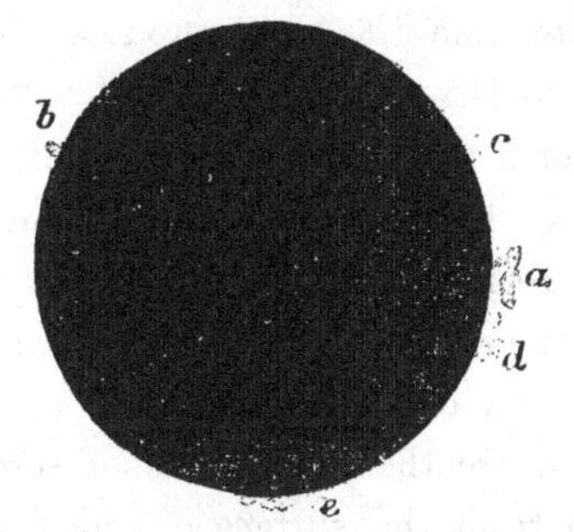

Fig. 18.—Prominences seen in Total Eclipse

be more fully considered later on. For the present it will be sufficient to observe that by the movement of the moon it may so happen that the moon completely hides the sun, and thus for a few minutes produces what we call a total eclipse. The few minutes during which a total eclipse lasts are of the most priceless value to the astronomer. Darkness reigns over the earth, and in that darkness rare and beautiful sights can be witnessed.

We have in Fig. 18, a view of a total eclipse, showing some of those remarkable objects known as prominences, (*a*, *b*, *c*, *d*, *e*), which project from the surface of the sun. Objects of this character surround the sun at other times as well as at eclipses, but their light is so faint that the great light of the sun renders them invisible. With the obscurity which surrounds the sun during a total eclipse as a background, the phenomenon starts into brilliancy.

It has been demonstrated that these very curious objects are, as their appearance indicates, really mighty glowing masses of gas; and a most beautiful arrangement has been discovered by which it has been made possible to view the prominences without waiting for the aid of an eclipse. It would be anticipating what we shall have to say in a future chapter were we at this point to give any detailed explanation of the ingenious contrivance by which these objects can be seen in the full blaze of sunlight. Suffice it now to observe that the principle of the method depends upon the peculiar character of the light from the prominences, which the spectroscope enables us to isolate from the glare produced by the ordinary solar beams. It gives to astronomers the great advantage of looking at the prominences for hours together, instead of being limited to the few minutes during which an eclipse lasts. The prominences appear to be merely protuberant portions of a layer of red incandescent gas surrounding the sun. This gas has been shown to consist of hydrogen and probably other substances.

Majestic indeed are the proportions of some of those mighty flames which leap from the surface of the sun; yet these flames flicker as do our terrestrial flames, when we allow them time comparable to their gigantic dimensions. Drawings of the same prominence often show great changes in a few hours, or even less. The magnitude of the changes could not be less than many thousands of miles, and the actual velocity with which such masses move is often not less than 100 miles a second. Still more violent are the solar convulsions which some observers have been so fortunate as to behold, when from the sun's surface, as from a mighty furnace, vast incandescent masses are projected upwards. All indications point to the surface of the sun as the seat of the most frightful storms and tempests, in which the winds sweep along incandescent vapours.

The remarkable power which the spectroscope places at our disposal of enabling the prominences to be seen without a total eclipse has been largely availed of in making drawings of these objects. Plate IV. gives a very beautiful view of a number of them as seen by Trouvelot with the great telescope at Cambridge, U.S. These drawings show the red colour of the flame-like objects, not very happily described as prominences; and they also show, in the different pictures, the wondrous variety of aspect which these objects assume. The dimensions of the prominences may be inferred from the scale appended to the plate. The largest of them is fully 80,000 miles high; but many observers have recorded prominences of much greater altitude. The rapid changes of these objects is well illustrated in the two sketches on the left of the lowest line, which were drawn on April 27th, 1872. These are both drawings of the same prominence taken at an interval no greater than twenty minutes. This mighty flame is so vast that its length is ten times as great as the diameter of the earth, yet in this brief period it has completely changed its aspect; the upper part of the flame has, indeed, broken away, and is now shown in that part of the drawing between the two figures on the line above. The drawings also show various instances of the remarkable spike-like prominences, taken at different times and on different parts of the sun. These spikes usually attain altitudes not greater than 20,000 miles, but sometimes they stretch up to stupendous distances. We may quote one special object of this kind, whose remarkable history has been chronicled by Professor Young,* the well-known authority in this department of astronomy. On October 7th, 1880, a prominence was seen, at about 10.30 a.m., on the south-east limb of the sun. It was then an object of no unusual appearance, being about 40,000 miles high, and attracted no special attention; but half an hour witnessed a marvellous transformation. During that brief interval the prominence became very brilliant, and doubled its length. For another hour the mighty flame still soared upwards, until it attained the unprecedented elevation of 350,000 miles—a distance more than one-third of the diameter of the sun. Here the energy

* During a visit to the United States in the autumn of 1884, the author was fortunate enough, by the kindness of Professor Young, to observe several solar prominences with the superb instruments at Princeton, New Jersey.

of the mighty outbreak seems to have expended itself: the flame broke up into filaments, and by 12.30—an interval of only two hours from the time when it was first noticed—the huge prominence had completely faded away.

The facts we have recorded give a surprising indication of the violence of those fiery storms by which the surface of the sun is occasionally disturbed. No doubt this vast prominence was exceptional in its magnitude, and in the vastness of the changes of which it was an indication; but we may, at all events, take it as the basis of an estimate of the maximum changes which the surface of the sun witnesses. The velocity must have been 200,000 miles an hour—a rate which must have more than averaged fifty miles a second. This mighty flame leaped up from the sun with a velocity more than 100 times as great as that of the swiftest bullet that was ever fired from a rifle.

The most striking feature of a total eclipse of the sun is unquestionably the Corona, or aureole of light which is then seen to surround the sun. On such an occasion, when the sky is clear, the moon appears of an inky darkness, not like a flat screen, but like the huge black ball that it really is. "From behind it (I quote Professor Young) stream out on all sides radiant filaments, beams, and sheets of pearly light, which reach to a distance sometimes of several degrees from the solar surface, forming an irregular stellate halo with the black globe of the moon in its apparent centre. The portion nearest the sun is of dazzling brightness, but still less brilliant than the prominences which blaze through it like carbuncles. Generally this inner corona has a pretty uniform height, forming a ring three or four minutes of arc in width, separated by a somewhat definite outline from the outer corona, which reaches to a much greater distance, and is far more irregular in form; usually there are several "rifts," as they have been called, like narrow beams of darkness, extending from the very edge of the sun to the outer night, and much resembling the cloud shadows which radiate from the sun before a thunder shower. But the edges of these rifts are frequently curved, showing them to be something else than real shadows; sometimes there are narrow bright streamers

PLATE IV.

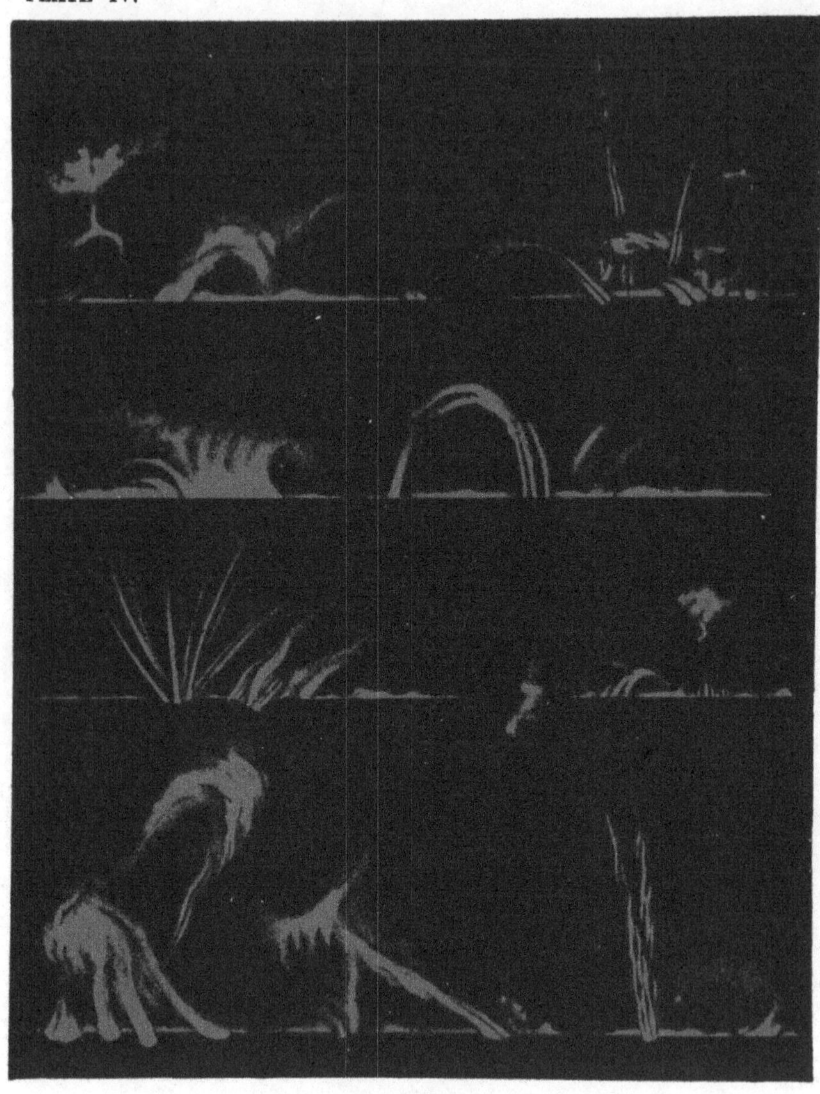

Scale of English Miles.

0 10 20 30 40 50 60 70 80 90 100,000

SOLAR PROMINENCES.

DRAWN BY TROUVELOT AT HARVARD COLLEGE, CAMBRIDGE, U.S., IN 1872.

as long as the rifts, or longer. These are often inclined, occasionally are even nearly tangential to the solar surface, and frequently curved. On the whole, the corona is usually less extensive and brilliant on the solar poles, and there is a recognisable tendency to accumulation above the middle latitudes or spot zones,

Fig. 19.—View of the Corona in a total eclipse.

so that roughly speaking the corona shows a disposition to assume the form of a quadrilateral or four-rayed star, though in almost every individual case this form is greatly modified by abnormal streamers at some point or other." Fig. 19 represents a view of the corona during a total eclipse.

We further present, in Plate V., the drawing made by Professor W. Harkness, which represents the corona as obtained from a comparison of a large number of photographs taken at different places in the United States during the total eclipse of July 29th, 1878.

As to the precise nature of this wonderful appendage, we must for the present be content to wait for further light. Probably when we understand the streamers of the aurora borealis and the tails of comets, we shall have learned something of those substances, well nigh spiritual in texture, which constitute the solar corona.

A remarkable appendage to the sun, which extends to a distance much greater than that of the corona, produces the phenomenon of the zodiacal light. A pearly glow is sometimes seen to spread over a part of the sky in the vicinity of the point where the sun has disappeared after sunset. The same spectacle may also be witnessed before sunrise, and it would seem as if the material producing the zodiacal light, whatever it may be, had a lens-shaped form with the sun in the centre. The nature of this object is still a matter of great uncertainty. We have represented in Fig. 20, a view of this mysterious phenomenon.

In all directions the sun pours forth, with the most prodigal liberality, its torrents of light and of heat. The greater part of that light and heat seems quite wasted in the depths of space. Our earth intercepts only the merest fraction, less than the 2,000,000,000th part of the whole. Our fellow planets and the moon also intercept a trifle; but what portion of the mighty flood can they utilise? The sip that a flying swallow takes from a river is as far from exhausting the water in the river as are the planets from using all the heat which streams from the sun. Were the radiation of the sun to be intercepted, all life on this earth must cease. An immovable atmosphere would brood over an ocean which, if not actually frozen, could be only disturbed by the sullen undulations of the tides, and the silence of death over the surface of the earth would only be broken by the occasional groans of a volcano.

We must postpone to a future chapter the important question of the source of the sun's heat. Let us simply terminate this chapter by a brief recital of what we at present enjoy by the benign influence of the sun. His gracious beams supply the magic power

Fig. 20.—The Zodiacal Light in 1874

that enables our corn to grow and ripen. It is the heat of the sun which raises water from the ocean in the form of vapour, and then sends down that vapour as rain to refresh the earth and to fill the rivers, which bear our ships down to the ocean. It is the heat of the sun beating on the large continents, which gives rise to the breezes and winds that waft our vessels across the deep; and when on a winter's evening we draw around the fire and feel its invigorating rays, we are really only enjoying sunbeams which shone on the earth countless ages ago. The heat in those ancient sunbeams developed the mighty vegetation of the coal epoch, and in the form of coal that heat has slumbered for millions of years, till we now call it again into activity. It is the power of the sun stored up in coal that urges on our steam-engines. It is the light of the sun stored up in coal that beams from every gas-light in our cities.

For our power to live and move, for the plenty with which we are surrounded, for the beauty with which nature is adorned, we are immediately indebted to one body in the countless hosts of space; and that body is the sun.

PLATE V.

TOTAL SOLAR ECLIPSE, JULY 29TH, 1878.

THE CORONA FROM THE PHOTOGRAPHS.

(HARKNESS.)

CHAPTER III.

THE MOON.

The Moon and the Tides—The Use of the Moon in Navigation—The Changes of the Moon—The Moon and the Poets—Whence the Light of the Moon?—Sizes of the Earth and the Moon—Weight of the Moon—Changes in Apparent Size—Variations in its Distance—Influence of the Earth on the Moon—The Path of the Moon—Explanation of the Moon's Phases—Lunar Eclipses—Eclipses of the Sun, how produced—Visibility of the Moon in a Total Eclipse—How Eclipses are Predicted—Uses of the Moon in finding Longitude—The Moon not connected with the Weather—Topography of the Moon—Nasmyth's Drawing of Triesnecker—Volcanoes on the Moon—Normal Lunar Crater—Plato—The Shadows of Lunar Mountains—The Micrometer—Lunar Heights—Former Activity on the Moon—Nasmyth's View of the Formation of Craters—Gravitation on the Moon—Varied Sizes of the Lunar Craters—Other features of the Moon—Is there Life on the Moon?—Absence of Water and of Air—Explanation of the Rugged Character of Lunar Scenery—Possibility of Life on Distant Bodies in Space.

If the moon were suddenly to be struck out of existence, we should be immediately apprised of the fact by a wail from every seaport in the kingdom. From London, from Liverpool, from Bristol, we should hear the same story—the rise and fall of the tide had almost ceased. The ships in dock could not get out; the ships outside could not get in; and the maritime commerce of the world would be thrown into dire confusion.

It is the moon which, principally, causes the daily ebb and flow of the tide, and this is the most important work which the moon has to do. Fleets of fishing boats around our coasts time their daily movements by the tide, and are largely indebted to the moon for bringing them in and out of harbour. Experienced sailors assure us that the tides are of the utmost service to navigation. The question how the moon causes the tides, must be postponed to a future chapter, where we shall also sketch the marvellous part which the tides seem to have played in the past history of our earth.

Who is there who has not watched, with admiration, the beautiful series of changes through which the moon passes every month? We first see her as an exquisite crescent of pale light in the western sky after sunset. Night after night she moves further and further to the east, until she becomes full, and rises about the same time that the sun sets. From the time of full moon the disc of light begins to diminish until the last quarter is reached. Then it is that the moon is seen high in the heavens in the morning. As the days pass by, the crescent shape is again assumed. The crescent wanes thinner and thinner as the moon draws closer to the sun. Finally she becomes lost in the overpowering light of the sun, again to emerge as the new moon, and again to go through the same cycle of changes.

The brilliancy of the moon arises solely from the light of the sun, which falls on the dark or not self-luminous substance of the moon. Out of the vast flood of light, which the sun pours forth with such prodigality into space, the dark body of the moon intercepts a little, and of that little it reflects a small fraction to illuminate the earth. The moon sheds so much light, and seems so bright, that it is often difficult at night to remember that the moon has no light except what falls on it from the sun. Nevertheless, the actual surface of the brightest full moon is perhaps not much brighter than the streets of London on a clear sunshiny day. A very simple observation will suffice to show that the moon's light is only sunlight. Look some morning at the moon in daylight, and compare the moon with the clouds. The brightness of the moon and of the clouds are directly comparable, and then it is seen plainly that the sun which illuminates the clouds has also illumined the moon. The attempt has been made to measure the relative brightness of the sun and the full moon. If 600,000 full moons were shining at once, their collective brilliancy would be equal to that of the sun.

The beautiful crescent moon has furnished a theme for many a poet. Indeed, if we may venture on the most gentle criticism, it would seem that some poets have forgotten that the moon is not to be seen every night. A poetical description of evening is almost certain to be associated with a description of the moon in some phase or other. We may cite one notable instance in which a poet, describing an historical event, has enshrined in exquisite verse a statement which cannot have been correct. Every child that speaks our language has been taught that the burial of Sir John Moore took place

"By the struggling moonbeams' misty light."

There is an appearance of detail in this statement which wears the garb of truth. We are not going to doubt that the night was really misty, and inquire whether the moonbeams really had to struggle into visibility, for the question is a much more fundamental one. We do not know who was the first to raise the point as to whether the moonbeams really shone on that memorable event at all or not; but the question having been raised, the Nautical Almanac immediately supplies an answer. From it we learn in language, whose truthfulness constitutes its only claim to be poetry, that the moon was new on the 16th January, 1809, at one o'clock in the morning of the day of the battle of Corunna. It is evidently implied in the ballad that the funeral took place on the night following the battle. We are therefore assured that the moon can hardly have been a day old when the hero was consigned to his grave. But the moon in such a case is practically invisible, and yields no appreciable moonbeams at all, misty or otherwise. Indeed, if the funeral took place at the "dead of night," as the poet tells us it did, the moon must have been far below the horizon at the time.*

In alluding to this and similar instances, Mr. Nasmyth gives a word of advice to authors or to artists who desire to bring the moon on a scene without knowing as a matter of fact that the moon was actually present. He recommends them to follow the example of Bottom in *Midsummer-Night's Dream,* and consult "a calendar,

* Some ungainly critic has observed that the poet himself seems to have felt a doubt on the matter, because he has supplemented the dubious moonbeams by the "lantern dimly burning." The more generous, if somewhat sanguine remark, has been also made, that "the time will come when the evidence of this poem will prevail over any astronomical calculations."

a calendar! look in the almanac; find out moonshine, find out moonshine!"

Among the countless host of celestial bodies—the sun, the moon, the planets, and the stars—the moon enjoys one special claim on our attention. The moon is our nearest neighbour. It is just possible that a comet may occasionally dart past the earth at a smaller distance than the moon, but with this exception the other

Fig. 21.—Comparative sizes of the Earth and Moon.

bodies are all hundreds or thousands, or even many millions of times as far off as the moon.

The moon is really one of the smallest objects visible to us which the heavens contain. Every one of the thousands of stars visible with the unaided eye is enormously larger than the moon. The brilliancy and apparent size of the moon arises from the fact that she is only 240,000 miles away, which is a distance almost immeasurably small when compared with the distances of the stars or other great bodies of the universe.

Fig. 21 exhibits the relative sizes of the earth and the moon. The small globe represents the moon, while the larger globe represents the earth. When we measure the actual diameters of the two globes, we find that of the earth to be 7,918 miles, and of the moon 2,160 miles, so that the diameter of the earth is nearly four times as great as the diameter of the moon. If the earth were cut into fifty pieces, all equally large, then one of these pieces rolled into a globe would equal the size of the moon. The superficial extent of the moon is equal to about one-thirteenth part of the surface of the earth. The hemisphere of the moon turned towards us exhibits at any moment an area equal to about one twenty-seventh part of the area of the earth. This, to speak approximately, is about half the area of Europe. The materials of the earth are, however, much heavier than those contained in the moon. It would take more than eighty globes, each as ponderous as the moon, to weigh down the earth.

Amid the incessant changes which the moon presents to us, one obvious fact stands forth prominently. Whether the moon be new or full, at first quarter or at last, whether it be high in the heavens or low near the horizon, whether it be in process of eclipse by the sun, or whether the sun himself is being eclipsed by the moon, one feature remains invariable—the apparent size of the moon is nearly constant. We can express the matter numerically. A globe one foot in diameter, at a distance of 110 feet from the observer, would under ordinary circumstances be just sufficient to hide the disk of the moon; occasionally, however, the globe would have to be brought in to a distance of only 101 feet, or occasionally it might have to be moved out to as much as 119 feet if the moon is to be exactly hidden. It is unusual for either of these limits to be approached, and the distance at which the globe must be situated so as to exactly cover the moon is usually more than 105 feet, and less than 115 feet. These fluctuations in the apparent size of the moon are contained within such narrow limits, that in the first glance at the subject they may be overlooked. It will be easily seen that the apparent size of the moon must be connected with its real distance from the earth. Suppose, for the sake of illustration, that the moon were to recede into space, its size would seem to dwindle, and long ere it had reached the distance of even the very nearest of the other celestial bodies, it would have shrunk into insignificance. On the other hand, if the moon were to come nearer to the earth, its apparent size would gradually increase, until, when close to the earth, it would seem like a mighty continent stretching over the sky. We find that the apparent size of the moon is nearly constant, and hence we infer that the real distance of the moon is also constant. The average value of that distance is 240,000 miles. It may approach under rare circumstances to a distance but little more than 220,000 miles; it may recede under rare circumstances to a distance hardly less than 260,000 miles, but the ordinary fluctuations do not exceed more than about 13,000 miles on either side of its mean value.

From the moon's incessant changes we perceive that she is in constant motion, and we now further see that whatever these movements may be, the earth and the moon must still always remain at nearly the same distance apart. If we further add that the path pursued by the moon around the heavens lies in a plane, then we are forced to the conclusion that the moon must be revolving in a nearly circular path around the earth at the centre. It can, indeed, be shown that the constant distance of the two bodies involves as a necessary condition the revolution of the moon around the earth. The attraction between the moon and the earth tends to bring the two bodies together. The only way by which such a catastrophe can be permanently avoided is by causing the moon to revolve around the earth. The attraction between the earth and the moon still exists, but its effect is not then shown in bringing the moon in towards the earth. The attraction has now to exert its whole power in keeping the moon in its circular path; were the attraction to cease the moon would start off in a straight line, and recede never to return.

The fact of the moon's revolution around the earth is easily demonstrated by observations of the stars. The rising and setting of the moon is of course due to the rotation of the earth, and this apparent diurnal movement the moon has in common with the sun and with the stars. It will, however, be noticed that the moon is continually changing its place among the stars. Even in the course of a single night, the displacement of the moon will be conspicuous to a careful observer, without the aid of a telescope. The moon completes its revolution in 27·3 days.

In Fig. 22 we have a view of the relative positions of the earth, the sun, and the moon, but it is to be observed that the

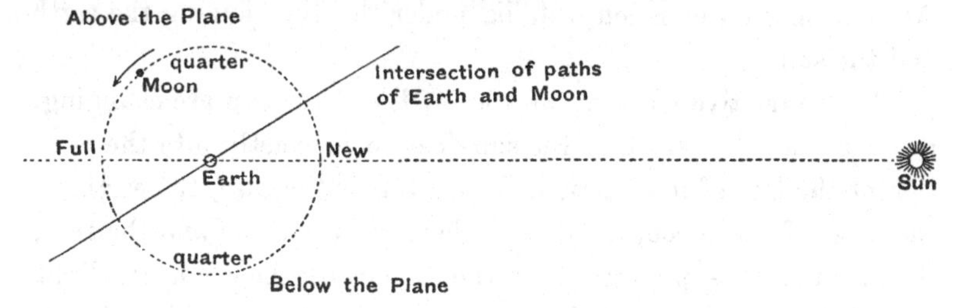

Fig. 22.—The Moon's path around the Earth.

distance of the sun is really much greater than can possibly be represented in the figure. That half of the moon which is turned towards the sun is brilliantly lighted up, and according as we see more or less of that brilliant half we say that the moon is more or less full, all the "phases" being visible in succession as shown by

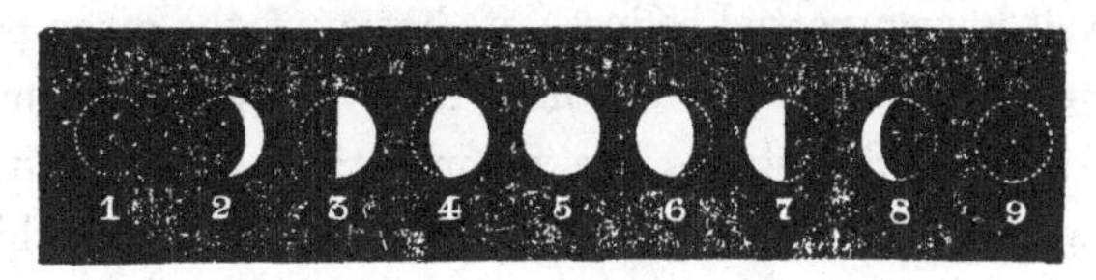

Fig. 23.—The Phases of the Moon.

the numbers in Fig. 23. A beginner sometimes finds a difficulty in understanding how a full moon at night is really lighted by the sun. "Is not," he will say, "the earth in the way? and must it not cut off the sunlight from every object on the other side of the earth to the sun?" A study of Fig. 22 will explain the difficulty. The plane in which the moon revolves does not coincide with the plane in which the earth revolves around the sun. The line in which the plane of the earth's motion is intersected by that of the moon divides the moon's path into two semicircles. We must imagine the moon's path to be tilted a little, so that the upper semicircle is somewhat above the plane of the paper, and the other semicircle below. It thus follows that when the moon is in the position marked full, under the circumstances shown in the figure, the moon will be just above the line joining the earth and sun; the sunlight will thus pass over the earth to the moon, and the moon will be illuminated. At new moon the moon will be under the line joining the earth and the sun.

As the relative positions of the earth and the sun are changing, it sometimes happens that the sun does come exactly into the position of the line of intersection. When this is the case, the earth, at the time of full moon, lies directly between the moon and the sun; the moon is thus plunged into the shadow of the earth, the light from the sun is intercepted, and we say that the moon is eclipsed. The moon sometimes only partially enters the sun's shadow, in which case the eclipse is a partial one. When, on the other hand, the sun is situated on the line of intersection at the time of new moon, the moon lies directly between the earth and the sun, and the dark body of the moon then cuts off the sunlight from the earth, producing a solar eclipse. Usually only a part of the sun is thus obscured, forming the well-known partial eclipse; if, however, the moon pass centrally over the sun, then we may have either of two very remarkable kinds of eclipse. Sometimes the moon entirely blots out the sun, and then we have the sublime spectacle of a total eclipse, which tells us so much as to the nature of the sun, and to which we have already referred in the last chapter. Occasionally, however, even when the moon is placed centrally over the sun, a thin rim of sunlight is seen round the margin of the moon. We then have what is known as an annular eclipse. It is very remarkable that the moon is sometimes able to completely hide the sun, and sometimes fails to do so. It happens, curiously enough, that the average apparent size of the moon is equal to the apparent size of the sun, but owing to the fluctuations in their distances, the actual apparent sizes of both bodies undergo certain changes. It may happen that the apparent size of the moon is greater than that of the sun. In this case a central passage produces a total eclipse; but it may also happen that the apparent size of the sun exceeds that of the moon, in which case a central passage can only produce an annular eclipse.

There are hardly any more interesting celestial phenomena than the different descriptions of eclipses. The almanac will always give timely notice of the occurrence, and the more striking features can be observed without a telescope. In an eclipse of the moon (Fig. 24) it is interesting to note the moment when the black shadow is first detected, to watch its gradual encroachment over the bright surface of the moon, to follow it, in case the eclipse is total, until there is only a thin crescent of moonlight left, and to watch the final extinction of that crescent when the whole moon is plunged into the shadow. But now a spectacle of great interest and beauty is often manifested; for though the moon is so hidden behind the earth that not a single direct ray of the sunlight could reach the surface, yet it is often found that the moon remains visible, and indeed actually glows with a copper hue bright enough to permit several of the markings on the surface to be seen. Whence this light? It is due to the sunbeams which have just grazed the edge of the earth. In doing so they have become bent by the refraction of the atmosphere, and thus turned inwards into the shadow. Such beams have passed through a prodigious thickness of the earth's atmosphere, and in this long journey through hundreds of miles of air they have become tinged with a ruddy or copper hue. Nor is this property of our atmosphere an unfamiliar one. Does not the sun at sunrise or at sunset glow with a light much more ruddy than the beams it dispenses at noonday? But

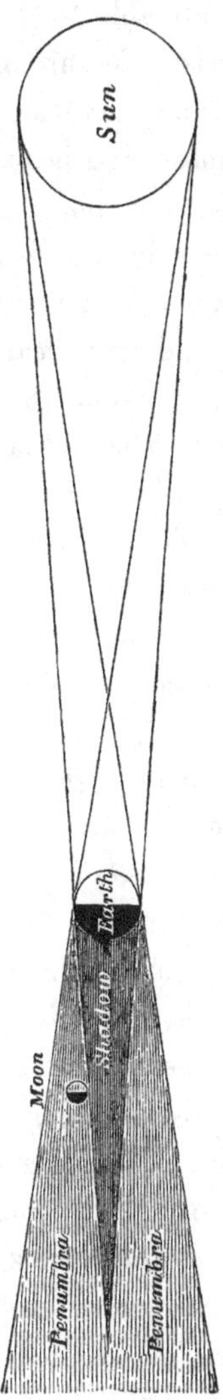

Fig. 24.—Form of the Earth's Shadow, showing the Penumbra or partially-shaded region. Within the Penumbra the Moon is visible; in the shadow it is nearly invisible.

at sunset or at sunrise the rays have a vastly greater mass of atmosphere to penetrate than they have at noon, and accordingly they are able to give to the rays the ruddy colour which is a characteristic feature of sunset. In the case of the eclipsed moon, the sunbeams have an atmospheric journey double as great as that at sunset, and hence the ruddy glow of the moon is accounted for.

The almanacs give the full particulars of each eclipse that happens in the corresponding year. They are able to do this because astronomers have been carefully observing the moon for ages, and have learned from these observations not only how the moon moves at the present, but also how it will move for ages to come. The actual calculations are very troublesome and complicated, but there is one leading principle about eclipses which is so simple that we must refer to it. The eclipses occurring this year have no very obvious relation to the eclipses that occurred last year, or to those that will occur next year. Yet, when we take a more extended view of the sequence of eclipses, a very definite principle becomes manifest. If we observe all the eclipses in a period of eighteen years, or nineteen years, then we can predict future eclipses for a long time. It is only necessary to recollect that in 6,585⅓ days after one eclipse, a nearly similar eclipse follows. For instance, a beautiful eclipse of the moon occurred on the 5th of December, 1881. If we count back 6,585 days from that date, or, that is, eighteen years and eleven days, we come to November 24th, 1863, and a similar eclipse of the moon took place then. Again, there were four eclipses in the year 1881. If we add 6,585⅓ days to the date of each eclipse, it will give the dates of all the four eclipses in the year 1899. It was this rule which enabled the ancient astronomers to predict the recurrence of eclipses, before they understood the motions of the moon nearly as well as we do now.

During a long voyage, and perhaps under critical circumstances, the moon will often render the sailor the most invaluable information. To navigate a ship, suppose from Liverpool to China, the captain must frequently determine the precise position which his ship then occupies. If he could not do this, he would never find his way across the trackless ocean. It is in the first place by observations of the sun that the place of the ship is found; but in addition to these observations, which tell him his local time, the captain requires to know the Greenwich time before he can place his finger at a point of the chart and say, "My ship is here." To ascertain the Greenwich time, the ship carries a chronometer, which has been carefully rated before starting, and, as a precaution, two or three or more chronometers are usually employed; for an unknown error in the chronometer will be a very serious or dangerous matter. Every minute that the chronometer is wrong may perhaps put the captain fifteen miles out of his reckoning. It is therefore sometimes of importance to have the means of testing the chronometer; and it would be a great convenience if every captain, when he wished, could actually consult some infallible standard of Greenwich time. We want, in fact, a Greenwich clock which may be visible over the whole globe. There is such a clock; and, like any other clock, it has a face on which certain marks are made, and a hand which travels round that face. The great clock at Westminster shrinks into insignificance when compared with the mighty clock which the captain uses for setting his chronometer. The face of this stupendous dial is the surface of the heavens. The numbers engraved on the face of a clock are replaced by the twinkling stars; while the hand which moves over the dial is the beautiful moon herself. When the captain desires to test his chronometer, he measures the distance of the moon from a neighbouring star. For example, he may see that the moon is three degrees from the star Regulus. In the Nautical Almanac he finds the Greenwich time at which the moon was three degrees from Regulus. Comparing this with the indications of the chronometer, he finds the required correction.

We owe much to the moon. We hope, indeed, in a subsequent chapter, to point out that we owe a great deal more to her than was formerly suspected; but there is one widely-credited myth about the moon which must be regarded as devoid of real foundation. The idea that the moon and the weather are connected has no doubt been entertained by high authority, but careful comparison has shown that there is no definite connection between the two.

We often notice large blank spaces on maps of Africa and of Australia, which indicate our ignorance of the interior of those

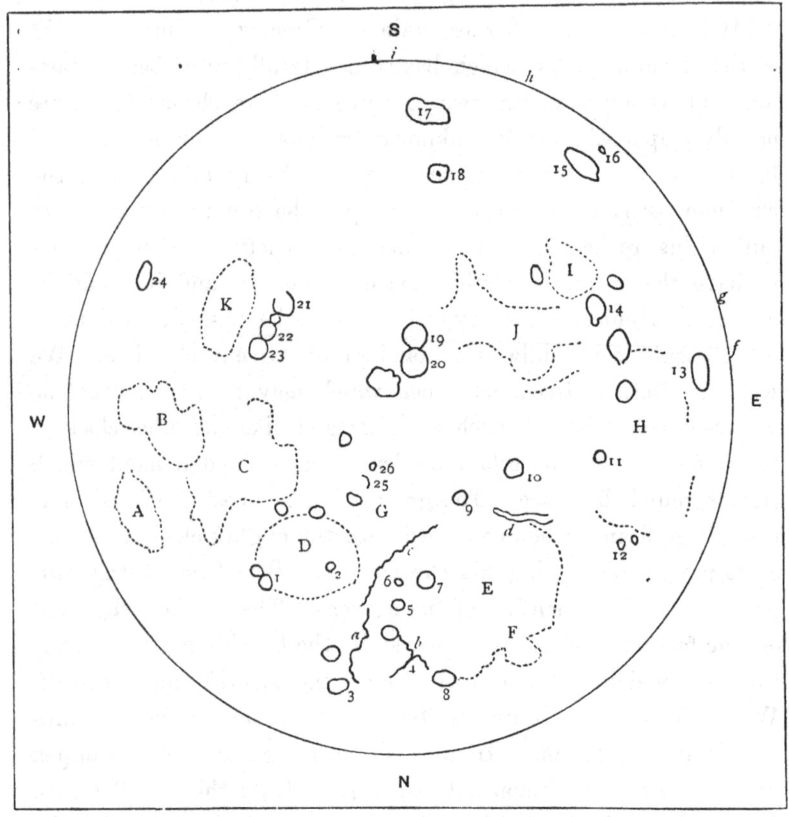

Fig. 25.—Key to Chart of the Moon (Plate VI.)

great continents. We can find no such blank spaces in the map of the moon. Astronomers know the surface of the moon better than the geographers know the interior of Africa. Every spot on the face of the moon which is as large as an English parish, has been surveyed and mapped, and in many cases even christened.

The map of the moon shown in Plate VI., and based upon drawings made with small telescopes, gives a general view of that side

PLATE VI.

CHART OF THE MOON'S SURFACE.

of the moon which is presented to the earth. We can see from it the most characteristic features of lunar scenery. Those dark regions so well known in the ordinary full moon are easily recognised on the map. They were originally thought to be seas, and indeed they still retain the name, though it is obvious that they contain no water. The map also shows certain ridges or elevated portions, and when we apply measurement to these objects we learn that they are mighty mountain ranges. But the most striking features in the map are those ring-like objects which are scattered over the surface in profusion. These are known as the lunar craters.

To facilitate reference to the different points of interest we have arranged an index map, which, upon comparison with the plate of the moon, will give a clue to the names of the different objects. The seas are represented by capital letters; thus, for instance, A is the Mare Crisium, and H the Oceanus Procellarum. The ranges of mountains are indicated by small letters; thus *a* on the index indicates the so-called Caucasus mountains, and the Apennines are denoted by *c*. The remarkable objects of such varied size known as craters are distinguished by numbers; thus the feature on the map lying under 1 on the index is the crater Posidonius.

A. Mare Crisium.
B. ,, Fœcunditatis.
C. ,, Tranquillitatis.
D. ,, Serenitatis.
E. ,, Imbrium.
F. ,, Sinus Iridium.
G. ,, Vaporum.
H. Oceanus Procellarum.
I. Mare Humorum.
J. ,, Nubium.
K. ,, Nectaris.

---

a. Caucasus mountains.
b. Alps ,,
c. Apennine ,,
d. Carpathian ,,
f. Cordilleras and D'Alembert mountains.
g. Rook mountains.
h. Dœrfel ,,
i. Leibnitz ,,

---

1. Posidonius.
2. Linné.
3. Aristotle.
4. Great Valley of the Alps.
5. Aristillus.
6. Autolycus.
7. Archimedes.
8. Plato.
9. Eratosthenes.
10. Copernicus.
11. Kepler.
12. Aristarchus.
13. Grimaldi.
14. Gassendi.
15. Schickard.
16. Wargentin.
17. Clavius.
18. Tycho.
19. Alphons.
20. Ptolemy.
21. Catharina.
22. Cyrillus.
23. Theophilus.
24 Petavius.
25. Hyginus.
26. Triesnecker.

Lunar objects can only be well seen when the sunlight falls upon them in such a manner as to exhibit strongly-contrasted light and shade. It is on this account impossible to observe the moon well at the time when it is full, for then no conspicuous shadows are cast. The best time for seeing any object usually coincides with the time when the boundary between light and shade passes through the neighbourhood, and then the features are brought out with exquisite distinctness.

Plate VII.* gives an illustration of the lunar scenery, the object represented being known to astronomers by the name Triesnecker. The district included is only a very small fraction of the entire surface of the moon, yet the actual area is very considerable, embracing as it does many hundreds of square miles. We see in it various ranges of lunar mountains, while the central object in the picture is one of those remarkable lunar craters which are the characteristic features of lunar scenery. This crater is about twenty miles in diameter, and it has a lofty mountain in the centre, the peak of which is just shown, illuminated by the rising sun.

A typical view of a lunar crater is shown in Plate VIII. This is no doubt a somewhat imaginary sketch. The point of view from which the artist is supposed to have taken the picture is one quite unattainable by terrestrial astronomers, yet there can be little doubt that it is a fair representation of the objects on the moon. We should, however, recollect the scale on which it is drawn. The vast crater must be many miles across, and the mountain at its centre must be thousands of feet high. The telescope will, even at its best, only show the moon as well as we could see it with the unaided eye if it were only about 250 miles away instead of being 240,000. We must not, therefore, expect to see any details on the moon even with the finest telescopes, unless they were coarse enough to be visible at a distance of 250 miles. A view of England from a distance of 250 miles would only show London as a sort of coloured spot, by contrast with the general surface of the country.

* This sketch has been copied by permission from the very beautiful view in Messrs. Nasmyth and Carpenter's book, of which it forms Plate XI. The other illustrations of lunar scenery in Plates VII., VIII., IX., are from the same. The photographs were obtained from models carefully constructed to illustrate the features on the moon.

We return, however, from a somewhat fancy sketch to a more prosaic examination of what the telescope does actually reveal. Plate IX. represents a large crater which is well known to every user of a telescope, and which has been christened Plato. The floor of the crater of Plato is very nearly flat, and in this case the central mountain, so often seen in other craters, is wanting.

The mountain peaks on the moon throw long, sharply defined shadows. These shadows are indeed characterised by a definiteness which we do not find in terrestrial shadows, the difference no doubt arising from the absence of air from the moon. The air on the earth diffuses a certain amount of light, which mitigates the blackness of terrestrial shadows, and tends to soften their outline. No such influences are at work on the moon, and the sharpness of the shadows is taken advantage of in our attempts to measure the heights of the lunar mountains.

It is often easy to compute the altitude of a church steeple, a lofty chimney, or a similar object, by measuring the length of its shadow. The simplest and the most accurate process is to measure at noon the number of feet from the base of the object to the end of the shadow. The elevation of the sun on the day in question can be obtained from the almanac, and then the length of the shadow follows by a simple calculation. Indeed, if the observations were made on the 4th of March or the 6th of September at London, the calculation would then be unnecessary, for the noonday length of the shadow is equal to the altitude of the object. In summer the length of the shadow is less than the altitude; in winter the length of the shadow exceeds the altitude. At sunrise or sunset the shadows are, of course, much longer than at noon, and it is shadows of this kind that are measured on the moon. These measures are made by that most indispensable adjunct to the equatorial telescope known as the micrometer.

The micrometer, in its most ordinary form, is a small piece of apparatus which can be screwed on to the eye end of the telescope. The word denotes an instrument for measuring *small* distances. In one sense the term is not a very happy one. The objects to which the astronomer applies the micrometer are usually anything but small objects. They are, in fact, often objects of the most transcendent dimensions, vastly exceeding the moon or the sun, or even our whole system, in bulk. Still, the name is not altogether inappropriate, for vast though the objects may be, they yet seem minute, even in the telescope, on account of their great distance. We require, therefore, for such measures an instrument capable of the very greatest nicety. Here, again, we invoke the aid of the spider, to whose assistance in another department we have already referred. In the filar micrometer two spider lines are parallel, and one intersects them at right angles. One or both of the parallel lines can be moved by means of a screw. The distance through which the line has been moved is accurately indicated by noting the number of revolutions and parts of a revolution of the screw. Suppose, then, the two lines be first brought into coincidence, and then moved apart until the apparent length of the shadow of the mountain be equal to the distance of the lines: we then know the number of revolutions of the micrometer screw, which is equivalent to the length of the shadow. The value of the screw is known by other observations, and hence the length of the shadow can be determined, and its equivalent in miles can be ascertained. The elevation of the sun, at the moment when the measures were made, is also found, and hence the actual elevation of the mountain can be calculated. By measures of this kind the height of the rampart surrounding a crater on the moon can be learned.

The beauty and interest of the moon as a telescopic object induces us to give to the student a somewhat detailed account of the more remarkable features which it presents. Most of the objects we are to describe can be effectively seen with only a very moderate telescopic power. It is, however, to be observed that all of them cannot be well seen at one time. The student must remember that the region most distinctly shown is the boundary between light and darkness. He will, therefore, select for observation such objects as may happen to lie on or near that boundary at the time when he is observing. We shall here follow the numbering on the key to the map.

1. *Posidonius.*—The diameter of this large crater is nearly 60 miles. Although its surrounding wall is comparatively narrow, it is so distinctly marked as to make the object very conspicuous. As so frequently happens in lunar volcanoes, the bottom of the crater is below the level of the surrounding plain, in the present instance to the extent of nearly 2,500 feet. Towards the close of last century, Schroeter, the industrious Hanoverian observer, fancied he saw traces of activity in the little crater on the floor of Posidonius. It is remarkable that he described it as being only grey in the interior, at a time when it ought to have been full of black shadow. This is particularly interesting in connection with what has been observed in the next object.

2. *Linné.*—This small crater lies right out in the Mare Serenitatis. Fifty or sixty years ago it was described as about 6½ miles in diameter, and otherwise so conspicuous as to be used by two astronomers as a fundamental point in the survey of the moon. In 1866 Schmidt, of Athens, announced that the crater was obscured, apparently by cloud. Since then an exceedingly small crater has been visible, but the whole object is now so inconspicuous that it would hardly be chosen as a landmark. On the whole, this is the most clearly proved case of present change in a lunar object.

3. *Aristotle.*—This great philosopher's name is attached to a grand crater 50 miles in diameter, the interior of which, although very hilly, shows no decidedly marked central cone. But the lofty wall of the crater, exceeding 10,500 feet in height, overshadows the bottom so long that its irregularities are never seen to advantage.

4. *The Great Valley of the Alps.*—Right through the lunar Alps runs this wonderfully straight valley, with a width ranging from 3½ to 6 miles. It is, according to Mädler, at least 11,500 feet deep, and over 80 miles in length. A few low ridges run parallel to its sides, possibly the result of landslips.

5. *Aristillus.*—Under favourable conditions Lord Rosse's great telescope has shown the exterior of this magnificent crater to be marked over with deep gullies, radiating from its centre. It is about 34 miles wide, and about 10,000 feet in depth.

6. *Autolycus* is somewhat smaller than the foregoing, to which it forms a companion in accordance with what Mädler thought a well defined relation amongst lunar craters, by which they frequently occurred in pairs, with the smaller one more usually to the south. This does indeed occur repeatedly, but towards the edge the arrangement is generally rather apparent than real, and is merely a result of foreshortening.

7. *Archimedes.*—This large plain, about 50 miles in diameter, has its vast smooth interior divided into seven distinct zones running east and west. There is no central mountain or other obvious internal sign of former activity, but its irregular wall rises into abrupt towers, and is marked outside by decided terraces.

8. *Plato.*—In the northern part of the moon this extensive "steel-grey" plain is noticeable with the smallest telescope. It is admirably represented in Plate IX. The average height of the rampart is about 3,800 feet on the eastern side; the western side is somewhat lower, but there is one peak rising to the height of nearly 7,300 feet. The plain girdled by this vast rampart is of ample proportions. It is a somewhat irregular circle, about sixty miles in diameter, and containing an area of 2,700 square miles. On its floor the shadows of the western wall are beautifully shown, as also three small craters, of which a large number have been detected by persevering observers. The narrow sharp line leading from the crater to the left is one of those remarkable "clefts" which traverse the moon in so many directions. Another may be seen further to the left. Above Plato are several detached mountains, the loftiest of which is Pico, about 8,000 feet in height. Its long and pointed shadow would at first sight lead one to suppose that it must be very steep; but Schmidt, who specially studied the inclination of lunar slopes, is of opinion that it is not nearly so steep as many of the Swiss mountains that are frequently ascended. To give some idea of Schmidt's amazing industry in lunar researches, it may be mentioned that in six years he made nearly 57,000 individual settings of his micrometer in the measurement of lunar altitudes. His great chart of the mountains in the moon is based on no less than 2,731

drawings and sketches, if those are counted twice that may have been used for two divisions of the map.

9. *Eratosthenes.*—This profound crater, upwards of 37 miles in diameter, lies at the end of the gigantic range of the Apennines, and not improbably, in accordance with Mädler's suggestion, it once formed the outlet of the stupendous forces that elevated those comparatively craterless peaks.

10. *Copernicus.*—Of all the lunar craters this is one of the grandest and best known. It is particularly well known through Sir John Herschel's drawing, so beautifully reproduced in the many editions of the "Outlines of Astronomy." The region to the west is dotted over with innumerable minute craterlets. It has a central many-peaked mountain about 2,400 feet in height. There is good reason to believe that the terracing shown in its interior is mainly due to the repeated alternate rise, partial congealation and retreat of a vast sea of lava. At full moon it is surrounded by radiating streaks.

11. *Kepler.*—Although the internal depth of this crater is scarcely less than 10,000 feet, it has but a very low surrounding wall, which is remarkable for being covered with the same glistening substance that also forms a system of bright rays, not unlike those surrounding the last object. It has been pointed out that this is the only instance in which these mysterious bright rays occur in a level part of the moon.

12. *Aristarchus* is the most brilliant crater in the moon, being specially vivid with a low power in a large telescope. So bright is it, indeed, that it has often been seen in the dark side of the moon, just after new moon, and has thus given rise to marvellous stories of active lunar volcanoes. To the south-east is another smaller crater, Herodotus, north of which is a narrow deep valley, nowhere more than 2½ miles broad, which makes a remarkable zigzag. It is one of the largest of the lunar "clefts."

13. *Grimaldi* calls for notice as the darkest object of its size in the moon. Under very exceptional circumstances it has been seen with the naked eye, and as its area has been estimated at nearly 14,000 square miles, it gives an idea of how little unaided vision

can discern in the moon; but it must be added that we always see Grimaldi considerably foreshortened.

14. The great crater *Gassendi* has been very frequently mapped on account of its very elaborate system of "clefts." At its northern end it communicates with a smaller but much deeper crater, that is often still filled with black shadow, after the whole floor of Gàssendi is illuminated.

15. *Schickard* is another large crater being very little less in area than the foregoing. Within its vast expanse Mädler detected 23 minor craters. It was with regard to this crater that Chacornac first pointed out that owing to the curvature of the surface of the moon, a spectator at the centre of the floor "would think himself in a boundless desert," the surrounding wall, although in part over 10,000 feet high, being entirely beneath the horizon.

16. Close to the foregoing is *Wargentin*. There can be little doubt that this is really a huge crater filled almost to the very brim with congealed lava.

17. *Clavius.*—Near the 60th parallel of lunar south latitude, lies this enormous enclosure, whose area is not less than 16,500 square miles. Both in its interior and on its walls are many peaks and secondary craters. The telescopic view of a sunrise upon the surface of Clavius is truly said by Mädler to be indescribably magnificent. One of the peaks rises not less than 24,000 feet above the bottom of one of the included craters. Mädler even expressed the opinion that in this wild neighbourhood there are craters so profound that no ray of sunlight ever penetrated their lowest depths, while, as if to compensate for this, there are peaks whose summits enjoy a mean day almost twice as long as their night.

18. If the full moon be viewed through even an opera-glass or the smallest hand-telescope, one crater is immediately seen to be conspicuous beyond all others, by reason of the brilliant rays or streaks that radiate from it. This is the majestic *Tycho*, 17,000 feet in depth and 50 miles in diameter (Plate X.). A peak 6,000 feet in height rises in the centre of its floor, while a series of terraces diversify its interior slopes, but it is the wonderful and profoundly mysterious

PLATE VII.

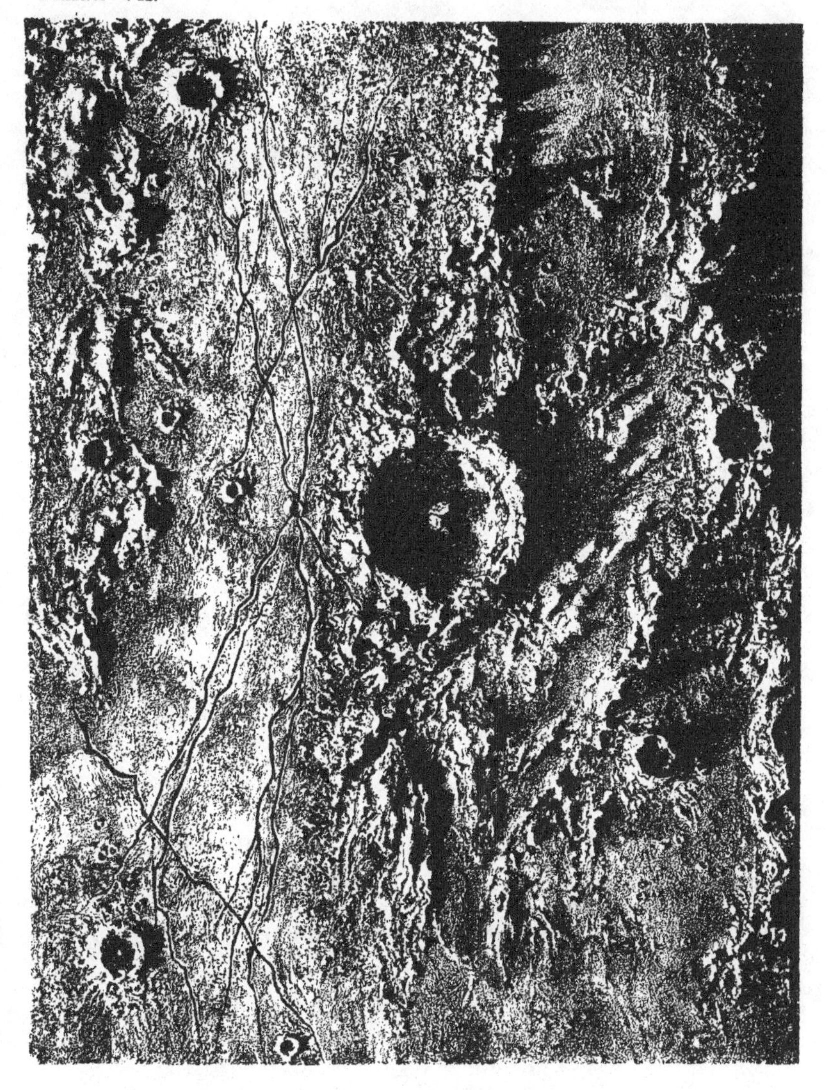

TRIESNECKER.

(AFTER NASMYTH.)

bright rays that chiefly call forth our admiration and surprise. When the sun rises on Tycho, these streaks are utterly invisible in its neighbourhood, so much so indeed that it then requires a practised eye to recognise Tycho amidst its mountainous surroundings. But as soon as the sun has attained a height of about 25° to 30° above the horizon, the rays emerge from their obscurity, and gradually increase in brightness until full moon, when they become the most conspicuous objects on her surface. As yet no satisfactory explanation has been given of the origin of these objects. They are of all lengths, from a few hundred to two, or in one instance nearly three thousand miles in length. They extend themselves with superb indifference across vast plains, into the deepest craters, or over the highest opposing ridges. We know of nothing on our earth to which they can be compared.

Near the centre of the moon's disc is a fine range of craters, always full open to our view under all illuminations, of these two may be mentioned—*Alphons* (19), the floor of which is strangely marked by two bright and several dark markings not due to irregularities in the surface. *Ptolemy* (20). Besides several small inclosed craters, its floor is crossed by very numerous low ridges, visible when the sun is rising or setting.

21, 22, 23.—When the moon is five or six days old a beautiful group of three craters will be readily found on the boundary line between night and day. These are *Catharina, Cyrillus,* and *Theophilus.* Catharina, the most southerly of the group, is more than 16,000 feet deep, and connected with Cyrillus by a wide valley; but between Cyrillus and Theophilus there is no such connection. Indeed Cyrillus looks as if its huge surrounding ramparts, as high as Mount Blanc, had been completely finished when the volcanic forces commenced the formation of Theophilus, the rampart of which encroaches considerably on its older neighbour. Theophilus stands as a well-defined round crater about 64 miles in diameter, with an internal depth of 14,000 to 18,000 feet, and a beautiful central group of mountains, one-third of that height on its floor. This proves that the last eruptive efforts in this part of the moon fully equalled in intensity those that had preceded them. Although Theophilus is on the whole the deepest crater we can see in the moon, it has received little or no deformation by secondary eruptions, while the floor and wall of Catharina show complete sequences of lesser craters of various sizes that have broken in upon and partly destroyed each other. In the spring of the year somewhat before the first quarter, this instructive group of extinct volcanoes can be seen to great advantage at a convenient hour in the evening.

24. *Petavius* is remarkable not only for its great size, but also for the rare feature of having a double rampart. It is a beautiful object in the very new moon, or just after full moon, but disappears absolutely when the sun is more than 45° above its horizon. The crater floor is remarkably convex, culminating in a central group of hills intersected by a deep cleft.

25. *Hyginus* is a small crater near the neighbourhood just mentioned. One of the largest of the lunar chasms passes right through it, making an abrupt turn as it does so. It is not difficult to find under any illumination.

26. *Triesnecker.*—This fine crater has been already described on p. 62, but is again alluded to in order to draw attention to the elaborate system of chasms so conspicuously shown in Plate VII. That these chasms are depressions is abundantly evident by the shadows inside them. Very often their margins are appreciably raised. They seem to be fractures in the moon's surface.

Of the various mountains that are occasionally seen as projections on the actual edge of the moon, those called after Leibnitz (*i*) seem to be the highest. Schmidt found the highest peak upwards of 41,900 feet above a neighbouring valley. In comparing these altitudes with those of mountains on our earth, we must for the latter add the depth of the sea to the height of the land. Reckoned in this way, our highest mountains are still higher than any we know of in the moon.

We must now discuss the very important question as to the origin of these remarkable features on the surface of the moon. We shall admit at the outset that our evidence on this subject is only indirect. To establish by unimpeachable evidence the volcanic origin of the remarkable lunar craters, it would be almost necessary that volcanic outbursts should have been witnessed on the moon through our telescopes, and that such outbursts should have been seen to result in the formation of the well-known ring, with or without the mountain rising from the centre of the plain that the ring surrounds. Have any such phenomena been witnessed by astronomers? To say that nothing of the kind has ever been witnessed would be rather too emphatic a statement. On certain occasions careful observers have reported the occurrence of minute local changes on the moon. As we have already remarked, a crater named Linné, of dimensions respectable, no doubt, to a lunar inhabitant, but forming a very inconsiderable telescopic object, was thought to have undergone some change. On another occasion a minute crater was thought to have arisen near to the well-known crater named Hyginus; but the mere enumeration of such instances gives real emphasis to the statement that there is at the present time no appreciable source of disturbance of the moon's surface. Even were these trifling cases of suspected change really established —and this is perhaps rather farther than most astronomers would be willing to go—they still are utterly insignificant when compared with the mighty phenomena that gave rise to the host of great craters which cover so large a portion of the moon's surface.

We are led inevitably to the conclusion that the surface of the moon once possessed much greater activity than is at present the case. We can also give a reasonable, or, at all events, a plausible, explanation of the cessation of that activity in recent times. Let us glance at two other bodies of our system, the earth and the sun, and compare them with the moon. Of the three bodies, the sun is enormously the greatest, while the moon is much less than the earth. We have also seen that the sun is heated to an enormous temperature, and further on in this work the reasons will be given for believing that the sun is gradually parting with its heat. The surface of the earth, formed as it is of solid rocks and clay, or covered as it is in great part by the vast expanse of ocean, bears but few obvious traces of a high temperature. Nevertheless, it is highly probable from ordinary volcanic phenomena that the interior of the earth still possesses an enormously high temperature. The argument is, then, as follows. A large body takes a longer time to cool than a small body. A large iron casting will take days to cool; a small casting will become cold in a few hours. Whatever may have been the original source of heat in our system—a question which need not now be discussed—it seems sufficiently plain that the different bodies in the system were all originally heated, and have now for ages been gradually cooling. The sun is so vast that he has not yet had time to cool; the earth, of intermediate bulk, has had time to cool on the outside, while still retaining vast stores of internal heat; while the moon, least of all, has had time to cool to such an extent that changes of importance on its surface can no longer be originated by internal fires.

We are thus led to refer the origin of the lunar craters to some ancient epoch in the moon's history. How ancient that epoch is to be we have no means of knowing, but it is possible to form a surmise that, in all probability, the antiquity of the lunar craters is enormously great. At the time when the moon was sufficiently heated to have these vast volcanic convulsions, of which the mighty craters are the survivals, the earth must have been very much hotter than it is at present. It is not, indeed, at all unreasonable to believe that when the moon was hot enough for its volcanoes to be active the earth was so hot that life was impossible on its surface. This supposition would point to an antiquity for the moon's craters far too great to be estimated by the centuries and the thousands of years which are adequate for the lapse of time as recognised by the history of human events. It seems not unlikely that millions of years may have elapsed since the mighty craters of Plato or of Copernicus consolidated into their present form.

It will now be possible for us to attempt to account for the formation of the lunar craters. The most probable views on the subject are certainly those adopted by Mr. Nasmyth, though it must be admitted that they are by no means free from difficulty. We can explain the way in which the rampart around the lunar crater is formed, and the great mountain which so often adorns the centre of the plain. The view in Fig. 26 contains an imaginary

PLATE VIII.

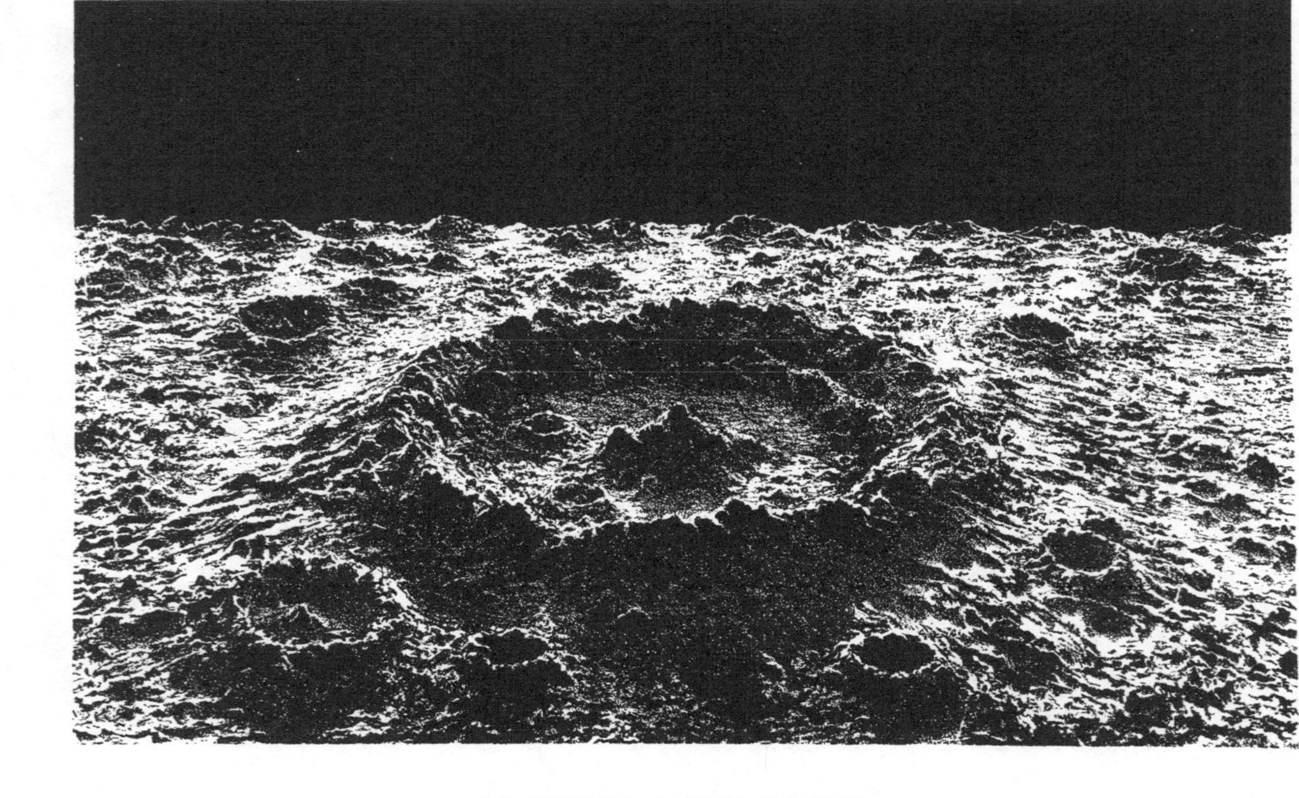

A NORMAL LUNAR CRATER.

(AFTER NASMYTH.)

sketch of a volcanic vent on the moon in the days when the craters were active. The eruption is here in the full flush of its energy,

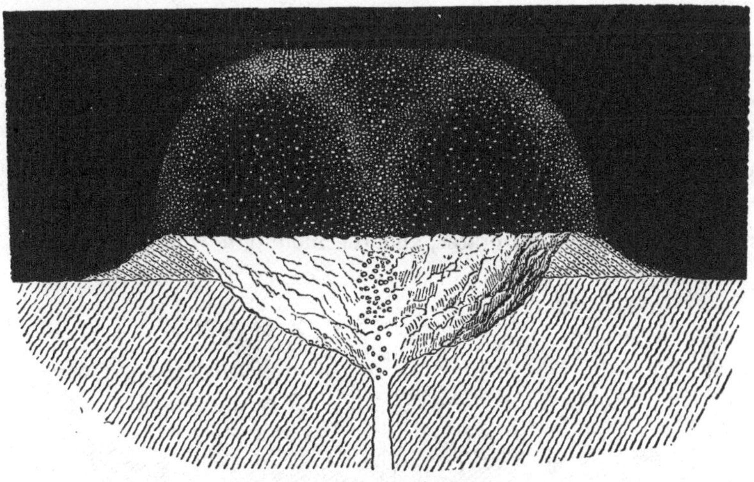

Fig. 26.—Volcano in Activity.

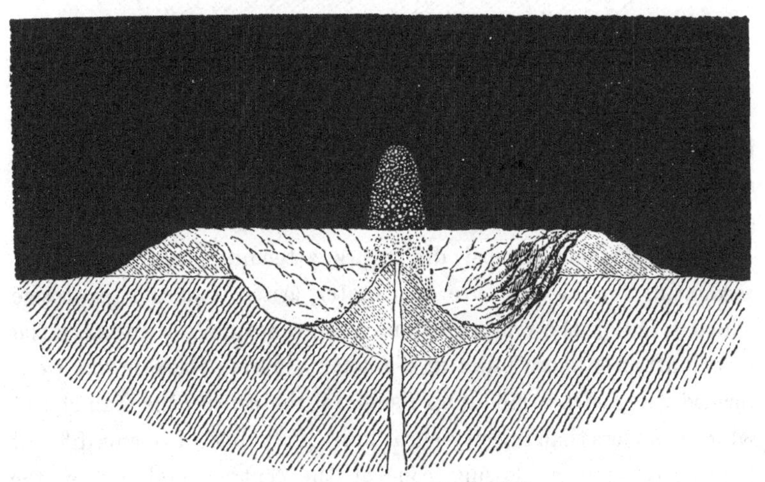

Fig. 27.—Subsequent Feeble Activity.

when the internal forces are hurling forth a fountain of ashes or stones, which fall down at a considerable distance from the vent; and these accumulations constitute the rampart surrounding the crater.

The second picture depicts the crater in a later stage of its history. The prodigious explosive power has now been exhausted, and has perhaps been intermitted for some time. Again, the volcano bursts into activity, but this time with only a small part of its original energy. A feeble jet now issues from the same vent and deposits the materials close around the orifice, and thus gradually raises a mountain in the centre. Finally, when all the activity has subsided, and the volcano is silent and still, we find the remains of the early energy testified to by the rampart which surrounds the ancient crater, and the mountain which rises in its

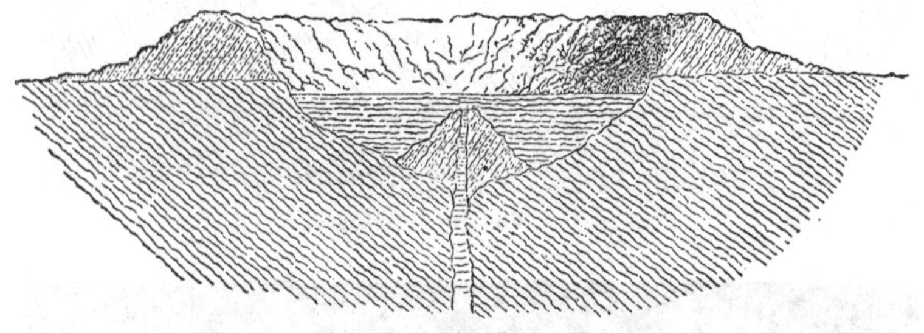

Fig. 28.—Formation of the Level Floor by Lava.

interior. The flat floor which is found in some of the craters may not improbably be due to an outflow of lava which has afterwards consolidated. A sketch of this is shown in Fig. 28.

One of the principal difficulties attending this method of accounting for the structure of a crater, arises from the great size which some of the craters attain. There are craters on the moon forty, fifty, or more miles in diameter; indeed, there is one well-formed ring, with a mountain rising in the centre, the diameter of which is no less than seventy-eight miles (Petavius). It seems difficult to conceive how a blowing cone at the centre could convey the materials to such a distance as the thirty-nine miles between the centre of Petavius and the rampart. The difficulty is, however, greatly smoothed away when it is borne in mind that the force of gravitation is much less on the moon than on the earth. Have we not already seen that the moon is so much smaller than the earth, that eighty moons rolled into one would not weigh as much as the earth? On the earth an ounce weighs an ounce and a pound weighs a pound; but owing to the small size of the moon its gravitation is much less than that on the surface of the earth. A weight of six ounces here would only weigh one ounce on the moon, and a weight of six pounds here would only weigh one pound on the moon. A labourer who can carry one sack of corn on the earth could, with the same exertion, carry six sacks of corn in the moon. A cricketer who can throw a ball 100 yards on the earth could with precisely the same exertion throw the same ball 600 yards on the moon. Hiawatha could shoot ten arrows into the air one after the other before the first reached the ground; on the moon he might have shot his whole quiver full. The volcano, which on the moon drove projectiles to the distance of thirty-nine miles, need only possess the same explosive power as would have been sufficient to drive the missiles six or seven miles on the earth. A modern cannon properly elevated would achieve this feat.

It must also be borne in mind that there are innumerable craters on the moon of the same general type, but of the most varied dimensions; from a tiny telescopic object two or three miles in diameter, we can proceed in gradually ascending stages until we reach the mighty Petavius just considered. With regard to the smaller craters, there is obviously little or no difficulty in attributing them to a volcanic source, and as the continuity from the smallest to the largest craters is unbroken, it seems quite reasonable to suppose that even the greatest has had a similar origin.

But we must not linger any longer over these most interesting features of the moon. Attractive and beautiful as is the spectacle of our nearest neighbour in the heavens, we have yet so long a task to accomplish that we must dismiss any further discussions of the subject. Even a single spot on the moon is capable of affording a diligent observer materials for many a long night's study. All such details we must pass over, and can only devote the brief space which remains to a few features of more general interest.

Why are these lunar landscapes so excessively weird and so

rugged? Why do they always remind us of sterile deserts, and why do we see no grassy plains or green forests such as we are familiar with on the earth? In some respects the moon is not very differently circumstanced to the earth. Like it, the moon has the pleasing alternations of day and night, though the day in the moon is fourteen of our days, and the night of the moon is as long as fourteen of our nights. We are warmed by the rays of the sun by day, so is the moon; but, whatever may be the temperature during the long day on the moon, it seems certain that the cold of the lunar night would transcend that known in the bleakest regions of our earth.

Even our largest telescopes can tell nothing directly as to whether life can exist on the moon. The mammoth trees of California might be growing on the lunar mountains, and elephants might be walking about on the plains, but our telescopes could not show them. The smallest object that we can see on the moon must be about as large as a good sized cathedral, so that organised beings, if they existed, could not make themselves visible as telescopic objects.

We are therefore compelled to resort to indirect evidence as to whether life would be possible on the moon. We may say at once that astronomers believe that life, as we know it, could not exist. Among the necessary conditions of life water is one of the first. Take every form of vegetable life, from the lichen which grows on the rock up to the giant tree of the forest, and we find the substance of every plant contains water, and could not exist without water. Nor is water less necessary for the existence of animal life. Without water animal life, the life of man himself, is inconceivable.

Unless, therefore, water be present in the moon we shall be bound to conclude that life, as we know it, is impossible. The great question then is, can we tell whether there is water in the moon or not? If any one stationed on the moon were to look at the earth through a telescope, would he be able to see any water here? Most undoubtedly he would. He would see the clouds and he would notice their incessant changes, and the clouds alone would be almost conclusive evidence of the existence of water. An astronomer on the moon would also see our oceans as smooth coloured surfaces, remarkably contrasted with the land. In fact, considering that much more than half of our globe is covered with oceans, and that most of the remainder is liable to be obscured by clouds, the lunar astronomer in looking at our earth would often see hardly anything but water in one form or other, and very likely he would come to the conclusion that our globe was quite unfitted to be a residence except for amphibious animals.

But when we look at the moon with our telescopes, do we see any traces of water? There are no doubt many large districts which at a first glance seem like oceans, and were indeed termed "seas" by the old astronomers, a name which they still absurdly retain. Closer inspection shows that the so-called lunar seas are deserts, often marked over with small craters and with rocks. The telescope reveals no seas and no oceans, no lakes and no rivers. Nor is the grandeur of the moon's scenery ever impaired by clouds over her surface. Whenever the moon is above the horizon, and terrestrial clouds are out of the way, we can see the features of her surface with distinctness. There are no clouds in the moon; there are not even the mists or the vapours which invariably arise wherever water is present, and therefore astronomers have been led to the conclusion that our satellite is a sterile and a waterless desert.

There is another equally essential element of life absent from the moon. Our globe is surrounded with a deep clothing of air resting on the surface, and extending above our heads to the height of about 200 or 300 miles. We need hardly say how necessary air is to life, and therefore we turn with interest to the question as to whether the moon is surrounded with an atmosphere. Let us clearly understand the problem we are about to consider. Imagine that a traveller started from the earth on a journey to the moon; as he proceeded the air would gradually get rarer and rarer, until at length, when he was a few hundred miles above the earth's surface, he would have left the last perceptible traces of the earth's atmosphere behind him; but even by the time he was completely free from the earth's atmosphere he would only have advanced a very small fraction of the whole journey of 240,000 miles, and there is still a vast void to be traversed before the moon is reached. If the moon were furnished with an atmosphere like the earth, as the traveller got near the end of his journey, and within a few hundred miles of the moon's surface, he would expect to meet again with traces of an atmosphere, which would gradually increase in density until he arrived at the moon's surface. The traveller would thus have passed through one atmosphere at the beginning of his journey, and another at the end, while the main portion of the journey would have been through space more void than the receiver of an air-pump.

Such would be the case if the moon were coated with an atmosphere like that surrounding our earth. But what are the facts? The traveller as he drew near the moon would look in vain for an atmosphere at all resembling ours. It is possible—indeed it is probable—that when he was close to the surface he might meet with faint traces of some gaseous covering surrounding the moon, but it would not be a fractional part of the ample clothing which the earth now enjoys. For all purposes of respiration, as we understand the term, we may say that there is no air on the moon, and an inhabitant of our earth transferred to the moon would be as certainly suffocated there as he would be in the middle of space.

It may, however, be asked how we learn this. Is not air transparent, and how, therefore, could our telescopes be expected to show whether the moon really possessed such an envelope? The fact is that it is by indirect methods of observation that we learn the nakedness of the moon. There are various arguments that can be adduced; but the most conclusive is that obtained on the occurrence of what is called the "occultation" of a star. It sometimes happens that the moon comes directly between the earth and a star, and the temporary extinction of the star is an "occultation." We can observe the moment when the occultation takes place, and the suddenness of the extinction of the star is extremely remarkable. If the moon had a copious atmosphere, the gradual interposition of this atmosphere by the movement of the moon would produce a gradual extinction of the star, and not the sudden phenomenon usually observed.

The absence of air and of water from the moon, explains the peculiar and weird ruggedness of the lunar scenery. We know that on the earth the action of wind, and of rain, of frost, and of snow, is constantly tending to wear down our mountains and reduce their asperities, but no such agents are at work on the moon. Volcanoes sculptured the moon into its present condition, and, though the volcanoes have been silent for ages, the traces of their handiwork seem nearly as fresh to-day as they were when the mighty fires were extinguished.

"The cloud-capped towers, the gorgeous palaces, the solemn temples" have but a brief career on earth. It is chiefly the incessant action of water and of air that makes them vanish like the "baseless fabric of a vision." On the moon these causes of disintegration and of decay are all absent, though perhaps the changes of temperature in the transition from lunar day to lunar night would be attended with expansions and contractions that might compensate in some degree for the absence of more potent agents of dissolution.

It seems probable that a building on the moon would remain for century after century just as it was left by the builders. There need be no glass in the windows, for there is no wind and no rain to keep out. There need not be fireplaces in the rooms, for fuel cannot burn without air. Dwellers in a city in the moon would find that no dust can rise, no odours be perceived, no sounds be heard.

Man is a creature adapted for life in circumstances which are very narrowly limited. A few degrees of temperature, more or less; a slight variation in the composition of air; the precise suitability of food; make all the difference between health and sickness, between life and death. Looking beyond the moon, into the length and breadth of the universe, we find countless celestial globes with every conceivable variety of temperature and of constitution. Amid this vast number of worlds with which space teems, are there any inhabited by living beings? To this great question science can make no response, save this; we cannot tell. Yet it is impossible to resist a conjecture. We find our earth teeming with life in every part. We find life under the most varied conditions that can be conceived. We have life under the burning heat of the tropics; we have life in the everlasting frost at the poles. We have life in caves where not a ray of light ever penetrates. We have life in the depths of the ocean, at the pressure of tons on the square inch. Whatever be the external circumstances, some form of life can generally be found to which those circumstances are congenial.

It is not at all likely that even among the million spheres of the universe there is a single one exactly like our earth; like it in the possession of air and of water, like it in size and in composition. It does not seem probable that a man could live for one hour on any body in the universe except the earth, or that an oak-tree could live in any other sphere for a single season. Men can dwell on the earth, and oak-trees can thrive there, because the constitutions of the man and of the oak are specially adapted to the particular circumstances of the earth.

Could we obtain a closer view of some of the celestial bodies, we should probably find that they, too, teem with life, but with life specially adapted to the environment. Life in forms strange and weird; life far stranger to us than Columbus found it to be in the New World when he first landed there. Life, it may be, stranger than ever Dante described, or Doré drew. Intelligence may yet have a home among those spheres no less than on the earth. There are globes greater and globes less—atmospheres greater and atmospheres less. And the truest philosophy on this subject is crystallised in the exquisite language of Tennyson :—

> "This truth within thy mind rehearse,
> That in a boundless universe
> Is boundless better, boundless worse.
>
> Think you this mould of hopes and fears
> Could find no statelier than his peers
> In yonder hundred million spheres?"

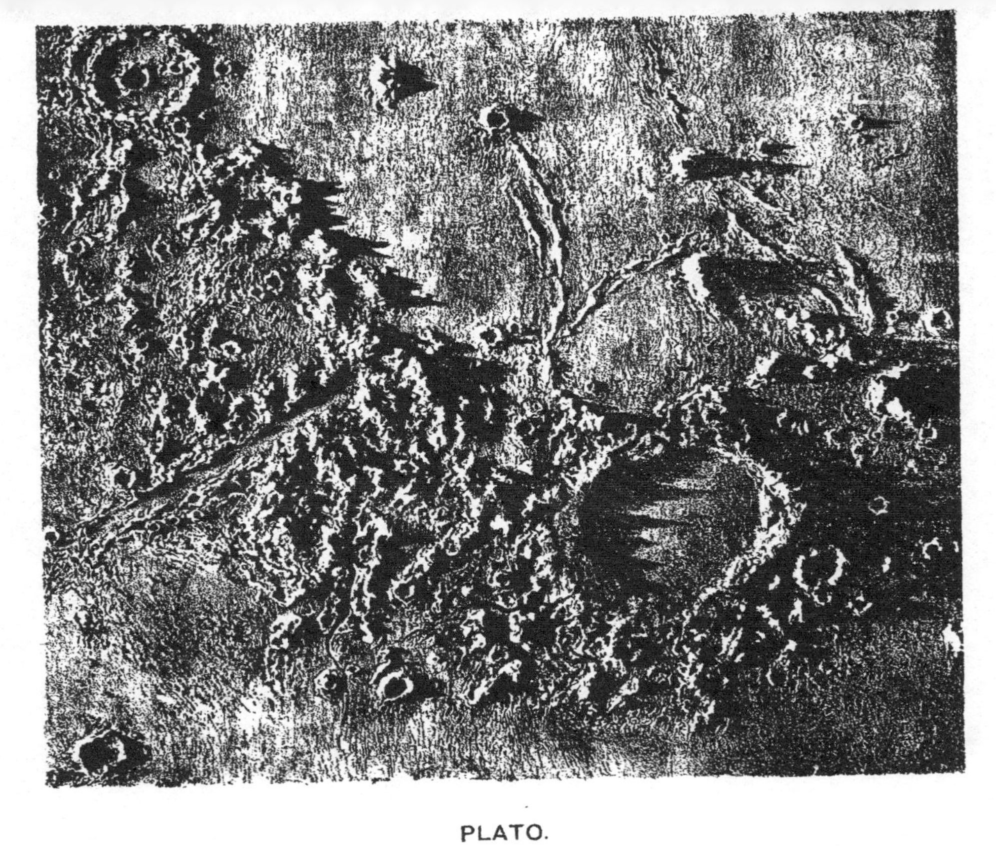

PLATO.

(AFTER NASMYTH)

CHAPTER IV.

THE SOLAR SYSTEM.

Exceptional Importance of the Sun and Moon—The Course to be pursued—The Order of Distance—The Neighbouring Orbs—How are they to be discriminated?—The Planets Venus and Jupiter attract notice by their brilliancy—Sirius not a neighbour—The Planets Saturn and Mercury—Telescopic Planets—The Criterion as to whether a Body is to be ranked as a neighbour—Meaning of the word *Planet*—Uranus and Neptune—Comets—The Planets are Illuminated by the Sun—The Stars are not—Ths Earth is really a Planet—The four Inner Planets, Mercury, Venus, the Earth, and Mars—Velocity of the Earth—The Outer Planets, Jupiter, Saturn, Uranus, Neptune—Light and Heat received by the Planets from the Sun—Comparative Sizes of the Planets—The Minor Planets—The Planets all Revolve in the same direction—The Solar System—An Island Group in Space.

In the two preceding chapters of this work we have endeavoured to describe the heavenly bodies in the order of their relative importance to mankind. Passing on from the preliminary chapters, which principally discussed the means by which we were enabled to observe the heavenly bodies, we then proceeded to describe those objects themselves. Could we hesitate for a moment as to which of the bodies in the universe should be the first to receive our attention? We do not now allude to the intrinsic significance of the sun when compared with other bodies or groups of bodies scattered through space. It may be that the sun has many orbs rivalling it in real splendour, in bulk, and in mass. We shall, in fact, show later on in this volume that this is the case; and we shall afterwards attempt to indicate the true rank of the sun amid the countless host of heaven. But whatever may be the importance of the sun, viewed merely as one of the bodies which teem through space, there can be no hesitation for a moment as to how immeasurably his influence on the earth surpasses that of all these bodies in the universe put together. It was therefore natural—indeed inevitable—that our first excursion into the abyss of space should be to explore that mighty body which is the source of our life itself.

Nor could there be very much hesitation as to the second step which ought to be taken. The intrinsic importance of the moon, when compared with other celestial bodies, may be small; it is, indeed, we shall afterwards see, almost infinitesimal. But in the economy of our earth the moon still plays, and has played, a part second only in importance to that of the sun himself. The moon is so close to us that her brilliant rays pale to invisibility countless orbs of a size and an intrinsic splendour incomparably greater. The moon also occupies quite an exceptional position in the history of astronomy; for the greatest discovery that science has yet witnessed was accomplished by means of her motion. It was therefore natural that an early chapter in our Story of the Heavens should be devoted to a body whose interest approximated so closely to that of the sun himself.

But the sun and the moon partly described (we shall afterwards find it necessary to return to both of them), some hesitation is natural in the choice of the next step. The two great luminaries being abstracted from our view, there is no other celestial body of such exceptional interest and significance as to make it quite clear what course to pursue, in the attempt to unfold the story of the heavens in the most natural manner. Did we attempt to describe the celestial bodies in the order of their actual magnitude, our ignorance must at once pronounce the attempt to be impossible. We cannot even make a conjecture as to which body in the heavens is to stand first on the list. Even if that mightiest body be within reach of our telescopes (in itself a highly improbable supposition), we have not the least idea in what part of the heavens it is to be sought. And even if this were possible—if we were able to arrange all the visible bodies rank by rank in the order of their magnitude and their splendour—still the scheme would be impracticable, for of most of them we know little or nothing.

We are therefore compelled to adopt a different method of procedure, and the simplest, as well as the most natural, will be to follow as far as possible the order of distance of the different bodies. We have already spoken of the moon as the nearest neighbour to the earth, we shall next consider some of the other celestial bodies which are comparatively near to us; then, as the work proceeds, we shall discuss the objects further and further away, until towards the close of the volume we shall be engaged in considering the most distant bodies in the universe which the telescope has yet revealed to us.

Even when we have decided on this principle, our course is still not free from doubt. Many of the bodies in the heavens are moving so that their distances from the earth change, but this is a difficulty which need not really detain us. We shall make no attempt to adhere closely to the principle in all its details. It will be sufficient if we first describe those great bodies—not a very numerous class—which are, comparatively speaking, in our vicinity, though still at vast and varied distances; and then we shall pass on to the almost infinitely numerous bodies which are separated from us by distances so stupendous that the imagination is baffled in the attempt to realise them.

Let us, then, scan the heavens, to choose therefrom those neighbouring orbs which are to form the material for our earliest consideration. The sun has set, the moon has not risen; a cloudless sky discloses a heaven glittering with countless gems of light. Some are grouped together into well-marked constellations; others scattered promiscuously hither and thither, with every degree of lustre, from the very brightest down to the faintest point that the eye can just discover. Amid all this host of objects, how are we to identify those which are nearest to the earth? Look to the west: the sun has but lately set, and over the spot where his departing beams still linger we see the lovely evening star shining forth. This is the planet Venus—a beauteous orb, twin-sister to the earth. The brilliancy of this planet, its rapid changes both in position and in lustre, would suggest at once that it was much nearer to the earth than other star-like objects. This presumption has been amply confirmed by careful measurements, and therefore Venus is to be included in the list of the orbs which constitute our neighbours.

Another conspicuous planet—almost rivalling Venus in lustre, and vastly surpassing Venus in the magnificence of its proportions and its retinue—has borne from all antiquity the majestic name of Jupiter. No doubt Jupiter is much more distant from us than Venus. Indeed, he is always at least twice as far away, while he is sometimes ten times as far. But still we must include Jupiter among our neighbours. Compared with the host of stars with which the heavens glitter, Jupiter must be regarded as quite near. The distance of Jupiter can only be expressed in hundreds of millions of miles; yet, vast as that distance is, it would have to be multiplied by tens of thousands, or hundreds of thousands, before it would be large enough to span the stupendous abyss which intervenes between the earth and the stars which form the constellations.

Venus and Jupiter have invited our attention by their far superior brilliancy. To them must now be added a few other bodies, in which brilliancy can hardly any longer be a reliable indication. An observer unacquainted with astronomy might not improbably point to the Dog Star—or Sirius, as astronomers more generally know it—as an object whose exceptional lustre showed it to be one of our neighbours, but this is not the case. We shall afterwards have occasion to refer to this gem of our northern skies with considerable detail, and then it will appear that Sirius is really a mighty globe far transcending in splendour our own sun, but plunged into the depths of space to such an appalling distance that his enfeebled rays, when they reach the earth, give us the impression, not of a mighty sun, but only of a brilliant star.

The principle of selection which we are to adopt will be explained presently; in the meantime, it will be sufficient to observe that our list is to be augmented first, by the addition of the unique object known as Saturn, whose brightness is, however, far surpassed by that of Sirius, as well as by a few other stars. Then we add Mars, an object which occasionally draws in close to the earth, and shines with a fiery radiance which would hardly prepare us for the truth that Mars is intrinsically one of the smallest of the celestial bodies. Besides the objects we have mentioned, the ancient astronomers had detected a fifth, known as Mercury—a planet which is usually invisible amid the light surrounding the sun. Mercury, however, occasionally wanders far enough from the sun to be seen before sunrise or after sunset, and was thus discovered ages ago. These five—Mercury, Venus, Mars, Jupiter, and Saturn —exhaust the list of planets so far as the ancient astronomers knew them.

We can, however, now extend the list somewhat further, by adding to it the telescopic objects which have in modern times been found to be justly entitled to be included among our neighbours. Here we must no longer postpone the introduction of the criterion by which we are to decide whether a body is near the earth or not. The planets can be to some extent recognised by the steady radiance of their light as contrasted with the incessant twinkling of the stars. A very small amount of attention devoted to any of the bodies we have named will, however, point out a more definite contrast between them and the stars. We observe, for instance, Jupiter, on any clear night when the stars can be well seen, and note carefully his position with regard to the stars in his neighbourhood—how he is to the right of this star, or to the left of that; directly between this pair of stars, or directly pointed to by that. We then either mark down the place of Jupiter on a map or a globe, or make a sketch of the stars in the neighbourhood showing the position of the planet. After a month or two, when the observations are to be repeated, the place of Jupiter is again to be compared with those stars by which it was before defined. It will be found that, while the stars have preserved their mutual positions, the place of Jupiter has changed. Hence it is that Jupiter is with propriety called a *planet*, or a wanderer, incessantly moving from one part of the starry heavens to another. By similar comparisons it can be shown that the other bodies we have mentioned—Venus and Mercury, Saturn and Mars—are also wanderers, and belong to that group of heavenly bodies known as planets. Here, then, we have the simple criterion by which the earth's neighbours are to be readily discriminated from the stars. Each of the bodies near the earth is a planet, or a wanderer, and the mere fact that a body is a wanderer is alone sufficient to prove it to be one of the class which we are now studying.

Provided with this test, we can at once make a considerable addition to our list. Amid the myriad orbs which the telescope reveals, here and there is found one which is a wanderer. Two other mighty planets, known as Uranus and as Neptune, must thus be added to the five already mentioned, making in all a group of seven great planets. A vastly greater number may also be reckoned when we admit to our view bodies which not only seem minute telescopic objects, but really are minute globes when compared with the mighty bulk of our earth. These lesser planets, to the number of between two and three hundred, are also among the earth's neighbours.

We should, at this point, remark that a class of heavenly bodies widely differing from the planets must also be included among our neighbours. The bodies we refer to are the comets, and, indeed, it may happen that a comet will sometimes draw nearer to the earth than even the closest approach ever made by a planet. These remarkable and mysterious visitors will necessarily engage a good deal of our attention later on; but for the present we confine our attention to those more substantial globes, whether large or small, which are always termed planets.

In some of the lighthouses, surrounding our coasts, which exhibit flashing or intermittent lights, the cutting off the light is accomplished by two opaque semi-cylinders which clasp the central flame at stated intervals, and thus produce the obscurity. Imagine for a moment that some similar obscuring apparatus could be clasped around our sun so that all his beams were intercepted. That our earth would be plunged into the darkness of midnight is of course an obvious consequence. A moment's reflection will show that the moon, shining as it does only with the reflected rays of the sun, would become totally invisible. But would this extinction of the sunlight have any other effect? Would it, for instance, have any effect on the countless brilliant points that stud the heavens at midnight? On a winter's night, as the sun is far below the horizon, no beams could in any case illuminate the sky, and consequently, whether the sun be really obscured or not, could make no difference in the darkness of a moonless winter night. But yet the sudden obscurity of the sun would produce a remarkable effect, which a little attention would disclose. The stars, no doubt, would not exhibit the slightest change in brilliancy. Each star shines by its own light and is not indebted to the sun. The constellations would thus twinkle on as before, but a wonderful change would come over the planets. Were the sun to be obscured, instantly the planets would disappear from view. The midnight sky would thus witness the sudden extinction of the planets, while all the stars would remain unaltered. It seems hard at first to understand how the brilliancy of Venus, or the lustre of Jupiter can be really only due to the beams which fall upon these bodies from the distant sun. The evidence is, however, conclusive on the question; and it will be placed before the reader more fully when we come to discuss the several planets in detail.

Another objection may be here anticipated. Suppose that we are looking at Jupiter high in mid-heavens on a winter's night, it might be reasonably contended that the earth lies between Jupiter and the sun, and that sunbeams therefore could not fall on Jupiter. This is, perhaps, not an unnatural view for an inhabitant of this earth to adopt until he has become acquainted with the sizes of the various bodies concerned, and with the distances by which those bodies are separated. But the question would appear in a widely different form to an inhabitant of the planet Jupiter. If such a being were asked whether he suffered much inconvenience by the intrusion of the earth between himself and the sun, his answer would be something of this kind :—"No doubt such an event as the passage of the earth between me and the sun is possible, and has occurred on rare occasions separated by long intervals; but so far from it being the cause of any inconvenience, the whole earth, of which you think so much, is really so minute, as compared with the sun, that when the earth did come in front of the sun it was merely seen as a minute telescopic point, and the amount of sunlight which it intercepted was quite inappreciable."

The fact that the planets shine by the sun's light, points at once to the similarity between them and our earth. We are thus led to regard our sun as a central glowing globe associated with a number of other much smaller globes, each of which, being a dark body, is indebted to the sun for its light and its heat.

That was, indeed, a grand step in astronomy which demonstrated the nature of the solar system. The discovery that our earth, of

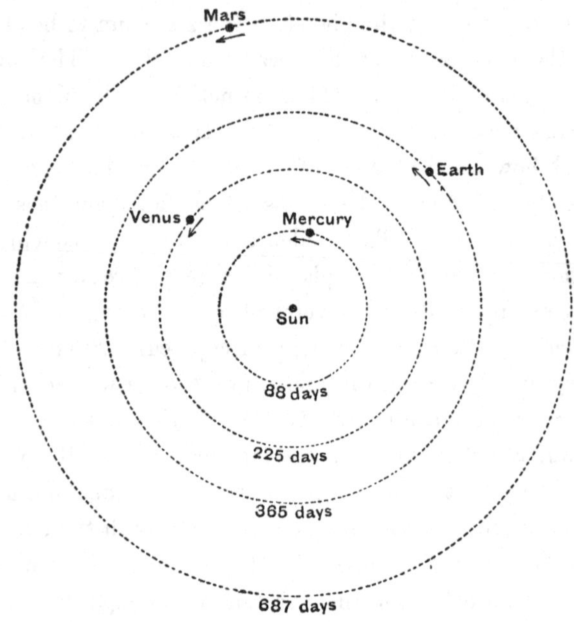

Fig. 29.—The Orbits of the four interior Planets.

whose stupendous bulk we are so conscious, really formed a globe poised freely in space, was in itself a mighty exertion of human intellect; but when it came to be recognised that this globe was but one of a whole group of globes, some smaller, no doubt, but others very much larger, and that this mighty series was subordinated to the supreme control of the sun, we have a chain of discoveries of an importance and a magnificence which knows no parallel in nature.

We thus see that the sun presides over a numerous family. The members of that family are dependent upon the sun, and have

PLATE X.

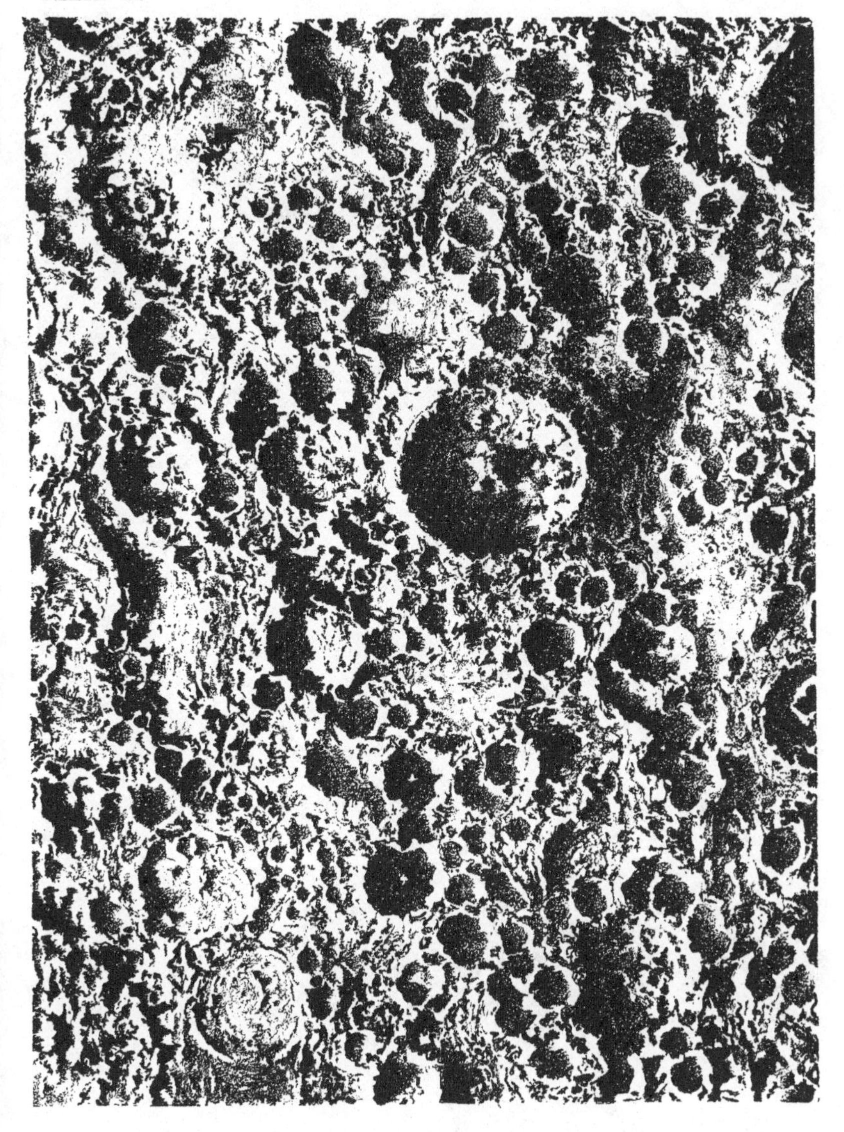

TYCHO AND ITS SURROUNDINGS.

(AFTER NASMYTH.)

a size suitably proportioned to their subordinate position. Even Jupiter, the largest member of that family, does not contain one-thousandth part of the material which forms the vast bulk of the sun. Yet Jupiter alone exceeds all the rest of the planets together.

In the centre of the diagram in Fig. 29 we have represented the sun—not, indeed, in his true proportions as to bulk, for it would be impossible to maintain the true proportions in a figure of reasonable dimensions. Around this sun we have four nearly circular paths, indicated by dotted lines to denote the orbits in which the different bodies move. The innermost of these four paths is the orbit of the planet Mercury. The planet moves around the sun in this path, and regains the place from which it started in eighty-eight days.

The next orbit, proceeding outwards from the sun, is that of the planet Venus, which we have already referred to as the well-known Evening Star. Venus moves round its path in a period of 225 days. One step further from the sun and we come to the orbit of another planet. This planet is almost the same size as Venus—it is much larger than Mercury. The planet now under consideration accomplishes its circle in a period of 365 days. This period sounds familiar to our ears. It is the length of the year; and the planet is the earth on which we stand. There is an impressive way in which to realise the enormous course which the earth pursues in each annual journey. Everyone knows that the circumference of a circle is about three and one-seventh times the diameter of the circle; so that, taking the distance from the earth to the sun to be 92,700,000 miles, the diameter of the circle which the earth describes around the sun will be 185,400,000 miles, and consequently the circumference of the mighty circle in which the earth moves round the sun is about 583,000,000 miles. Over this distance the earth has to travel in a year. It will be merely a sum in division to find from this how far the earth must travel each second in order to accomplish this long journey within a year. It will appear that the earth must actually move about eighteen miles every second, as otherwise it would not complete its journey in the allotted time.

Pause for a moment to think what a velocity of eighteen miles a second really implies. Can we realise a speed so tremendous? Let us compare it with our ordinary types of rapid movement. Look at that express train how it crashes under the bridge, how, in another moment, it is lost to view! Can any velocity be greater than that? Let us try it by figures. The train moves a mile a minute; multiply that velocity by eighteen and it becomes eighteen miles a *minute*, but we must further multiply it by sixty to make it eighteen miles a *second*. The velocity of the express train is not even the thousandth part of the velocity of the earth. Let us take another illustration. We stand at the rifle ranges to see a rifle fired at a target 1,000 feet away, and we find that a second or two is sufficient to carry the bullet over that distance. Can any

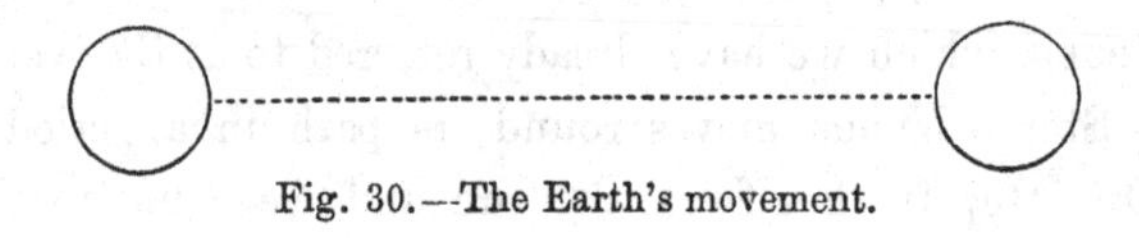

Fig. 30.—The Earth's movement.

moving body exceed in speed the rifle-bullet? The earth moves nearly one hundred times as fast. Yet, viewed in another way, the stupendous speed of the earth does not seem so inconceivable. The earth is a mighty globe, so great indeed that even moving at this speed it takes almost eight minutes to pass over its own diameter. The speed is thus not so enormously great when the great bulk of the earth is considered. To illustrate this we show here a view of the progress made by the earth. The distance between the centres of these circles is about six times the diameter; and, accordingly, if these two circles represent the earth, the time taken to pass from one position to the other is about forty-eight minutes.

Still one more step from the earth, and we come to the orbit of the fourth planet Mars, which requires 687 days, or nearly two years, to complete its circuit of the heavens. With our arrival at Mars we have gained the limit to the inner portion of the solar system.

The four planets we have mentioned form a group in themselves, distinguished by their comparative nearness to the sun, and also because they are bodies of moderate dimensions. In our system Venus and the Earth are globes of about the same size. Mercury and Mars are smaller globes intermediate between the bulk of the earth and that of the moon. The four planets which come nearest to the sun are, however, vastly surpassed in bulk and weight by the giant bodies of our system—the stately group of Jupiter and Saturn, Uranus and Neptune.

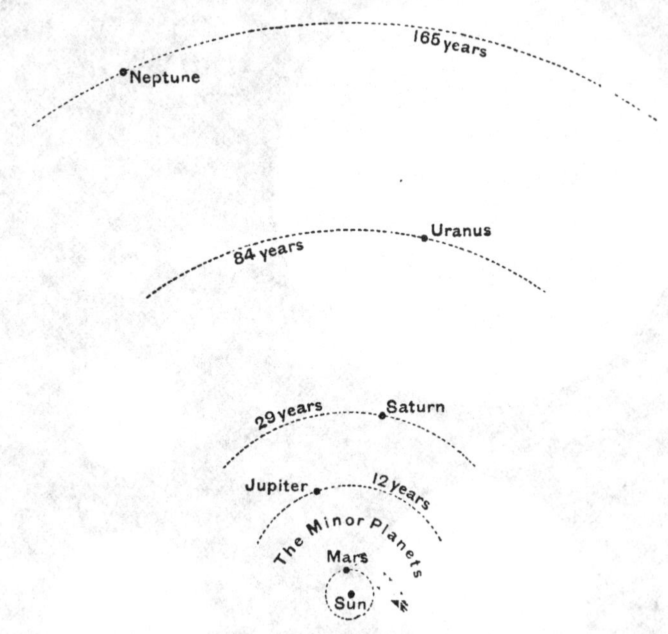

Fig. 31.—The Orbits of the four giant Planets.

These giant planets enjoy the sun's guidance equally with their weaker brethren. On the above diagram parts of the orbits of the great outer planets are represented. The sun, as before, presides at the centre, but the inner planets would now be so close to the sun that it is only possible to represent the orbit of Mars. After the orbit of Mars comes a considerable interval, not, however, devoid of planetary activity, and then follow the orbits of Jupiter and of Saturn; further still, we have Uranus, a great globe on the verge of unassisted vision; and, lastly, the whole system is bounded by the grand orbit of Neptune—a planet of which we shall have a marvellous story to narrate.

The various circles in Fig. 32 show the apparent sizes of the sun as seen from the different planets. Taking the circle corres-

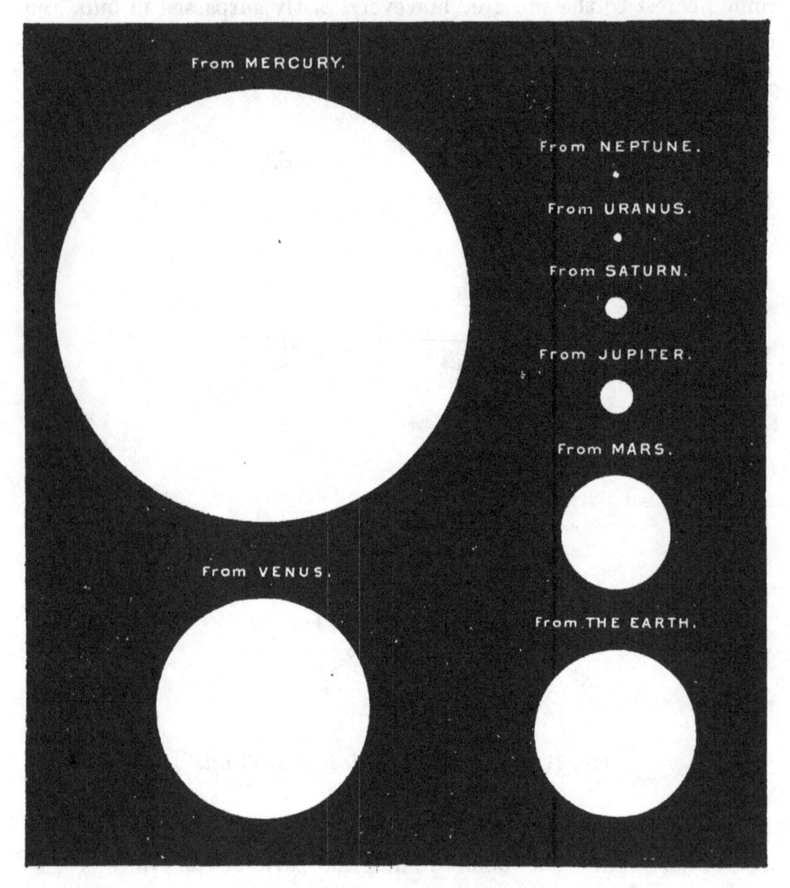

Fig. 32.—Apparent size of the Sun, as seen from the various Planets.

ponding to the earth to represent the amount of heat and light which the earth derives from the sun, then the other circles denote the heat and the light enjoyed by the corresponding planets. The next outer planet to the earth is Mars, whose share of solar blessings is not so very inferior to that of the earth; but when we look at Jupiter, at Uranus, or at Neptune, we fail to see how bodies so remote from the sun can enjoy climates at all comparable with those of the planets which are more favourably situated.

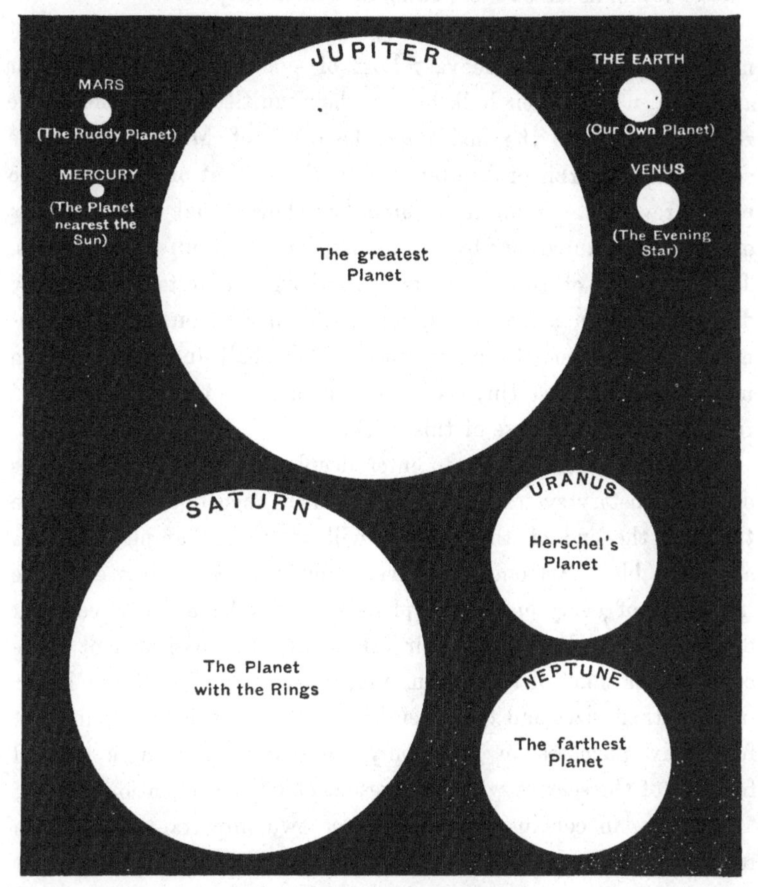

Fig. 33.—Comparative sizes of the Planets.

Fig. 33 shows a picture of the whole family of planets surrounding the sun—represented on the same scale, for exhibiting their comparative sizes. Jupiter, it will be seen, is not only the greatest planet, but he is much greater than all the other planets put together. Measured by bulk, Jupiter is more than 1,200 times as great as the earth, so that it would take at least

1,200 earths rolled into one to form a globe equal to the globe of Jupiter. Measured by weight, the disparity between the earth and Jupiter, though still enormous, is not quite so great; but this is a matter to be discussed more fully in a later chapter.

Even in this preliminary survey of the solar system we must not omit to refer to the vast host of planets which attract our attention, not by their bulk but by their multitude. In the ample zone, bounded on the inside by the orbit of Mars, and on the outside by the orbit of Jupiter, it was thought at one time that no planet revolved. Modern research has shown that this vast area of space is tenanted, not by one planet, but by hundreds of planets. The discovery of these planets is a charge undertaken by many diligent modern astronomers, while the discussion of their movements affords labour to many others. We shall find much to learn in the study of these tiny bodies, to which a chapter will be devoted further on in the course of this work.

But we do not propose to enter deeply into the mere statistics of the planetary system at present. Were such our intention, the tables at the end of this volume will show that ample materials are available. Astronomers have taken a more or less complete inventory of every one of the planets. They have measured their distances, the shapes of their orbits and the positions of those orbits, their times of revolution, and, in the case of all the larger planets, their sizes and their weights. Such results are of interest for many purposes in astronomy; but it is the more general features of the science which at present claim our attention.

Let us, in conclusion, note one or two important truths with reference to this beautiful planetary system. We have seen that the planets all revolve in nearly circular paths around the sun. We have now to supplement this by another statement of very great importance. Each of the planets pursues its path in the same direction. It may thus happen that one planet may overtake another, but it can never happen that two planets pass by each other as do the trains on adjacent lines of railway. We shall subsequently find that the whole welfare of our system, nay, its continuous existence, is dependent upon this remarkable uniformity.

Such is our solar system; a mighty organised group of planets circulating round a common sun, poised in space, and completely isolated from all external interference. No star, no constellation, has any appreciable influence on our solar system. We constitute a little island group, separated from the nearest stars by the most appalling distances. It may be that as the other stars are suns, so they too may have systems of planets circulating around them; but of this we know nothing. Of the stars we can only say that they are points of light, and if they had hosts of planets those planets must for ever remain invisible to us, even if they were many times as large as Jupiter.

At this limitation to our possible knowledge we need not repine, for just as we find in the solar system all that is necessary for our daily bodily wants, so shall we find ample occupation for whatever faculties we may possess, in endeavouring to understand those mysteries of the heavens which lie within our reach.

CHAPTER V.

THE LAW OF GRAVITATION.

Gravitation—The Falling of a Stone to the Ground—All Bodies fall Equally—Sixteen Feet in a Second—Is this True at Great Heights?—Fall of a Body at a height of a Quarter of a Million Miles—How Newton obtained an Answer from the Moon—His great Discovery—Statement of the Law of Gravitation—Illustrations of the Law—How is it that all the Bodies in the Universe do not rush together?—The Effect of Motion—How a Circular Path can be produced by Attraction—General Account of the Moon's Motion—Is Gravitation a Force of great Intensity?—Two Weights of 50 lbs.—Two Iron Globes, 53 yards in diameter, and a mile apart, attract with a force of 1 lb.—Characteristics of Gravitation—Orbits of the Planets not strictly Circles—The Discoveries of Kepler—Construction of an Ellipse—Kepler's First Law—Does a Planet move uniformly?—Law of the Changes of Velocity—Kepler's Second Law—The Relation between the Distances and the Periodic Times—Kepler's Third Law—Kepler's Laws and the Law of Gravitation—Movement in a straight line—A Body unacted on by disturbing Forces would move in a straight line with constant Velocity—Application to the Earth and the Planets—The Law of Gravitation deduced from Kepler's Laws—Universal Gravitation.

OUR narrative of the heavenly bodies must undergo a slight and temporary interruption, while we now enunciate and describe with appropriate detail the extremely important principle known as the law of gravitation, which underlies the whole of astronomy. To the law of gravitation must be ascribed the movements of the moon around the earth, and of the planets around the sun. Those movements can be completely accounted for when once the law of gravitation has been admitted. It is accordingly incumbent upon us to explain that law, before we proceed to the more detailed account of the separate planets. We shall find, too, that the law of gravitation opens up vast chapters in the history of the stars situated at the most stupendous distances in space, while it also affords the key by which we are enabled to cast a retrospective glance into the vistas of time past, and trace with plausibility, if not with certainty, early phases in the history of our system.

The sun and the moon, the planets and the comets, the stars and the nebulæ, all alike are subject to this universal law, which is now to engage our attention.

What is more common than the fact that when a stone is dropped it will fall to the ground? so common is it that no one at first thinks it worthy of remark. People are often surprised at seeing a piece of iron drawn to a magnet. Yet the fall of a stone to the ground is the manifestation of a force quite as interesting as the force of magnetism. It is the earth which draws the stone, just as the magnet draws the iron. In each case the force is one of attraction; but while the magnetic attraction is confined to a few substances, and is of comparatively limited importance, the attraction of gravitation extends far and wide throughout the whole universe.

Let us commence with a few very simple experiments upon the force of gravitation. Take in the hand a small piece of lead, and allow it to fall upon a cushion. The lead requires a certain time to move from the fingers to the cushion, but that time is always the same when the height is the same. Take now another and larger piece of lead, and hold one in each hand at the same height. If both are released at the same moment, they will both reach the cushion at the same moment. It might have been thought that the heavy body would fall more quickly than the light body; but when the experiment is tried it is seen that this is not the case. Repeat the experiment with various other substances. Try an ordinary marble. It also will fall in the same time as the piece of lead. With a piece of cork we again try the experiment, and again obtain the same result. At first it seems to fail when we compare a feather with the piece of lead; but that is solely on account of the air, which resists the feather more than it resists the lead. If, however, the feather be placed upon the top of a penny, and the penny be horizontal when dropped, it will clear the air out of the way of the feather in its descent, and then the feather will fall as quickly as the penny, as quickly as the marble, or as quickly as the piece of lead.

If the observer were in a gallery when trying these experiments, and if the cushion were sixteen feet below his hands, then the time the marble would take to fall through the sixteen feet would be one second. The time occupied by the cork or by the lead would be the same, and even the feather itself would fall through sixteen feet in one second, if it could be screened from the interference of the air. Try this experiment where we like, in London, or in any other city, in any island or continent, on board a ship at sea, at the North Pole, or the South Pole, or the equator, it will always be found that any body of any size, or any material, will fall about sixteen feet in one second of time.

Lest any erroneous impression should arise, we may just mention that the distance traversed in one second does vary slightly at different parts of the earth, but from causes which need not at this moment detain us. We shall for the present regard sixteen feet as the distance through which any body, free from interference, would fall in one second at any part of the earth's surface. But now let us extend our view above the earth's surface, and inquire how far this law of sixteen feet in a second may find obedience elsewhere. Let us, for instance, ascend to the top of a mountain and try the experiment there. It would be found that at the top of the mountain a marble would take a little longer to fall through sixteen feet than the same marble would if let fall at its base. The difference would be very small; but yet it would be measurable, and sufficient to show that the power of the earth to pull the marble down to the ground was somewhat weakened when we ascend high above the earth's surface. Yet no matter how high we ascend, either to the top of a high mountain, or to the still greater heights that have been reached in balloon ascents, we shall find that the tendency of bodies to fall to the ground remains, though no doubt the higher we go the more is that tendency weakened. It would be of the greatest interest to find how far this power of the earth to draw bodies towards it extends. We cannot get more than about five or six miles above the earth's surface in a balloon; yet we want to know what would happen if we could go 500 miles, or 5,000 miles, or still further, into the depths of space.

Conceive that a traveller were endowed with some means of soaring aloft for miles and thousands of miles, still up and up, until at length he had attained the awful height of nearly a quarter of a million of miles above the earth. Glancing down at the surface of that earth, which is at such a stupendous depth beneath, he would be able to see the whole world, as it were, in a bird's-eye view. He would lose, no doubt, the details of towns and villages; the features in such a landscape would be whole continents and whole oceans, in so far as the openings of the clouds far beneath would permit the earth's surface to be exposed.

At this stupendous height he could try one of the most interesting experiments that was ever in the power of a philosopher. He could test whether the earth's attraction attained to such a height, and he could measure the amount of that attraction. Take for the experiment a cork, a marble, or any other object, large or small; hold it between the fingers, and let it go. Every one knows what would happen in such a case down here; but it required Sir Isaac Newton to tell us what would happen in such a case up there. Newton says that the power of the earth to draw bodies towards it extends even up to this great height, and that the marble will fall. This is the doctrine that we can now test. We are ready for the experiment. The marble is released, and, lo, our first exclamation is one of wonder. The moment the marble was released, instead of dropping instantly, it appears to be suspended. We are on the point of exclaiming that we are beyond the earth's attraction, and that Newton is wrong, when our attention is arrested; the marble is beginning to move, so slowly that at first we have to watch it carefully. Then it moves more rapidly, so that its motion is beyond all doubt until, gradually acquiring more and more velocity, the marble speeds on its long journey of a quarter of a million of miles to the earth.

But surely, it will be said, such an experiment must be entirely impossible; and no doubt it cannot be performed in the way described. The bold idea occurred to Newton of making use of the moon itself, which is almost a quarter of a million of miles above the earth, for the purpose of answering the question. Never was the moon put to such noble use before. The moon is actually at each moment falling in towards the earth. We can calculate how much the moon is deflected in each second towards the earth, and thus obtain the measure of the earth's attractive power. From the moon Newton was able to learn that a body released at the distance of 240,000 miles above the surface of the earth would still be attracted by the earth, that in virtue of the attraction the body would commence to move off towards the earth—not, indeed, with the velocity with which a body falls in experiments on the surface —but with a very much lesser speed. At the distance of the moon a body, so far from falling a distance of sixteen feet in a second of time, would commence its long journey so slowly that a *minute*, instead of a *second*, would have elapsed before the distance of sixteen feet had been accomplished.*

It was by pondering on information thus won from the moon that Newton made his immortal discovery. The gravitation of the earth is a force which extends far and wide through space. The more distant the body, the weaker is the gravitation; here Newton found the means of determining the great problem as to the law according to which the intensity of the gravitation decreased. The information derived from the moon, that a body 240,000 miles away requires a minute to fall through a space equal to that through which it would fall in a second down here, was of unspeakable importance. In the first place, it shows that the attractive power of the earth, by which it draws all bodies earthward, becomes weaker at a distance. This might, indeed, have been anticipated. It is as reasonable to suppose that as we retreated further and further into the depths of space the power of attraction should diminish, as that the lustre of light should diminish as we recede from it; and it is remarkable that the law according to which the attraction of gravitation decreases with the increase of distance, is precisely the same as the law according to which the brilliancy of a light decreases as its distance increases. The law of nature, stated in its simplest form, asserts that the intensity of gravitation varies inversely as the

* The space described by a falling body is proportional to the product of the force and the square of the time. The force varies inversely as the square of the distance from the earth, so that the space will vary as the square of the time, and inversely as the square of the distance. If, therefore, the distance be increased sixty-fold, the time must also be increased sixty-fold, if the space fallen through is to remain the same.

square of the distance. Let us endeavour to elucidate this somewhat abstract statement by one or two simple illustrations. Suppose a body were raised above the surface of the earth to a height of nearly 4,000 miles, so as to be at an altitude equal to the radius of the earth. In other words, a body so situated would be twice as far from the centre of the earth as a body which was on the surface. The law of gravitation says that the intensity of the attraction is then to be decreased to one-fourth part, so that the pull of the earth on a body, 4,000 miles up, is only one quarter of the pull of the earth on that body so long as it lies on the surface. We may imagine the effect of this pull to be shown in different ways. Allow the body to fall, and in the interval of one second it will only drop through four feet, a mere quarter of the distance that gravity accomplishes near the earth's surface. To put the matter in another way; suppose that the pull of the earth is to be measured by one of those little weighing machines known as a spring balance. If a weight of four pounds be hung on a spring balance, at the earth's surface, the index of course shows a weight of four pounds; but conceive this balance still bearing the weight appended thereto were to be carried up and up, the *indicated* strain would become less and less, until, by the time the balance reached 4,000 miles high, where it was *twice* as far away from the earth's centre as at first, the indicated strain would be reduced to the *fourth* part, and the balance would only show one pound. If we could imagine the mighty voyage prolonged still further into the depths of space, the reading on the scale of the balance would still continue to decrease. By the time the apparatus had reached a distance of 8,000 miles high, being then *three* times as far from the earth's centre as at first, the law of gravitation tells us that the attraction must have decreased to one-ninth part. The strain thus shown on the balance would be only the ninth part of four pounds, or less than half a pound. But let the voyage be once again resumed, and let not a halt be made this time until the balance and its four-pound weight has retreated to that orbit which the moon traverses in its monthly course around the earth. The distance thus attained is about sixty times the radius of the earth, and consequently the attraction of gravitation is diminished in the proportion of one to the square of sixty; the balance will then only be strained by the inappreciable fraction of 1-3,600th part of four pounds; and we also see that a weight which on the earth weighed a ton and a half would, if raised 240,000 miles, weigh less than a pound. But even at this mighty distance we are not to halt; imagine that we retreat still further and further, the strain shown by the balance will ever decrease, but it will still exist no matter how far we go. Astronomy appears to teach us that the attraction of gravitation can extend, with suitably enfeebled intensity, across the most profound gulfs of space.

The principle of gravitation is of far wider scope than we have yet indicated. We have spoken merely of the attraction of the earth, and we have stated that its attraction extends throughout space. But the law of gravitation is not so limited. Not only does the earth attract every other body, and every other body attract the earth, but each of these bodies attracts each other; so that in its more complete shape the law of gravitation announces that "every body in the universe attracts every other body with a force which varies inversely as the square of the distance."

It is impossible for us to over-estimate the importance of this law. It supplies the clue by which we can unravel the complicated movements of the planets. It has led to marvellous discoveries, in which the law of gravitation has enabled us to anticipate the telescope, and, indeed, actually to feel the existence of bodies before those bodies have even been seen.

An objection which may be raised at this point must first be dealt with. The objection is, indeed, a plausible one. If the earth attracts the moon, why does not the moon tumble down on the earth? If the earth is attracted by the sun, why does it not tumble into the sun? If the sun is attracted by other stars, why do they not rush together with a frightful collision? It may not unreasonably be urged that if all these bodies in the heavens are attracting each other, it would seem that they must all rush together in consequence of that attraction, and thus weld the whole material universe into a single mighty mass. We know, as a matter of fact, that these collisions do not often happen, and that there is extremely little likelihood of their taking place. We see that although our earth is said to have been attracted by the sun for countless ages, yet that the earth is just as far from the sun as ever it was. Is not this in conflict with the doctrine of universal gravitation? In the early days of astronomy such objections would be regarded, and doubtless were regarded, as well nigh insuperable; even still we occasionally hear them raised, and it is therefore the more incumbent on us to explain how it happens that the solar system has been able to escape from the catastrophe by which it seems to have been threatened.

There can be no doubt that if the moon and the earth had been initially placed *at rest,* they would have been drawn together by the attraction. So, too, if the system of planets surrounding the sun had been left initially *at rest* they would have dashed into the sun and the system would have been annihilated. It is the fact that the planets are *moving,* and that the moon is *moving,* which has enabled these bodies successfully to resist the attraction in so far, at least, as that they are not drawn thereby to total destruction.

It is so desirable that the student should understand clearly how a central attraction is compatible with revolution in a nearly circular path, that we give an illustration to show how the moon pursues its monthly orbit under the guidance and the control of the attracting earth.

The imaginary sketch in Fig. 34 denotes a section of the earth with a high mountain thereon. If a cannon were stationed on the top of the mountain at C, and if the cannon-ball were fired off in the direction C E with a small charge of powder, the ball would move down along the first curved path. If it be fired a second time with a still greater charge, the path will be along the second curved line, and the ball would again fall into the sea. But let us try next time with a stronger charge, and, indeed, with a far stronger cannon than any piece of ordnance ever yet made. The velocity of the projectile must now be assumed to be some miles per second, but we can conceive that the velocity shall be so adjusted that the ball shall move along the path C D, always at the same height above the sea, though still curving, as every projectile must curve, from the horizontal line in which it moved at the first moment. Arrived at D, the ball will still be at the same height above the surface, and its velocity will be unabated. It will therefore directly start off again and move round another arc of the circle without getting nearer to the surface. In this manner the projectile will travel completely round the whole globe, coming back again to C and then taking another start in the same path. If we could then

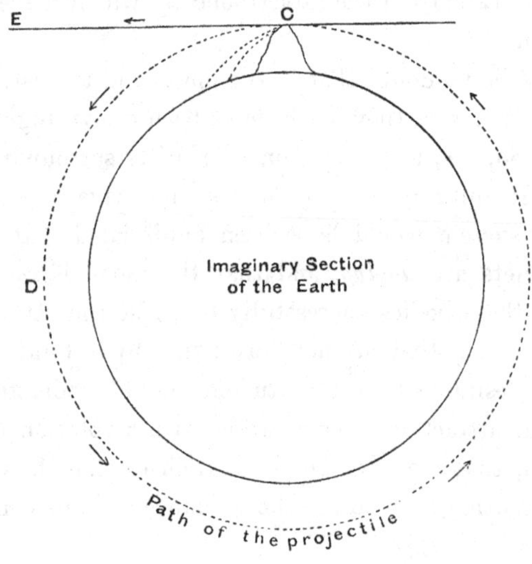

Fig. 34.—Illustration of the Moon's motion.

abolish the mountain and the cannon at the top, we should have a body revolving for ever around the earth in consequence of the attraction of gravitation.

Make now a bold stretch of the imagination. Conceive a mighty cannon which could fire off a round bullet no less than 2,000 miles in diameter. Discharge this enormous bullet with a velocity of about 3,000 feet per second, which is two or three times as great as the velocity actually attainable in modern artillery. Let this mighty bullet be fired horizontally from some station nearly a quarter of a million miles above the surface of the earth. That stupendous bullet would sweep right round the earth in a nearly circular orbit, and return to where it started in about four weeks.

It would then commence another revolution, four weeks more would find it again at the starting point, and this motion would go on for ages.

Do not suppose that we are entirely romancing. We cannot show the mighty cannon, but we can show the mighty bullet. We see it every month ; it is the beautiful moon herself. No one asserts that the moon was ever shot from such a cannon; but it must be admitted that she moves as if she were. In a later chapter we shall inquire into the history of the moon, and show how she came to revolve in this wonderful manner.

As with the moon around the earth, so with the earth around the sun. The illustration shows that a circular or nearly circular motion harmonises with the conception of the law of universal gravitation.

We are accustomed to regard gravitation as a force of stupendous magnitude. Does not gravitation control the moon in its revolution around the earth? Is not even the mighty earth itself retained in its path around the sun by the surpassing power of the sun's attraction? No doubt the actual force which keeps the earth in its path, as well as that which retains the moon in our neighbourhood, is of vast intensity, but that is because gravitation is in such cases associated with bodies of most stupendous mass. No one can deny that all bodies accessible to our observation appear to attract each other in accordance with the law of gravitation; but it must be confessed that, unless one or both of the attracting bodies possess enormously great mass, the intensity is almost immeasurably small. Let us attempt to illustrate how feeble is the gravitation between masses of easily manageable dimensions. Take for instance two iron weights, each weighing about 50lb., and separated by a distance of one foot from centre to centre. There is a certain attraction of gravitation between these weights. The two weights are drawn together, yet they do not move. The attraction between them, though it certainly exists, is an extremely minute force. We are not here dealing with a force at all comparable in its intensity with magnetic attraction. Every one knows that a magnet will draw a piece of iron with considerable force, but the intensity of gravitation is very much less on masses of equal amount. The attraction between these two 50lb. weights is less than the one-ten-millionth part of a single pound. Such a force is utterly infinitesimal in comparison with the friction between the weights and the table on which they stand, and hence there is no response to the attraction by even the slightest movement. Yet, if we can conceive each of these weights mounted on wheels absolutely devoid of friction, and running on absolutely perfect horizontal rails, then there is no doubt that the weights would slowly commence to draw together, and in the course of time would arrive in actual contact.

If we desire to conceive gravitation as a force of measurable intensity, we must employ masses immensely more ponderous than those 50lb. weights. Imagine a pair of globes, each composed of 417,000 tons of cast iron, and each, if solid, being about 53 yards in diameter. Imagine these globes placed at a distance of one mile apart. Each globe attracts the other by the force of gravitation. It does not matter that buildings and obstacles of every description intervene; gravitation will pass through such impediments as easily as light passes through glass. No screen can be devised dense enough to intercept the passage of gravitation. Each of these iron globes will therefore under all circumstances attract the other; but, notwithstanding their ample proportions, the intensity of that attraction is still very small, though appreciable. The attraction between these two globes is not more than a force equal to the pressure exerted by a single pound weight. A child could hold back one of these massive globes from its attraction by the other. Suppose that all was clear, and that friction could be so neutralised as to permit the globes to follow the impulse of their mutual attractions. The two globes will then commence to approach, but the masses are so large, while the attraction is so small, that the speed will be accelerated very slowly. A microscope would be necessary to show when the motion has actually commenced. An hour and a half must elapse before the distance is diminished by a single foot; and although the pace improves subsequently, yet three or four days must elapse before the two globes will come together.

The most remarkable characteristic of the force of gravitation must be here specially explained. The intensity appears to depend only on the magnitudes of the masses, and not at all on the nature of the substances of which these masses are composed. We have described the two globes as made of cast iron, but if either or both were composed of lead or copper, of wood or stone, of air or water, the attractive power would still be the same, provided only that the masses remained unaltered. In this we observe a profound difference between the attraction of gravitation and magnetic attraction. In the latter case the attraction is not perceptible at all in the great majority of substances, and is only considerable in the case of iron.

In our account of the solar system we have represented the moon as revolving around the earth in a *nearly* circular path, and the planets as revolving around the sun in orbits which are also approximately circular. It is now our duty to give a more minute description of these remarkable paths; and, instead of dismissing them as being *nearly* circles, we must ascertain precisely in what respects they differ from circles.

If a planet revolved around the sun in a truly circular path, of which the sun was always at the centre, it would then be obvious that the distance from the sun to the planet, being always equal to the radius of the circle, will be of constant magnitude. Now, there can be no doubt that the distance from the sun to each planet is approximately constant; but when accurate observations are made, it becomes clear that the distance is not absolutely constant. The variations in distance may amount to many millions of miles, but, even in extreme cases, the variation in the distance of the planet is only a small fraction—usually a very small fraction—of the total amount of that distance. The circumstances vary in the case of each of the planets. The orbit of the earth itself is such that the distance from the earth to the sun departs but little from its mean value. Venus makes even a closer approach to perfectly circular movement; while, on the other hand, the path of Mars, and much more the path of Mercury, show considerable relative fluctuations in the distance from the planet to the sun.

It has often been noticed that many of the great discoveries in science have their origin in the nice observation and explanation of minute departures from some law approximately true. We have in this department of astronomy an excellent illustration of this principle. The orbits of the planets are nearly circles, but they are not exactly circles. Now, why is this? There must be some natural reason. That reason has been ascertained, and it has led to several of the grandest discoveries that the mind of man has ever achieved in the realms of Nature.

In the first place, let us see what we are to infer from the fact that the distance of a planet from the sun is not constant. The motion in a circle is one of such beauty and simplicity, that we are reluctant to abandon it, unless the necessity for doing so be made clearly apparent. Can we not devise any way by which the circular motion might be preserved, and yet be compatible with the fluctuations in the distance from the planet to the sun? This is clearly impossible with the sun at the centre of the circle. But suppose the sun were not at the centre of the circle, while the planet, as before, revolved around the sun. The distance between the sun and the planet would then fluctuate. The more eccentric the position of the sun, the larger would be the proportionate fluctuation in the distance of the planet when at the different parts of its orbit. It might further be supposed that by placing a series of circles around the sun the various planetary orbits could be thus accounted for. The centre of the circle belonging to Venus is to coincide very nearly with the centre of the sun, and the centres of the orbits of all the other planets are to be placed at such suitable distances from the sun, as will render a satisfactory explanation of the gradual increase and decrease of the distance between the two bodies.

There can be no doubt that the movements of the moon and of the planets would be, to a large extent, explained by such a system of circular orbits; but the spirit of astronomical inquiry is not satisfied with approximate results. Again and again the planets are observed, and again and again are the observations compared with the places which the planets would have if they moved in accordance with the system here indicated. The centres of the circles are moved hither and thither, their radii are adjusted with greater care; but it is all of no avail. The observations of the planets are minutely examined to see if they can be in error; but of errors there are none at all sufficient to account for the discrepancies. The conclusion was thus inevitable—astronomers were forced to abandon the circular motion, which was thought to possess such unrivalled symmetry and beauty, and were compelled to admit that the orbits of the planets were not circular.

Then if these orbits be not circles, what are they? Such was the great problem which Kepler proposed to solve, and which, to his immortal glory, he succeeded in solving and in proving to demonstration. The great discovery of the true shape of the planetary orbits stands out as one of the most conspicuous events in the history of astronomy. It may, in fact, be doubted whether any other discovery in the whole range of science has led to results of such far-reaching interest.

We must here adventure for a while into the field of science known as geometry, and study therein the nature of that curve which the discovery of Kepler has raised to such unparalleled importance. The subject, no doubt, is a difficult one, and to pursue it with any detail would involve us in many abstruse calculations which would be out of place in this volume; but a general sketch of the subject is indispensable, and we must attempt to render it such justice as may be compatible with our limits.

The curve which represents with perfect fidelity the movements of a planet in its revolution around the sun, belongs to that well-known group of curves which mathematicians describe as the conic sections. The particular form of conic section which denotes the orbit of a planet is known by the name of the *ellipse:* it is spoken of somewhat less accurately as an oval. The ellipse is a curve which can be readily constructed. There is no simpler method of doing so than that which is familiar to draughtsmen, and which we shall here briefly describe.

We represent on the next page (Fig. 35) two pins passing through a sheet of paper. A loop of twine passes over the two pins in the manner here indicated, and is stretched by the point of a pencil. With a little care the pencil can be guided so as to keep the string stretched, and its point will then describe a curve completely around the pins, returning to the point from which it started. We thus produce that celebrated geometrical curve which is called an ellipse. It will be very instructive to draw a number of ellipses, varying in each case the circumstances under which they are formed. If, for instance, the pins remain placed as before, while the length of the loop is increased, so that the pencil is farther away from the pins, then it will be

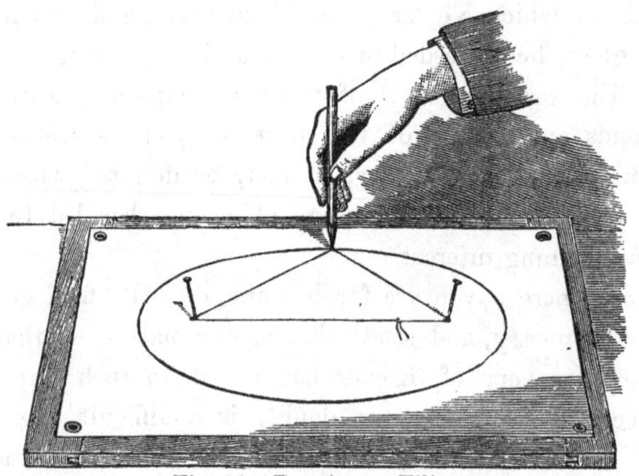

Fig. 35.—Drawing an Ellipse.

observed that the ellipse has lost some of its elongation, and approaches more closely to a circle. On the other hand, if the length of the cord in the loop be lessened, while the pins remain as before, the ellipse will be found more oval, or, as a mathematician would say, its *eccentricity* is increased. It is also useful to study the changes which the form of the ellipse undergoes when one of the pins is altered, while the length of the loop remains unchanged. If the two pins be brought nearer together the eccentricity will decrease, and the ellipse will approximate more closely to the shape of a circle. If the pins be separated more widely the eccentricity of the ellipse will be increased. That the circle is really a form of ellipse will be evident, if we suppose the two pins to draw in so close together that they become coincident; the point will then simply trace out a circle, as the pencil moves round the figure.

It will be obvious that the points marked by the pins possess very remarkable relations with respect to the curve. Each one of these points is called a *focus* of the ellipse, and an ellipse can only have one pair of foci. There is in fact only one pair of positions possible for the two pins, when an ellipse of specified size and shape is to be constructed.

The ellipse is thus a curve differing principally from a circle, inasmuch as it possesses variety of form. We can have large and small ellipses just as we can have large and small circles, but we can also have ellipses of greater or less eccentricity. If the ellipse has not the perfect simplicity of the circle, it has at least the charm of variety which the circle has not; while the ellipse has also the beauty which can be derived from an outline of perfect grace, and an association with the most noble conceptions.

The ancient geometricians had studied the ellipse; they had noticed its foci; they had become acquainted with its properties; and thus Kepler was familiar with the ellipse at the time when he undertook his ever-memorable researches on the movements of the planets. He had found, as we have already indicated, that the movements of the planets could not be reconciled with circular orbits. What shape of orbit should next be tried? The ellipse was ready to hand, its properties were known, and the comparison could be made; memorable, indeed, was the consequence of this comparison. Kepler found that the movement of the planets could be explained, by supposing that the path in which each one revolved was an ellipse. This in itself was a discovery of the most commanding importance. On the one hand, it reduced to order the movements of the great globes which circulate round the sun; while on the other, it took that beautiful class of curves which had exercised the highest geometrical powers of the ancients, and at once gave to those curves the dignity of defining the great highways of the universe.

But we have as yet only partly enunciated the first discovery of Kepler. We have seen that a planet revolves in an ellipse around the sun, and that the sun is therefore at some point in the interior of the ellipse—but at what point? Interesting indeed is the answer to this question; for have we not pointed out how the foci possess a significance which no other points enjoy? Kepler showed that the sun must be situated in one of the foci in the ellipse in which each planet revolves. We can thus enunciate the first of Kepler's laws of planetary motion in the following words:—

Each planet revolves around the sun in an elliptic path, having the sun at one of the foci.

We are now enabled to form a clear picture of the orbits of the planets, be they ever so numerous, as they revolve around the sun. In the first place we observe that the ellipse is eminently a plane curve; that is to say, each planet must, in the course of its long journey, confine its movements to one plane. Each planet has thus a certain plane appropriated to it. It is true that all these planes are very nearly coincident, at least in so far as the great planets are concerned; but still they are distinct, and the only feature in which they all agree, is that each one of them passes through the sun. All the elliptic orbits of the planets have one focus in common, and that focus lies at the centre of the sun.

It is well to illustrate this remarkable law by considering the circumstances of two or three different planets. Take first the case of the earth, where the path, though really an ellipse, is very nearly circular. In fact, if it were drawn accurately to scale on a sheet of paper, the difference between the elliptic orbit and the circle would hardly be detected without careful measurement. In the case of Venus the ellipse is still more nearly a circle, and the two foci of the ellipse are very nearly coincident with the centre of the circle. On the other hand, in the case of Mercury, we have an ellipse which departs from the circle to a very marked extent, while in the orbits of some of the lesser planets the eccentricity is still greater. It is extremely remarkable that every planet, no matter how far from the sun, should be found to move in an ellipse of some shape or other. We shall presently show that necessity compels each planet to pursue an elliptic path, and that no other form of path is possible.

Started on its elliptic path, the planet pursues its stately course, and after a certain interval known as the *periodic time*, regains the position from which it started. Again the planet traces out anew the elliptic path which it formerly pursued, and thus, revolution after revolution, the same track is followed around the sun. Let us now attempt to follow the planet in its course, and observe the history of its motion during the time requisite for a complete circuit of the sun. The dimensions of a planetary orbit are so stupendous that the planet must run its course very rapidly in order to complete the journey within the allotted time. The earth, as we have already seen, would have to move about eighteen miles

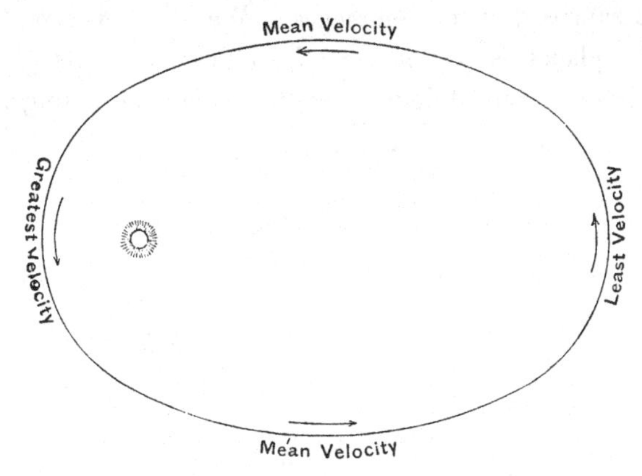

Fig. 36.—Varying velocity of elliptic motion.

a second to complete a voyage round the sun in the course of $365\frac{1}{4}$ days. The question then arises, as to whether the rate at which a planet moves is uniform or not. Does the earth, for instance, actually move at all times with the velocity of eighteen miles a second, or does our planet sometimes move more rapidly and sometimes more slowly, so that the average of eighteen miles a second is still maintained? This is a question of very great importance, and we are able to answer it in the clearest and most emphatic manner. The velocity of a planet is *not* uniform, and the variations of that velocity can be explained by the adjoining figure. Let us imagine the planet first of all to be situated at that part of its path most distant from the sun towards the right of the figure. In this position the velocity is at its lowest value; as the planet begins to approach the sun the speed gradually improves until it attains its mean value. After this point has been passed, and the planet is now rapidly hurrying on towards the sun, the velocity with which it moves becomes gradually greater and greater, until at length, as it dashes round the sun, its velocity reaches the highest point. After passing the sun the distance of the planet from the sun increases, and the velocity of the motion begins to abate; gradually it sinks down until the mean value is again reached, and then it falls still lower, until the planet recedes to its greatest distance from the sun, by which time the velocity has abated to the point from which we supposed it to commence. We thus observe that the nearer the planet is to the sun the quicker it moves. We can, however, give numerical definiteness to the law, according to which

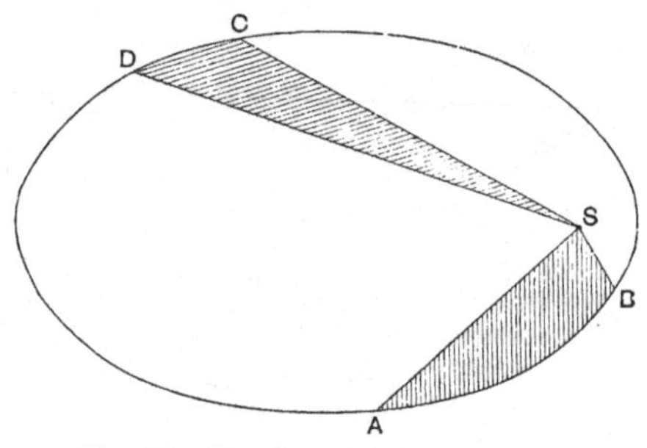

Fig. 37.—Equal areas in equal times.

the velocity of the planet varies. The adjoining figure gives a planetary orbit with, of course, the sun at the focus, S. We have taken two portions, A B and C D, round the ellipse; we have joined their extremities to the focus; and we have marked the two nearly triangular areas by shading. Then Kepler's second law may be stated in these words:—

"Every planet moves round the sun with such a velocity at every point, that a straight line drawn from it to the sun passes over equal areas in equal times."

If, therefore, the two shaded portions, A B S and D C S, are equal in area, then the times occupied by the planet in travelling over the portions of the ellipse, A B and C D, are equal. If the one area be greater than the other, then the time occupied is greater in the proportion of the areas.

This law being admitted, the reason of the increase in the planet's velocity when it approaches the sun is at once apparent. To accomplish a definite area when near the sun, a larger arc is obviously necessary than at other parts of the path; while at the opposite extremity of the path, a small arc suffices for a large area, and the velocity is accordingly less.

These two laws completely enunciate the motion of a planet round the sun. The first defines the path which the planet pursues; the second describes how the velocity of the planet varies at different points along its path. But Kepler added to these a third law, which enables us to compare the movements of two different planets revolving round the same sun. Before stating this law, it is necessary to explain exactly what is meant by the *mean* distance of a planet. In its elliptic path the distance from the sun to the planet is constantly changing; but it is nevertheless easy to attach a distinct meaning to that distance which is an average of all the distances. This average is called the mean distance. The simplest way of finding the mean distance is to add the greatest distance to the least distance, and take half the sum. We have already defined the periodic time of the planet; it is the number of days which the planet requires for the completion of a journey round its path. Kepler's third law establishes a relation between the mean distance and the periodic time. That relation is stated in the following words :—

"*The squares of the periodic times are proportional to the cubes of the mean distances.*"

Kepler saw that the different planets had different periodic times; he also saw that the greater the mean distance of the planet the greater was its periodic time, and he was determined to find out the connection between the two. It was easily seen, that it would not be true to say that the periodic time is merely proportional to the mean distance. Were this the case, if one planet had a distance twice as great as another, the periodic time in the former case would have been double that in the latter; but observations showed that though the periodic time in the one case exceeded that in the other, yet that it was less than twice as much.

By repeated trials, which would have exhausted the patience of one less confident in himself, and less assured of the accuracy of the observations which he sought to interpret, Kepler at length discovered the true law, and expressed it in the form we have stated.

To illustrate the nature of this law, we shall take for comparison the earth and the planet Venus. If we denote the mean distance of the earth from the sun by unity, then the mean distance of Venus from the sun is 0·7233. If we omit decimals beyond the first place, we can represent the periodic time of the earth as 365·3 days, and the periodic time of Venus as 224·7 days. Now the law which Kepler asserts is that the square of 365·3 is to the square of 224·7 in the same proportion as unity is to the cube of 0·7233. The reader can easily verify the truth of this identity by actual multiplication. It is, however, to be remembered that, as only four figures have been retained in the expressions of the periodic times, so only four figures are to be considered significant in making the calculations. Perhaps, however, the most striking manner of making the verification will be to regard the time of the revolution of Venus as an unknown quantity, and deduce it from the known revolution of the earth and the mean distance of Venus. In this way, by assuming Kepler's law, we deduce the cube of the periodic time by a simple proportion, and the resulting value of 224·7 days can then be obtained. As a matter of fact, in the calculations of astronomy, the distances of the planets are usually ascertained from Kepler's law. The periodic time of the planet is an element which can be measured with great accuracy; and once it is known, then the square of the mean distance, and consequently the mean distance itself, is determined.

Such are the three celebrated laws of Planetary Motion, which have always been associated with the name of their discoverer. The profound skill by which these laws were elicited from the masses of observations, the intrinsic beauty of the laws themselves and their absolute truthfulness, their wide-spread generality, and the bond of union which they have established between the various members of the solar system, have given these laws quite an exceptional position in astronomy.

As established by Kepler, these planetary laws were merely the results of observation. It was found, as a matter of fact, that the planets did move in ellipses, but Kepler assigned no reason why they should move in ellipses rather than in any other curve. Still less was he able to give a reason why they should sweep over equal areas in equal times, or why that third law was invariably obeyed. The laws as they came from Kepler's hands stood out as three independent truths; thoroughly established, but wholly unsupported by any explanations as to why these movements rather than any other movements should be those appropriate for the revolutions of the planets.

It was the crowning triumph of the great law of universal gravitation to remove from Kepler's laws this empirical character. Newton's grand discovery bound together the three isolated laws of Kepler into one beautiful doctrine. He showed not only that those laws are true, but he showed why they must be true, and why no other laws could have been true. He proved to demonstration in his famous work, the "Principia," that the true explanation of the laws of Kepler was to be sought in the attraction of gravitation. He showed that in the sun resided a power of attraction, and how it was a necessary consequence of that attraction that every planet should revolve in an elliptic orbit round the sun, having the sun as one focus; that the radius of the planet's orbit should sweep over equal areas in equal times; and that in comparing the movements of two planets, it became necessary to have the squares of the periodic times proportional to the cubes of the mean distances.

As this is not a mathematical treatise, it will be impossible for us to discuss the proofs which Newton has given, and which have commanded the immediate and universal acquiescence of all who have taken the trouble to understand them. We must here only confine ourselves to a very brief and general survey of the subject, which will indicate the character of the reasoning employed, without introducing details of a technical character.

Let us, in the first place, endeavour to think of a globe freely poised in space and completely isolated from the influence of every other body in the universe. Let us imagine that this globe is set in motion by some impulse which starts it forward like a mighty bullet through the realms of space. When the impulse ceases the globe is in motion, and it continues to move onwards. But what will be the path which it will pursue? We are so accustomed to see a stone thrown into the air moving in a curved path, that we might naturally think a body projected into free space will also move in a curve. A little consideration will, however, show that the cases are very different. In the realms of free space we find no conception of upwards or downwards; all paths are alike; there is no reason why the body should swerve to the right or to the left; and hence we are led to surmise that under these circumstances a body, once started and freed from all interference, would move in a straight line. It is true that this statement is one which can never be submitted to the test of direct experiment. Circumstanced as we are on the surface of the earth, we have no means of isolating a body from external forces. The resistance of the air, as well as friction in various other forms, no less than the gravitation towards the earth itself, interfere with our experiments. A stone thrown along a sheet of ice will be exposed to but little interference, and in this case we see that the stone will take a straight course along the frozen surface. A stone similarly cast into empty space would pursue a course absolutely rectilinear. This we demonstrate, not by any attempts at an experiment which would necessarily be futile, but by indirect reasoning. The truth of this principle can never for a moment be doubted by one who has duly weighed the arguments which have been produced in its behalf.

Admitting, then, the rectilinear path of the body, the next question which arises relates to the velocity with which that movement is performed. The stone gliding over the smooth ice on a frozen lake will, as everyone has observed, travel a long distance before it comes to rest. There is but little friction between the ice and the stone, but there is some friction; and as that friction always tends to stop the motion, it will at length happen that the stone is brought to rest. In a voyage through the solitudes of space a body experiences no friction; there is no tendency for the velocity to be reduced, and consequently we believe that the body could journey on for ever with unabated speed. No doubt such a statement seems at variance with our ordinary experience. A sailing ship comes to rest on the sea when the wind dies away. A train will gradually lose its velocity when the steam has been turned off. A humming-top will slowly expend its rotation and come to rest. In these instances we seem to have proof that when the force which has imparted motion has ceased, the motion itself will gradually wane and ultimately cease entirely. But in all these cases it will be found, on reflection, that the decline of the motion is to be attributed to the action of resisting forces. The sailing ship is retarded by the rubbing of the water on its sides; the train is retarded by the friction of the wheels, and by the fact that it has to force its way through the air; and the resistance of the air is mainly the cause of the stopping of the humming-top, for if the air be withdrawn, by making the experiment in a vacuum, the top will continue to spin for a greatly lengthened period. When we duly weigh these considerations we shall find it possible to admit that a body, once projected freely in space and acted upon by no external resistance, will continue to move on for ever in a straight line, and will preserve unabated to the end of time the velocity with which it originally started. This principle is known as the *first law of motion*.

Let us apply this great principle to the important question of the movement of the planets. Take, for instance, the case of our earth, and discuss the consequences of the first law of motion. Our earth is every moment moving with a velocity of about eighteen miles a second, and the first law of motion assures us that if the earth were submitted to no external force, it would for ever pursue a straight track through the universe, and never depart from the precise velocity which it has at the present moment. But is the earth moving in this manner? Obviously not. We have already found that the earth is moving round the sun, and the beautiful laws of Kepler have given to that motion the most perfect distinctness and precision. The consequence is irresistible. The earth cannot be free from external force. Some potent influence on the earth must be in ceaseless action. That influence, whatever it may be, constantly deflects the earth from the rectilinear path which it tends to pursue, and constrains the earth to trace out an ellipse instead of a straight line.

The great problem to be solved is now easily stated. There must be some constant influence on the earth. What is that influence, from whence does it proceed, and to what law is it submitted? Nor is the question confined to the earth alone. Mercury and Venus, Mars, Jupiter, and Saturn proclaim aloud that, as they are not moving in rectilinear paths, they must be exposed to some force. What is this force which guides the planets in their paths? Before the time of Newton this question might have been asked in vain. It was the mighty genius of Newton which supplied the answer, and thus revolutionised the whole of modern science.

Where lie the data from which the answer to the question is to be elicited? We have here no problem which can be solved by mere mathematical meditation. Mathematics is no doubt a useful, indeed an indispensable instrument in the inquiry; but we must not attribute to mathematics a potency which it does not possess. In a case of this kind, all that mathematics can do is to interpret the results obtained by observation. The data, then, from which Newton proceeded, were the observed facts in the movement of the earth and the other planets. Those facts had found a most beautiful expression by the aid of Kepler's laws. It was, accordingly, the laws of Kepler which Newton took as the basis of his labours, and it was for the interpretation of Kepler's laws that Newton invoked the aid of that celebrated mathematical reasoning which he created.

The question is then to be approached in this way: A planet being subject to *some* external influence, we have to determine what that influence is, from our knowledge that the path of each planet is an ellipse, and that each planet sweeps round the sun over equal areas in equal times. The influence on each planet is what a mathematician would call a force, and a force must have a line of direction. The most simple conception of a force is that of a pull communicated along a rope, and the direction of the rope is in this case the direction of the force. Let us imagine that the force exerted on each planet is imparted by an invisible rope. What do Kepler's laws tell us with regard to the direction of this rope and to the intensity of the strain which is transmitted along it?

The mathematical analysis of Kepler's laws would be beyond the scope of this volume. We must, therefore, confine ourselves to the results obtained by them, passing by the details of the reasoning. Newton first took the law which asserted that the planet moved over equal areas in equal times, and he showed by unimpeachable logic that this at once gave the direction in which the force acted on the planet. If for the sake of illustration we regard as before the force to be exerted by the medium of a rope, Newton showed that that rope must be invariably directed towards the sun. In other words, that the force exerted on each planet was at all times directed exactly from the planet towards the sun.

It still remained to explain the intensity of the force, and to show how the intensity of that force varied when the planet was at different points of its path. Kepler's first law enables this question to be answered. If the planet's path be elliptic, and if the force be always directed towards the sun at one focus of that ellipse, then mathematical analysis obliges us to say, that the intensity of the force must vary inversely as the square of the distance from the planet to the sun.

The movements of the planets in conformity with Kepler's laws would thus be accounted for even in their minutest details, if we admit that an attractive power draws the planet towards the sun, and that the intensity of this attraction varies inversely as the square of the distance. Can we hesitate to say that such an attraction does exist? We have seen how the earth attracts a falling body; we have seen how the earth's attraction extends to the moon, and explains the revolution of the moon around the earth. We have now learned that the movement of the planets round the sun can be also explained to be the consequence of this law of attraction. But the evidence in support of the law of universal gravitation is, in truth, much stronger than any we have yet presented. We shall have occasion to dwell on this matter further on. We shall show not only how the sun attracts the planets, but how the planets attract each other; and we shall find how this mutual attraction of the planets has led to remarkable discoveries which have raised the truth of the law of gravitation beyond the possibility of doubt.

Admitting the law of gravitation, we can then show that the planets must revolve around the sun in elliptic paths with the sun in the focus. We can show that they must sweep over equal areas in equal times. We can prove that the squares of the periodic times must be proportional to the cubes of their mean distances. Still further we can show how the mysterious movements of comets can be accounted for. By the same great law we can explain the revolutions of the satellites. We can account for the tides, and for numerous other details throughout the Solar System. Finally, we shall show that when we extend our view beyond the limits of our Solar System to the beautiful starry systems scattered through space we find even there evidence of the great law of universal gravitation.

CHAPTER VI.

THE PLANET OF ROMANCE.

Outline of the Subject—Is Mercury the Planet nearest the Sun?—Transit of an Interior Planet across the Sun—Has a Transit of Vulcan ever been seen?—Visibility of Planets during a Total Eclipse of the Sun—Professor Watson's Researches in 1878.

PROVIDED with a general survey of the Solar System, and with such an outline of the law of universal gravitation as the last chapter has afforded us, we commence the more detailed examination of the planets and their satellites. We shall begin with the planets nearest to the sun, and then we shall gradually proceed outwards to one planet after another, until we reach the confines of the system. We shall find much to occupy our attention. Each planet is itself a globe, and it will be our duty to describe what is known of that globe. The satellites by which so many of the planets are accompanied possess many points of interest. The circumstances of their discovery, their sizes, their movements, and their distances, must all be duly considered. Then, too, it will be found that the movements of the planets present much matter for reflection and examination. We shall have occasion to show how the planets mutually disturb each other, and what remarkable consequences have arisen from these disturbances. We must also occasionally refer to the important problems of celestial measuring and celestial weighing. We must show how the sizes, the weights, and the distances of the various members of our system are to be discovered. A great part of our task will lead us over ground which is thoroughly certain, and where the results have been confirmed by frequent observation. It happens, however, that at the very outset of our course we are obliged to deal with observations which are far from certain. The existence of a planet much closer than those hitherto known has been asserted by competent authority. The question is still quite unsettled, and the planet cannot with certainty be pointed out. Hence it is that we have called the subject of this chapter, The Planet of Romance.

It had often been thought that Mercury, long supposed to be the nearest planet to the sun, was perhaps not really the body entitled to that distinction. Mercury revolves round the sun at an average distance of about 36,000,000 miles. In the interval between it and the sun there might have been one or many other planets. There might have been one revolving at ten million miles, another at fifteen, and so on. But did such planets exist? Did even one planet revolve inside the orbit of Mercury? There were certain reasons for believing in such a planet. In the movements of Mercury indications were perceptible of an influence that could have been accounted for by the supposition of an interior planet. But there was necessarily a great difficulty about seeing this object. It must always be close to the sun, and even in the best telescope it is generally impossible to see a starlike point in that position. Nor could such a planet be seen after sunset, for at the best it would set almost immediately after the sun, and a like difficulty would make it invisible at sunrise.

Our ordinary means of observing a planet have therefore completely failed. We are compelled to resort to extraordinary methods, if we would seek to settle the great question as to the existence of the intra-Mercurial planets. There are at least two extraordinary methods of observation which might be expected occasionally to answer our purpose.

The first of these methods would arise when it so happened that the unknown planet came directly between the earth and the sun. In the adjoining diagram we have the sun at the centre; the internal orbit denotes that of the unknown planet, which has received the name of Vulcan before even its very existence has been at all satisfactorily established. The outer orbit denotes that of the earth. As Vulcan moves more rapidly than the earth, it will frequently happen that the planet will overtake the earth, so that the three bodies will have the positions represented in the diagram. It would not, however, necessarily follow that Vulcan was exactly between the earth and the sun. The path of the planet may be tilted up so that, as seen from the earth, Vulcan would be over the sun, or under the sun, according to circumstances. If, however, Vulcan really do exist, we can be assured that sometimes the three bodies will be directly in line, and this would then give the desired opportunity of making the telescopic discovery of the planet. We

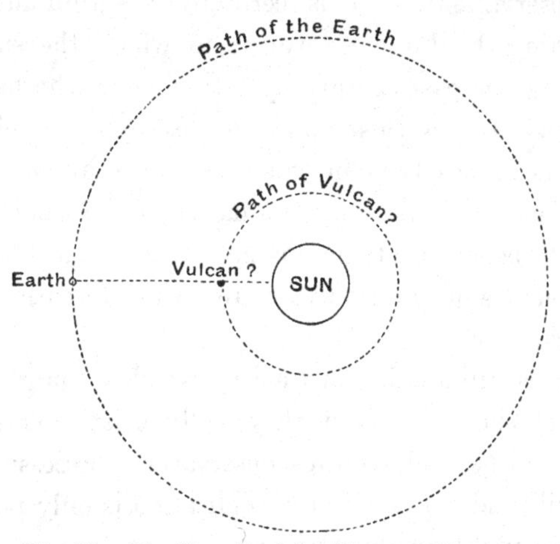

Fig. 38.—The Transit of the Planet of Romance.

should expect on such an occasion to see the planet as a dark spot, moving slowly across the face of the sun. The two other planets interior to the earth, namely, Mercury and Venus, are occasionally seen in the act of transit; and there cannot be a doubt that if Vulcan exist, its transits across the sun must be more numerous than those of Mercury, and far more numerous than those of Venus. On the other hand, it may reasonably be anticipated that Vulcan is a small globe, and as it will be much more distant from us than even Mercury at the time of transit, we could not expect that the transit of Vulcan would be a spectacle at all comparable in importance with the transit of either of the two other planets.

The question then arises, as to whether telescopic research has ever shown anything which can be regarded as a transit of Vulcan.

On this point it is not possible to speak with any degree of certainty. It has, on more than one occasion, been asserted by observers that a spot has been seen rapidly traversing the sun, and from its shape and general appearance they have presumed it to have been an intra-Mercurial planet. But a close examination of the circumstances under which such observations have been made has not tended to increase confidence in this presumption. Such discoveries have usually been made by persons little familiar with telescopic observations. It is certainly a significant fact that, notwithstanding the diligent scrutiny to which the sun has been exposed during the past century by astronomers who have specially devoted themselves to this branch of research, no telescopic discovery of Vulcan has been in this way made by any really experienced astronomer. From an examination of the whole subject, we are inclined to believe that there is not at this moment any reliable telescopic evidence of the transit of an intra-Mercurial planet over the face of the sun.

But there is still another method by which we might reasonably hope to detect such planets if they really exist. This method is one of great rarity, and requires observations possessing no small degree of skill and delicacy. Its application is only possible when the sun is obscured by a total eclipse.

When the moon is placed directly between the earth and the sun, the brightness of day is temporarily exchanged for the gloom of night. If the sky be free from clouds, the stars spring forth, and can be seen close up to where the corona shows the obscured sun to be situated. Even if a planet were quite close to the sun, it would be visible on such an occasion. Careful preparation is necessary when it is proposed to make a trial of this kind. The danger to be specially avoided is the risk of confounding the planet with the ordinary stars, which it will probably resemble. The late distinguished American astronomer, Professor Watson, specially prepared to devote himself to this research during the great total eclipse in 1878. The duration of this total eclipse was very brief; it lasted only two or three minutes, and all the work had to be compressed into this very short interval. Professor Watson had previously carefully studied the stars in the neighbourhood of the sun. When the eclipse occurred, the light of the sun vanished, and the stars burst forth. The practised eye of Professor Watson at once identified the stars, of which he had formed a map, not merely on paper, but also engraved on his memory; and among them he saw an object, which certainly seemed to be the long-sought inter-Mercurial planet. We should indeed add, that the same observer seems to have also seen a second planet on the same occasion. To a certain extent Mr. Watson's observation was confirmed by another observer, Mr. Swift. It is, however, right to add that Vulcan has not been observed, though specially looked for, during the eclipses which have occurred since 1878. We cannot, however, believe it possible that so experienced an astronomer as Mr. Watson was mistaken. He has been one of the most successful discoverers of minor planets, and, not improbably, posterity will have to admit, when the inter-Mercurial planet or planets become better known, that the first reliable observation on this subject was made by Watson.

CHAPTER VII.

MERCURY.

The Ancient Astronomical Discoveries—How Mercury was first found?—Not easily seen—Mercury was known in 265 B.C.—Skill necessary in the Discovery—The Distinction of Mercury from a Star—Mercury in the East and in the West—The Prediction—How to Observe Mercury—Its Telescopic Appearance—Difficulty of Observing its Appearance—Orbit of Mercury—Velocity of the Planet—Can there be Life on the Planet?—Changes in its Temperature—Transit of Mercury over the Sun—Gassendi's Observation—Atmosphere around Mercury—The Weight of Mercury.

LONG and glorious is the record of astronomical discovery. The discoveries of modern days have succeeded each other with such rapidity, they have so often dazzled our imaginations with their brilliancy, that we are sometimes apt to think that astronomical discovery is a purely modern product. But no error could be more fundamentally wrong. While we appreciate to the utmost the achievements of modern times, let us endeavour to do justice to the labours of the astronomers of antiquity.

And when we speak of the astronomers of antiquity, let us understand clearly what is meant. Astronomy is now growing so rapidly that each century witnesses a surprising advance; each generation, each decade, each year, has its own rewards for those diligent astronomers by whom the heavens are so carefully scanned. We must, however, project our glance to a remote epoch in time past, if we would view the memorable discovery of Mercury. Compared with it, the discoveries of Newton are to be regarded as very modern achievements; even the enunciation of the Copernican system of the heavens is itself a recent event in comparison with the discovery of the planet Mercury.

By whom was this great discovery made? Let us see if this question can be answered by examination of astronomical records.

At the close of the memorable life of the great Copernicus he was heard to express his sincere regret that his eyes had never shown him the planet Mercury. Often had he tried to see this planet, whose movements were in such a marked way illustrative of the great theory of the celestial motions which it was his immortal glory to have established; but he had never been successful. Mercury is not easily to be seen, and it may well have been that the vapours from the Vistula obscured the horizon at Frauenburg where Copernicus dwelt, and that thus his opportunities of seeing Mercury were probably even rarer than they are at other places. But one circumstance is plain. The existence of the planet was quite familiar to Copernicus, and therefore we must look to some earlier epoch for its discovery. In the scanty astronomical literature of the Middle Ages we find occasional references to the existence of this planet. We can trace observations of Mercury back through the early centuries to the commencement of our era. Back earlier still it can be followed, until at length we come on the first record of an observation which has descended to us, being one made in the year 265 before the Christian era. It is not pretended, however, that this observation records the *discovery* of the planet. It is merely an observation of an object already well-known, while the earlier observations and any account of the discovery seem to have totally perished. It does not appear in the least degree likely that the discovery was even then a recent one. It may have been that the planet was independently discovered in two or more localities, but all records of such discoveries are totally wanting; and we are ignorant alike of the name of the discoverer, of the nation to which he belonged, and of the epoch in which his great discovery was made.

Although this discovery is of such vast antiquity, although it was made before correct notions were entertained as to the true system of the universe, and, it is needless to add, long before the invention of the telescope, yet it must not be assumed that the discovery of Mercury was by any means a simple or obvious matter. This will be manifest when we try to conceive the manner in which the discovery must have been first accomplished.

Some primæval astronomer, long familiar with the heavens, had learned to recognise the various stars and constellations. Experience had impressed upon him the permanence of these objects; he had seen that Sirius invariably appeared at the same seasons of the year; and he had noticed how it was placed with regard to Orion and the other neighbouring constellations. In the same manner each of the other bright stars was to him a familiar object always to be found in a particular region of the heavens. He saw how the heavens, as a whole, rose and set in such a way, that though each star appeared to move, yet the relative positions of the stars were incapable of alteration. No doubt this ancient astronomer was acquainted with Venus; he knew it as the evening star; he knew it as the morning star; and he was accustomed to regard Venus as a body which oscillated from one side of the sun to the other. We can easily imagine how, in the clear skies of an Eastern desert, the discovery of Mercury was made. The sun has set, the brief twilight has almost ceased, when lo, near that part of the horizon where the glow of the setting sun still illuminates the sky, a bright star is seen. But surely, the careless observer might say, there is nothing wonderful here. Is not the whole heaven spangled with stars? why should there not be a star in this locality also? But the primæval astronomer will not accept this explanation. He knows that there is no bright star at this place in the heavens. If the object of his attention be not a star, what then can it be? Eager to examine this question, the heavens are watched next night, and there again, higher up in the heavens, and more brilliant still, is the object seen the night before. Each successive night the object grows more and more brilliant, until at length it becomes a conspicuous gem. Perhaps it will rise still higher and higher; perhaps it will increase till it attains the brilliancy of Venus itself. Such were the surmises not improbably made by those who first watched this object; but they were not realised. After a few nights of exceptional brilliancy the lustre of this mysterious orb declines. The planet again draws near the horizon at sunset until at length it sets so soon after the sun that it has become invisible. Is it lost for ever? Years may pass away before another good opportunity of observing the object after sunset happens; but then again it will be seen to run through the same series of changes, though, perhaps, under very different circumstances. The greatest height above the horizon, and the greatest brilliancy, both vary enormously. It was not until after long and careful observations had been made, that the primæval astronomer could assure himself that the various appearances could be all attributed to a single body. In the Eastern deserts the phenomena of sunrise must have been nearly as familiar as those of sunset, and in the clear skies, at the point where the sunbeams were commencing to dawn above the horizon, a bright starlike point might sometimes be perceived. Each successive day this object rose higher and higher above the horizon before the moment of sunrise, and its lustre increased with the distance; then again it would draw in towards the sun, and return for a while to invisibility. Such were the data which were presented to the mind of the primitive astronomer. One body was seen after sunset, another body was seen before sunrise. To us it may seem an obvious inference from the observed facts, that the two bodies were identical. The inference is a correct one, but it is in no sense an obvious one. Long and patient observation established the remarkable law that one of these bodies was never seen until the other had disappeared. Hence it was inferred that the phenomena, both at sunrise and at sunset, were due to the same body, which oscillated to and fro about the sun.

We can easily imagine that the announcement of the identity of these two bodies was one which would have to be carefully tested before it could be accepted. How are the tests to be applied in a case of this kind? There can hardly be a doubt that the most complete and convincing demonstration of scientific truth is found in the fulfilment of prediction. When Mercury had been observed for years, a certain regularity in the recurrence of its visibility was noticed. Once this regularity had been fully established, prediction became possible. The time when Mercury would be seen after sunset, the time when it would be seen before sunrise, could be foretold with accuracy! When it was found that these predictions were obeyed to the letter—that the planet was always seen when looked for in accordance with the predictions—it was impossible to refuse assent to the hypothesis on which these predictions were based. Underlying that hypothesis was the assumption that all the various appearances arose from the oscillations of a single body, and hence the discovery of Mercury was established on a basis as firm as the discovery of Jupiter or of Venus.

In the latitudes of the British Islands it is generally possible to see Mercury some time during the course of the year. It is not practicable to lay down, within reasonable limits, any general rule for finding the dates at which the search should be made; but the

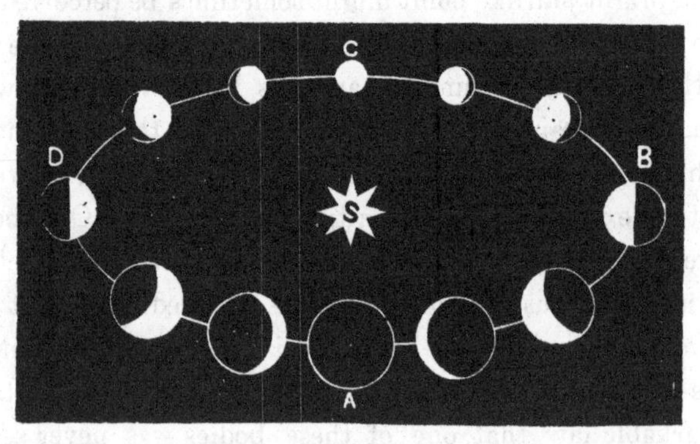

Fig. 39.—The Movement of Mercury, showing the variations in Phase and in apparent size.

student who is determined to see Mercury will generally succeed with a little patience. He must first consult an almanac which gives the planetary positions, and select an occasion when Mercury is stated to be an evening or a morning star. Such an occasion during the spring months is especially suitable, as the elevation of Mercury above the horizon is usually greater then than at other seasons; and in the evening twilight, about three-quarters of an hour after sunset, a view of this shy but beautiful object will reward the observer's attention.

To those astronomers who are provided with equatorial telescopes such instructions are unnecessary. To enjoy a telescopic view of Mercury, we first turn to the nautical almanac, and find the position in which the planet lies. If that position be above our horizon, we can at once direct the telescope to the planet, and even in broad daylight the planet will very often be seen. The telescopic appearance of Mercury is, however, not unfrequently disappointing. Though Mercury is really much larger than the moon, yet it is so very far off that its telescopic appearance is insignificant. There is, however, one feature in a telescopic view which would immediately attract attention. Mercury is then seen, not usually as a circular object, but more or less crescent-shaped, like a miniature moon. The phases of Mercury are also to be accounted for

Fig. 40.—Mercury as a Crescent.

on exactly the same principles as the phases of the moon. Mercury is a globe composed, like our earth, of materials possessing in themselves no source of illumination, but one-half of Mercury must always be turned towards the sun, and this half is accordingly lighted up brilliantly by the rays of the sun. When we look at Mercury we see nothing of the non-illuminated side, and the crescent is due to the fore-shortened view which we obtain of the illuminated half. Mercury is such a small object that, in the glitter of the naked-eye view, the *shape* of the luminous body cannot be defined. Indeed, even in the much larger crescent of Venus, the aid of the telescope has to be invoked before the crescent form can be observed. Beyond, however, the fact that Mercury is a crescent, and that it undergoes varying phases in correspondence with the changes in its relative position to the earth and the sun, we cannot see much of the planet. It is too small and too bright to admit of the delineation of details on its surface. No doubt attempts have been made, and observations have been recorded as to certain physical features on the planet. It has been supposed that traces of mighty mountains have been seen, and that the existence of an atmosphere surrounding the planet has been indicated. But such statements must be received with very great hesitation, if not with actual discredit, and beyond this mere allusion they need not further engage our attention.

The facts that are thoroughly established with regard to Mercury are mainly the numerical statements with regard to the path it describes around the sun. The time taken by the planet to complete a revolution is very nearly eighty-eight days. The average distance from the sun is about 36,000,000 miles, and the average velocity with which the planet moves is over twenty-nine miles a second. We have already alluded to the most characteristic and remarkable feature of the orbit of Mercury. That orbit differs from the paths of all the other large planets by its much greater departure from the circular form. In the majority of cases the planetary orbits are so little elliptic that a diagram of the orbit drawn accurately to scale would not be perceived to differ from a circle unless careful measures were made. In the case of Mercury the circumstances are different. The elliptic form of the path would be quite unmistakeable by the most casual observer. The distance from the sun to the planet fluctuates between very considerable limits. The lowest value it can attain is about 30,000,000 miles; the highest value is about 43,000,000 miles. In accordance with Kepler's second law, the velocity of the planet must exhibit corresponding changes. It must sweep rapidly around that part of his path near the sun, and more slowly round the remote parts of his path. The greatest velocity is about thirty-five miles a second, and the least is twenty-three miles a second.

For an adequate perception of the movements of Mercury, we ought not to dissociate the magnitude of the velocity from the vast dimensions of the body by which that velocity is performed. No doubt a velocity of twenty-nine miles in a second is enormously great when compared with ordinary velocities. The velocity of Mercury is not less than a hundred times as great as the velocity of the rifle-bullet. But when we compare the sizes of the bodies with their velocities the velocity of Mercury seems relatively much less than that of the bullet. A rifle-bullet traverses a distance equal to its own diameter many thousands of times in a second. But even though Mercury is moving so much faster, yet the diameter of the planet is so considerable that a period of two minutes will be required for it to move through a space equal to its diameter. Viewing the globe of the planet as a whole, the velocity of its movement is but a stately and dignified progress appropriate to its dimensions.

As we can learn little or nothing of the true surface of Mercury, it is utterly impossible for us to say whether life can exist on the surface of that planet. We may, however, not unreasonably conclude that there can hardly be life on Mercury at all analogous to the life which we know on the earth. The heat and the light of the sun beat down on Mercury with an intensity many-fold greater than we experience on the earth. When Mercury is at its longest distance from the sun, the intensity of solar heat is even then more than four times as great as the greatest heat which ever reaches the earth. But when Mercury, in the course of its remarkable changes of distance, draws in to the warmest part of its orbit, it is exposed to an appalling scorching. The intensity of the sun's heat must then be not less than nine times as great as the greatest radiation to which we are exposed. These changes succeed each other much more rapidly than the variations of our seasons. On Mercury the interval between midsummer to midwinter is only forty-four days, while the whole year is only eighty-eight days. These rapid and tremendous changes in solar heat must in themselves exercise a profound effect on the habitability of Mercury. Mr. Ledger well remarks, in his most interesting work,* that if there be inhabitants on Mercury the words "perihelion" and "aphelion," which are here often regarded as expressing ideas of an intricate or recondite character, must, on the

* "The Sun: its Planets, and their Satellites." London: 1882 (page 147).

surface of that planet, be familiar to everybody. The words imply, "near the sun," and "away from the sun;" but we do not associate these expressions with any obvious phenomena, because the changes in the distance from the earth to the sun are so inconsiderable. But in Mercury, where in six weeks the sun rises to more than double his apparent size, and gives more than double the quantity of light and of heat, such changes must be familiar to everyone. Perihelion and aphelion will embody ideas obviously and intimately connected with the whole economy of the planet.

It is nevertheless rash to found any inferences as to climate merely upon the proximity or the remoteness of the sun. Climate depends upon other matters besides the sun's distance. The atmosphere surrounding the earth has a profound influence on our climate, and if Mercury have an atmosphere—as has often been supposed—its climate may be thereby modified to an enormous extent. It seems, however, hardly possible to suppose that any atmosphere could form an adequate protection for the inhabitants from the violent and enormous fluctuations of solar radiation. All we can say is, that the problem of life in Mercury belongs to the class of unsolved, and perhaps unsolvable, mysteries.

It was in the year 1627 that Kepler made an important announcement of impending astronomical events. Kepler had been studying profoundly the movements of the planets. He had examined the former observations which had been made; and from his study of the past he had ventured to predict the future. Kepler announced that in the year 1631 the planets Venus and Mercury would both make a transit across the sun, and he assigned the dates to be November 7th for Mercury, and December 6th for Venus. This was at the time a very remarkable prediction. We are so accustomed to turn to our almanacs and learn from thence all the astronomical phenomena which are anticipated during the year, that we are apt to forget that in early times this was impossible. It has only been by slow degrees that astronomy has been rendered so perfect as to enable us to predict, with accuracy, the occurrence of the more delicate phenomena. The announcement of those transits by Kepler, some years before they occurred, was justly regarded at the time as a most remarkable achievement. The illustrious Gassendi prepared to apply the test of actual observation to the predictions of Kepler. We can now indicate the time of the transit accurately to a few minutes, but in those early attempts equal precision was not to be thought of. Gassendi considered it necessary to commence watching for the transit of Mercury two whole days before the time indicated by Kepler, and he had arranged a most ingenious plan for studying the transit. The light of the sun was admitted into a darkened room through a hole in the shutter, and by a lens an image of the sun was formed on a white screen. This is, indeed, an admirable and a very pleasing way of studying the surface of the sun, and even at the present day, with our best telescopes, one of the best-known methods of viewing the sun is founded on the same principle. Gassendi commenced his watch on the 5th of November, and carefully studied the sun's image at every available opportunity. It was not, however, until five hours after the time assigned by Kepler that the transit of Mercury actually commenced. Gassendi's preparations had been made with all the resources which he could command, but these resources seem very imperfect when compared with the appliances of our modern observatories. He was anxious to note the time when the planet appeared, and for this purpose he had stationed an assistant in the room beneath, who was to observe the altitude of the sun at the moment indicated by Gassendi. The signal to the assistant was to be conveyed by a very primitive apparatus. Gassendi was to stamp on the floor when the critical moment had arrived. In spite of the long delay, which exhausted the patience of the assistant, some valuable observations were obtained, and thus the first transit of a planet over the sun was observed.

The transits of Mercury are chiefly of importance on account of the accuracy which their observation infuses into our calculations of the movements of the planet. It has often been hoped that the observations during a transit would present to us reliable information as to the physical character of the globe of Mercury. To some extent—but not, unhappily, to any large extent—these hopes have been realised. Skilful observers have described the appearance of the planet in transit as that of a round, dark spot, surrounded by a luminous margin of a depth variously estimated on different occasions at from one-third to two-thirds of the planet's diameter. There can be hardly a doubt that a dense atmosphere surrounding Mercury would be capable of producing an appearance resembling that which has been described, and, therefore, the probability that such an atmosphere really exists must be admitted. On the other hand, the measurements of the intensity of light from Mercury seem to make the existence of an atmosphere somewhat doubtful.

Here we take leave of the planet Mercury—an interesting and beautiful object which stimulates our intellectual curiosity, while at the same time eluding all our attempts to obtain more complete knowledge. There is, however, one point of attainable knowledge which we have not touched on in this chapter. It is a difficult, but not by any means an impossible task to weigh Mercury in the celestial balance, and determine his mass in comparison with the other globes of our system. This is a delicate operation, but it leads us through some of the most interesting paths of astronomical discovery. The weight of the planet, as recently determined by Von Asten, is about one-twenty-fourth part of the weight of the earth.

CHAPTER VIII.

VENUS.

Interest attaching to this Planet—The Unexpectedness of its Appearance—The Evening Star—Visibility in Daylight—Only Lighted by the Sun—The Phases of Venus—Why the Crescent is not Visible to the Unaided Eye—Variations in the Apparent Size of the Planet—Resemblance of Venus to the Earth—The Transit of Venus—Why of such especial Interest—The Scale of the Solar System—Orbits of the Earth and Venus not in the same Plane—Recurrence of the Transits in Pairs—Appearance of Venus in Transit—Transits of 1874 and 1882—The Early Transits of 1631 and 1639—The Observations of Horrocks and Crabtree—The Announcement of Halley—How the Track of the Planet differs from different places—Illustrations of Parallax—Voyage to Otaheite—The result of Encke—Probable Value of the Sun's Distance—Observations of the recent Transit of Venus at Dunsink—The Question of an Atmosphere to Venus—Dr. Copeland's Observations—Utility of such Researches—Other Determinations of the Sun's Distance—Statistics about Venus.

It might, for one reason, be not inappropriate to have commenced our review of the planetary system by the description of the planet Venus. This planet is not especially remarkable for its size, for there are other planets hundreds of times larger. The orbit of Venus is no doubt larger than that of Mercury, but it is much smaller than that of the outer planets. Venus has not even the splendid retinue of minor attendants which give such dignity and such interest to the mighty planets of our system. Yet the fact still remains that Venus is peerless among the planetary host. We speak not now of spectacles only seen in the telescope, we refer to the ordinary observation which detected Venus ages before telescopes were invented. Who has not been delighted with the view of this glorious object? It is not to be seen at all times. For months together the beauties of Venus are hidden from mortal gaze. Its beauties are even enhanced by the caprice and the uncertainty that attends its appearance. We do not say that there is any caprice in the movements of Venus, as known to those who diligently consult their almanacs. The movements of the lovely planet are there prescribed with a prosaic detail hardly in correspondence with the character of the goddess of love. But to those who do not devote particular attention to the stars, the very unexpectedness of the appearance of Venus is one of its greatest charms. Venus has not been noticed, not been thought of, for many months. It is a beautiful spring evening; the sun has just set. The lover of nature turns to admire the sunset, as every lover of nature will. In the golden glory of the west a beauteous gem is seen to glitter; it is the evening star—the planet Venus. A week or two later another beautiful sunset is seen, and now the planet is no longer merely a glittering point low down; it has risen high above the horizon, and continues a brilliant object long after the shades of night have descended. Again, a little longer, and Venus has gained its full brilliancy and splendour. All the heavenly host—even Sirius and even Jupiter—must pale before the splendid lustre of Venus, the unrivalled queen of the firmament.

After weeks of splendour the height of Venus at sunset diminishes, and its lustre begins gradually to decline. It sinks to invisibility, and is forgotten by the great majority of mankind; but the capricious goddess has only moved from the west to the east. Ere the sun rises the morning star will be seen in the east. Its splendour gradually augments until it rivals the beauty of the evening star. Then again the planet draws near to the sun, and remains lost to view for many months, until the same cycle of changes recommences, after an interval of a year and seven months.

When Venus is at its brightest, it can be easily seen in broad daylight with the unaided eye. This striking spectacle proclaims in an unmistakable manner the unrivalled supremacy of Venus as compared with the other planets and with the fixed stars. Indeed, at this time Venus is from forty to sixty times as bright as the brightest star in the northern heavens.

The beautiful evening star is often such a very brilliant object, that it may seem difficult at first to realise that Venus is not self-luminous. Yet it is impossible to doubt that the planet is really only a dark globe, and to that extent resembling our own earth. The brilliancy of the planet is not so very much greater than the brilliancy of the earth on a sunshiny day, and the splendour of Venus entirely arises from the reflected light of the sun, in the manner already explained with respect to the moon.

We cannot, however, distinguish the beautiful crescent shape of the planet by the unaided eye, which merely shows a brilliant point too small to possess sensible form. This is to be explained on physiological grounds. The optical contrivances in the eye form an image of the planet on the retina; this image is very small. Even when Venus is nearest to the earth, the diameter of the planet subtends an angle not much more than one minute of arc. On the retina of the eye a picture of Venus is thus drawn about one six-thousandth part of an inch in diameter. Great as may be the delicacy of the retina, it is not adequate to the perception of form in a picture so minute. The nervous structure, which has been described as the source of vision, forms too coarse a canvas for the reception of the details of this tiny picture. Hence it is that to the unaided eye the brilliant Venus appears merely as a bright spot. The unaided eye cannot tell what shape it has; still less can it reveal the true beauty of the crescent. If the diameter of Venus were several times as great as it actually is; were Venus, for instance, as large as Jupiter or some of the other great planets, then its crescent could be readily discerned by the unaided eye. It is curious to speculate on what might have been the history of astronomy had Venus only been as large as Jupiter. Were every one able to see the crescent form without a telescope, it would then have been an elementary and almost obvious truth that Venus was a dark body revolving round the sun. The analogy between Venus and our earth would have been at once perceived; and the great theory which was left to be discovered by Copernicus in comparatively modern times, might not improbably have been handed down to us with the other discoveries which have come from the ancient nations of the East.

In Fig. 41 we have three views of Venus under different aspects. The planet is so much closer to the earth when the crescent is seen, that it appears to be part of a much larger circle than that made by Venus when more nearly full. This drawing shows the different aspects of the planet in their true relative proportions. It is very difficult to perceive distinctly any markings on the excessively brilliant surface. Sometimes observers have seen spots or other features, and occasionally the pointed extremities of the horns have been irregular, as if to show that the surface of Venus is not smooth. Attempts have even been made to prove from such

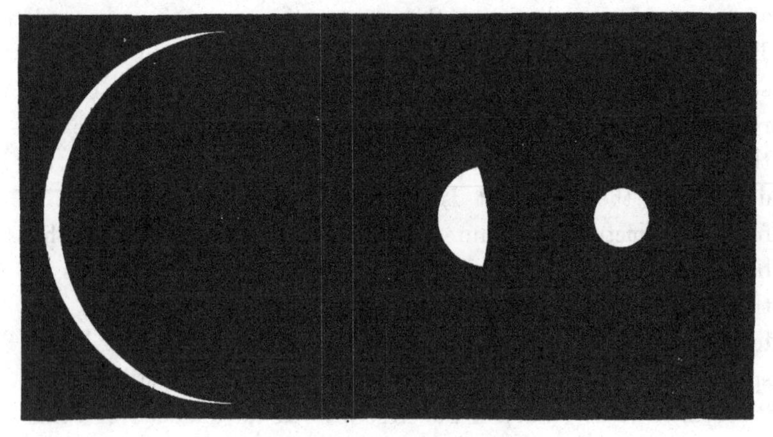

Fig. 41.—Different Aspects of Venus in the Telescope.

observations that there must be lofty mountains in Venus, but we cannot place much confidence in the results.

It so happens that our earth and Venus are very nearly equal in bulk. The difference is hardly perceptible, but the earth has a diameter a few miles greater than that of Venus. It is almost equally remarkable that the time of rotation of Venus on its axis seems to be very nearly equal to the time of rotation of the earth on its axis. The earth rotates once in a day, and Venus in about half an hour less than one of our days. There are also indications of the existence of an atmosphere around Venus, but we have no means of knowing at present what the gases may be of which that atmosphere is composed.

If there be oxygen in the atmosphere of Venus, then it would seem possible that there might be life on Venus which was not of a very different character from life on the earth. No doubt the sun's heat on Venus is greatly in excess of the sun's heat with which we are acquainted, but this is a difficulty not insuperable. We see at present on the earth, life in very hot regions and life in very cold regions. Indeed, as we go into the tropics we find life more and more exuberant, so that should water be present on the surface of Venus and oxygen in its atmosphere, we might expect to find in that planet a luxuriant tropical life, of a kind perhaps analogous in some respects to life on the earth.

In our account of the planet Mercury, as well as in our brief description of the hypothetical planet Vulcan, it has been necessary to allude to the phenomena presented by the transit of a planet over the face of the sun. This phenomenon, always of interest and always meriting the attention of astronomers, is especially noticeable in the case of Venus. The transit of Venus rises in fact, to, an importance hardly surpassed by any other phenomenon in our system, and hence it will necessarily engage our attention in the present chapter. We have in recent years had the opportunity of witnessing two of these rare occurrences. No future transit can occur till after this generation shall have passed away; not, indeed, until the June of A.D. 2004. It is hardly too much to assert that the recent transit of 1882 and the previous one of 1874 have received a degree of attention never before accorded to any astronomical phenomenon.

The transit of Venus is hardly to be described as a very striking or beautiful spectacle. It is not nearly so fine a sight as a great comet or a shower of shooting stars. Why is it, then, that the transit of Venus is regarded as of such great scientific importance? It is because the transit enables us to solve one of the greatest problems which has ever engaged the mind of man. It is by the transit of Venus that we attempt to determine the scale on which our solar system is constructed. Truly this is a noble problem. Let us dwell upon it for a moment. In the centre of our system we have the sun—a majestic globe more than a million times as large as the earth. Circling round the sun we have the planets, of which our earth is but one. There are hundreds of small planets. There are a few comparable with our earth; there are others vastly surpassing the earth. Besides the planets there are other bodies in our system. Many of the planets are accompanied by systems of revolving moons. There are hundreds, perhaps thousands, of comets, while the minor bodies of our system exist in countless millions. Each member of this stupendous host moves in a prescribed orbit around the sun, and collectively they form the solar system.

It is comparatively easy to learn the *shape* of this system, to measure the relative distances of the planets from the sun, and even the relative sizes of the planets themselves. Peculiar difficulties are, however, experienced when we seek to ascertain the actual *size* of the system, as well as its shape. It is this latter question which the transit of Venus enables us to solve.

Look, for instance, at an ordinary map of Europe. We see the various countries laid down with precision; we can tell the courses of the rivers; we can say that France is larger than England, and Russia larger than France; but no matter how perfectly the map may be constructed, something else is necessary before we can have a complete conception of the dimensions of the country. We must know *the scale on which the map is drawn*. The map contains a reference line with certain marks upon it. This line is to give the scale of the map. Its duty is to tell us that an inch on the map corresponds with so many miles on the actual surface. Without being supplemented by the scale, the map would for many purposes be quite useless. Suppose that we consulted a map in order to choose a route from London to Vienna, the map at once points out the direction to be taken and the various towns and countries to be traversed; but unless we consult the little scale in the corner, the map will not tell how many miles long the journey is to be.

A map of the solar system can be readily constructed. We can draw on it the orbits of some of the planets and of their satellites, and we can include many of the comets. We can give to the sun and to the planets their proper bulk. But to render the map quite efficient something more is necessary. We must have the scale which is to tell us how many millions of miles on the heavens will correspond to one inch of the map. It is at this point we encounter a difficulty. It is comparatively easy to have all the *relative* sizes of the orbits of the different bodies correct—very simple observations suffice for this purpose—but it is not at all easy to assign, with accuracy, the correct scale of the celestial map. There are, however, several ways of solving the problem, though they are all difficult and laborious. The most celebrated method is that presented on an occasion of the transit of Venus. Herein, then, lies the importance of the transit of Venus. It is one of the best known means of finding the actual scale on which our system is constructed. Observe the full importance of the problem. Once the transit of Venus has given us the scale, then all is known. We know the size of the sun; we know his distance; we know the bulk of Jupiter, and the distances at which his satellites revolve; we know the dimensions of the comets, and the number of miles to which they recede in their wanderings; we know the velocity of the shooting stars; and we learn the important lesson that our earth is but one of the minor members of the sun's majestic family.

As the path of Venus lies inside that of the earth, and as Venus moves more quickly than the earth, it follows that the earth is frequently passed by Venus, and just at the moment of passing it will sometimes happen that the earth, the planet, and the sun lie in the same straight line. We can then see Venus on the face of the sun, and this is the phenomenon which we call the transit of Venus. It is, indeed, quite plain that if the three bodies were exactly in a line, an observer on the earth, looking at the planet, will see it brought out vividly against the brilliant background of the sun.

Considering that the earth is overtaken by Venus once every nineteen months, it might be thought that the transits of Venus should occur with corresponding frequency. This is not the case; the transit of Venus is an exceedingly rare occurrence, and a hundred years or more will often elapse without a single transit taking place. This rarity of the transits arises from the fact that the path of the planet is inclined to the plane of the earth's orbit; so that while in half of its path Venus is above the plane of the earth's orbit, in the other half it is below. When Venus overtakes the earth, the line from the earth to Venus will therefore usually pass over or under the sun. If, however, it should happen that Venus overtakes the earth at or near either of the points in which the plane of the orbit of Venus passes through that of the earth, then the three bodies will be in line, and a transit of Venus will be the consequence. The rarity of the occurrence of a transit need no longer be a mystery. The earth passes through one of the critical parts every December, and through the other every June. If, therefore, it so happens that the conjunction of Venus occurs at, or close to, June 6th or December 7th, then a transit of Venus will occur at that conjunction, but under no other circumstances.

The most remarkable law with reference to the recurrence of the phenomenon is the well-known eight-year interval. The transits may be all grouped together into pairs; the two transits of any single pair being separated by an interval of eight years. For instance, a transit of Venus took place in 1761, and again in 1769. No further transits occurred until those recently witnessed in 1874 and in 1882. Then, again, comes a long interval, for another transit will not occur until 2004, but then it will be followed by another in 2012.

This recurrence of the transits in pairs admits of a very simple explanation. It so happens that the periodic time of Venus bears a remarkable relation to the periodic time of the earth. Venus accomplishes thirteen revolutions around the sun in very nearly the same time that the earth requires for eight revolutions. If, therefore, in 1874, Venus and the earth were in line with the sun, then in eight years more the earth will again be found in the same place; and so will Venus, for it has just been able to accomplish thirteen revolutions. A transit of Venus having occurred on the first occasion, a transit must also occur on the second.

It is not, however, to be supposed that every eight years the planets will again regain the same position with sufficient precision for a regular eight-year transit interval. For it is only approximately true that thirteen revolutions of Venus are coincident with eight revolutions of the earth. Each conjunction after an interval of eight years takes place at a slightly different position of the planets,

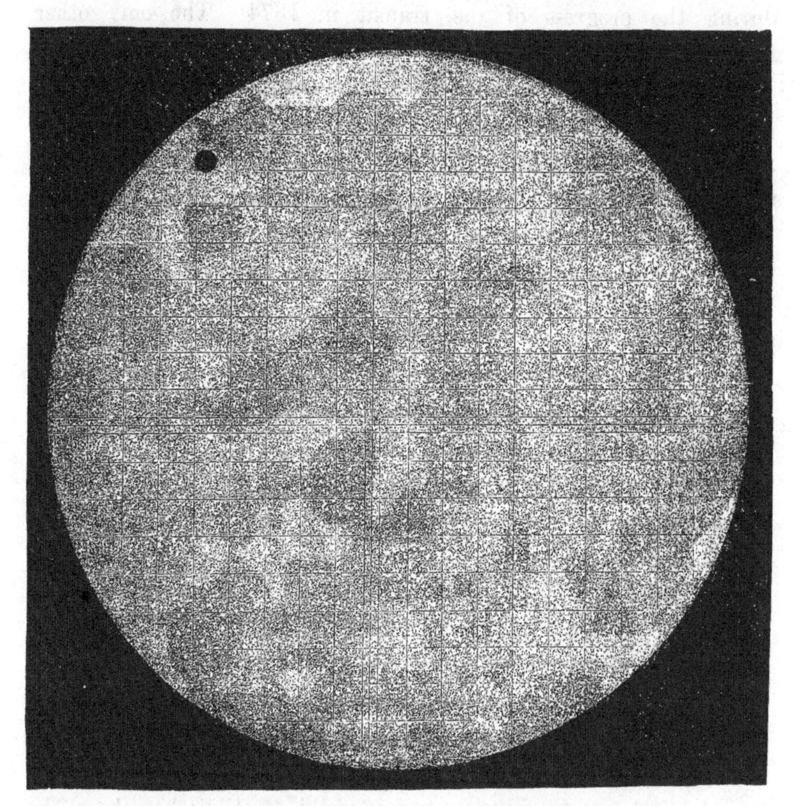

Fig. 42.—Venus on the Sun at the Transit of 1874.

so that when the two planets come together again in the year 1890 the point of conjunction will be so far removed from the critical point that the line from the earth to Venus will not intersect the sun, and thus, although Venus will pass very near the sun, yet no transit will take place.

Fig. 42 represents the transit of Venus in 1874. It is taken from a photograph obtained, during the occurrence of the transit, by M. Janssen. His telescope was directed towards the sun during the eventful minutes, and thus an image of the sun was depicted on the photographic plate placed in the telescope. The circular margin represents the disc of the sun. On that disc we see the round sharp image of the planet Venus, showing the appearance of the planet during the progress of the transit in 1874. The only other features to be noticed are a few spots on the sun rather dimly shown, and a network of lines which were stretched across the field of view of the telescope to facilitate the measurements. It might be supposed that the appearance of Venus in front of the sun could be mistaken for one of the spots in the sun, which are often large and round, and occasionally have even simulated the appearance of a planet. But this view will not bear examination. The occurrence of the transit at the predicted moment, and at the precise point of the sun's margin which the calculations had indicated, the sharpness of the planet's shape, and the circumstances of its motion, all discriminate the planet as something totally distinct from an ordinary sun spot.

The adjoining sketch exhibits the course which the planet pursued in its course across the sun on the two occasions in 1874 and 1882. Our generation has had the good fortune to witness the two occurrences indicated on this picture. The white circle denotes the disc of the sun; the planet enters on the white surface, and at first is like a bite out of the sun's margin. Gradually the black spot steals in front of the sun, until, after nearly half an hour, the black disc is entirely visible. Slowly the planet wends its way across, followed by hundreds of telescopes from every accessible part of the globe whence the phenomenon is visible, until at length, in the course of a few hours, the planet emerges from the other side.

It will be useful to take a brief retrospect of the different transits of Venus of which there is any historical record. They are not numerous. Doubtless hundreds of transits have occurred since man first came on the earth. It was not until the approach of the year 1631 that attention began to be directed to the matter, though the transit which undoubtedly occurred in that year was not, so far as we know, observed by any one.

The success of Gassendi, in observing the transit of Mercury, to which we referred in the last chapter, led him to hope that he would be equally fortunate in observing the transit of Venus, which Kepler had also foretold. Gassendi looked at the sun on the 4th, 5th, and 6th December. He looked at it again on the 7th, but he saw no sign of the planet. We now know the reason. The transit

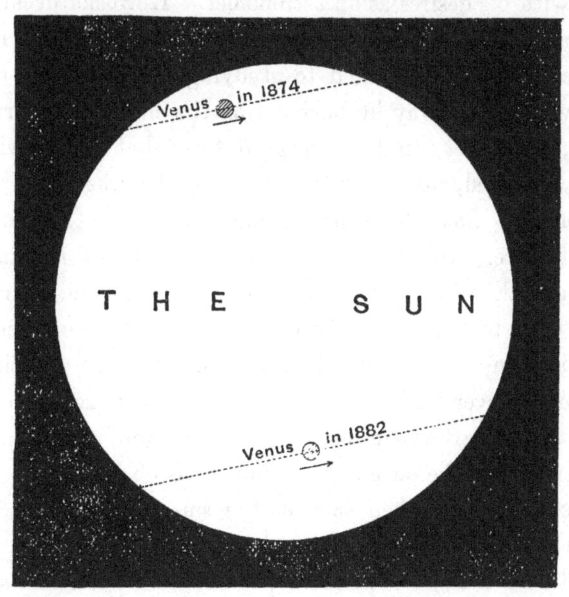

Fig. 43.—The path of Venus across the Sun in the Transits of 1874 and 1882.

of Venus took place during the night, between the 6th and the 7th, and would therefore have been invisible to European observers.

Kepler had supposed that, after the transit of 1631, there would not be another until 1761, but in this the usual acuteness of Kepler seems to have deserted him. He appears not to have fully appreciated the remarkable eight-year period, which necessitates that the transit of 1631 would be followed by another in 1639. This transit of 1639 is the one with which the history of the subject may be said to commence. It was the first occasion on which the transit was ever actually witnessed; nor was it then seen by many. So far as is known, that transit was only witnessed by two persons.

A young and ardent English astronomer, named Horrocks, had undertaken some computations about the motions of Venus. Horrocks made the discovery that the transit of Venus would be repeated in 1639, and he prepared to observe it. The sun rose on the morning of the eventful day—which happened to be a Sunday. The clerical profession, which Horrocks followed, here came into collision with his desires as an astronomer. Horrocks decided that his clerical duties must be performed, but that every spare moment of the day should be devoted to studying the sun. At nine, he says, he was called away by business of the highest importance—referring, no doubt, to his official duties; but the service was quickly performed, and a little before ten he was again on the watch, only to find the brilliant face of the sun without any unusual feature. It was marked with a spot, but nothing that could be mistaken for a planet. Again, at noon, came an interruption; he went to church, but he was back by one. Nor were these the only interruptions to his observations. The sun was also more or less clouded over during part of the day. However, at a quarter past three in the afternoon his clerical duties were over; the clouds had dispersed, and he once more resumed his observations. To his incredible delight he then saw on the sun the round dark spot, which he at once identified as the planet Venus. The observations could not last long; it was the depth of winter, and the sun was rapidly setting. Only half an hour was available, but he had made such careful preparations beforehand that half an hour sufficed to enable him to secure careful and exact measurements.

Horrocks had previously acquainted his friend, William Crabtree, with the impending occurrence. Crabtree was therefore on the watch, and succeeded in seeing the transit. But to no one else had Horrocks communicated the intelligence; as he says, "I hope to be excused for not informing other of my friends of the expected phenomenon, but most of them care little for trifles of this kind, rather preferring their hawks and hounds, to say no worse; and although England is not without votaries of astronomy, with some of whom I am acquainted, I was unable to convey to them the agreeable tidings, having myself had so little notice."

It was not till long afterwards that the full importance of the transit of Venus was perceived. Nearly a century had rolled away when the great astronomer, Halley (1656—1741), drew attention to the subject. The next transit was to occur in 1761, and forty-five years before that event Halley explained his celebrated method of finding the distance of the sun by means of the transit of Venus. Halley was then a man sixty years of age; he could have no expectation that he would live to witness the transit; but in noble language he commends the problem to the notice of the learned, and thus addresses the Royal Society of London:—"And this is what I am now desirous to lay before this illustrious Society, which I foretell will continue for ages, that I may explain beforehand to young astronomers, who may perhaps live to observe these things, a method by which the immense distance of the sun may be truly obtained. . . . I recommend it, therefore, again and again to those curious astronomers who, when I am dead, will have an opportunity of observing these things, that they would remember this my admonition, and diligently apply themselves with all their might in making the observations, and I earnestly wish them all imaginable success—in the first place, that they may not by the unseasonable obscurity of a cloudy sky be deprived of this most desirable sight, and then that, having ascertained with more exactness the magnitudes of the planetary orbits, it may redound to their immortal fame and glory." Halley lived to a good old age, but he died nineteen years before the transit occurred.

The student of astronomy who desires to learn how the transit of Venus will tell the distance from the sun must prepare to encounter a geometrical problem of no little complexity. We cannot give to the subject the detail that would be requisite for a full explanation. All we can attempt is to render a general account of the method, sufficient to enable the reader to see that the transit of Venus really does contain all the elements necessary for the solution of the problem.

We must first explain clearly the conception which is known to astronomers by the name of *parallax;* for it is by parallax that the distance of the sun, or, indeed, the distance of any other celestial body, must be determined. Let us take a simple illustration. Stand near a window from whence you can look at buildings, or the trees, the clouds, or any distant objects. Place on the glass a thin strip of paper vertically in the middle of one of the panes. Close the right eye, and note with the left eye the position of the strip of paper relatively to the objects in the background. Then, while still remaining in the same position, close the left eye and again observe the position of the strip of paper with the right eye. You will find that the position of the paper on the background has changed. As I sit in my study and look out of the window I see a strip of paper, with my right eye, in front of a certain bough on a tree a couple of hundred yards away; with my left eye the paper is no longer in front of that bough; it has moved to a position near the edge of the tree. This apparent displacement of the strip of paper, relatively to the distant background, is what is called parallax.

Move closer to the window, and repeat the observation, and you find that *the apparent displacement of the strip increases.* Move away from the window, and the displacement decreases. Move to the other side of the room, the displacement is much less, though probably still visible. We thus see that the change in the apparent place of the strip of paper, as viewed with the right eye or the left eye, varies in amount as the distance changes; but it varies in the opposite way to the distance, for as either becomes greater the other becomes less. We can thus associate with each particular distance a corresponding particular displacement. From this it will be easy to infer, that if we have the means of measuring the amount of displacement, then we have the means of calculating the distance from the observer to the window.

It is this principle, applied on a gigantic scale, which enables us to measure the distances of the heavenly bodies. Look, for instance, at the planet Venus; let this correspond to the strip of paper, and let the sun, on which Venus is seen in the act of transit, be the background. Instead of the two eyes of the observer, we now place two observatories in distant regions of the earth; we look at Venus from one observatory, we look at it from the other; we measure the amount of the displacement, and from that we calculate the distance of the planet. All depends, then, on the means which we have of measuring the displacement of Venus as viewed from the two different stations. There are various ways of accomplishing this, but the most simple is that originally proposed by Halley.

From the observatory at A Venus seems to pursue the upper of the two traeks shown in the adjoining figure. From the observatory at B it follows the lower track, and it is for us to measure the

Fig. 44.—To illustrate the observation of the Transit of Venus from two localities, A and B, on the Earth.

distance between the two tracks. This can be accomplished in several ways. Suppose the observer at A note the time that Venus has occupied in crossing the disc, and that similar observations be made at B. As the track seen from B is the larger, it must follow that the time observed at B will be greater than that at A. When the observers from the different hemispheres come together and compare their observations, the *times* observed will enable the lengths of the tracks to be calculated. The lengths being known, their places on the circular disc of the sun are determined, and hence the amount of displacement of Venus in transit is ascertained. Thus it is that the distance of Venus is measured, and thus the scale of the solar system is known.

The two transits to which Halley's memorable researches referred occurred in the years 1761 and 1769. The results of the first were not very successful, in spite of the arduous labours of those who undertook the observations; but the transit of 1769 will be for ever memorable, not only on account of the determination of the sun's distance, but as giving rise to the first of the celebrated voyages of Captain Cook. It was to see the transit of Venus, that Captain Cook was commissioned to sail to Otaheite, and there, on the 3rd of June, on a splendid day in that most exquisite climate, the transit of Venus was carefully observed and measured by different observers. Simultaneously with these observations others were obtained in Europe, and from the combination of the two the first accurate knowledge of the sun's distance was gained. The most complete discussion of these observations did not, however, take place for some time. It was not until the year 1824 that the illustrious Encke computed the distance of the sun, and gave as the definite result 95,000,000 miles.

For many years this number was invariably adopted, and most of the present generation will remember how they were taught in their school-days that the sun was 95,000,000 miles away. At length doubts began to be whispered as to the accuracy of this result. The doubts arose in different quarters, and were presented with different degrees of importance; but they all pointed in one direction, they all indicated that the distance of the sun was not really so great as the result which Encke had obtained. It must be remembered that there are several ways of finding the distance of the sun, and it will be our duty to allude to some other methods later on. It has been ascertained that the result obtained by Encke was too great, and that the distance of the sun may probably be now stated at 92,700,000 miles.

I venture to record our personal experience of the last transit of Venus, which we had the good fortune to view from Dunsink Observatory on the afternoon of the 6th of December, 1882.

The morning of the eventful day appeared to be about as

unfavourable for a grand astronomical spectacle as could well be imagined. Snow, a couple of inches thick, covered the ground, and more was falling, with but little intermission, all the forenoon. It seemed almost hopeless that a view of the great event could be obtained from this observatory; but it is well in such cases to bear in mind the injunction given to the observers on a celebrated eclipse expedition. They were instructed, no matter what the day should be like, that they were to make all their preparations precisely as they would have done were the sun shining in undimmed splendour. By this advice no doubt many observers have profited; and we acted upon it here with very considerable success.

We have at this observatory two equatorials, one of them an old, but tolerably good instrument, of about six inches aperture, the other the great South equatorial of twelve inches aperture already referred to. At eleven o'clock the day looked worse than ever; but we at once proceeded to make all ready. I stationed Mr. Rambaut at the small equatorial, while I myself took charge of the South instrument. The snow was still falling when the domes were opened: but, according to our pre-arranged scheme, the telescopes were directed, not indeed upon the sun, but to the place where we knew the sun was, and the clockwork was set in motion, which carried round the telescopes, still constantly pointing towards the invisible sun. The predicted time of the transit had not yet arrived. Mr. Hind, the distinguished superintendent of the "Nautical Almanac," had kindly sent us his computations, showing that, viewed from Dunsink, the transit ought to commence at 1 h. 35 min. 48 secs. mean time at Dublin, and that the point on the sun's disc where the planet would enter was 147° from the north point of the sun round by east. This timely intimation was of twofold advantage. It told us, in the first place, the precise moment when the event was to be expected; and, what was perhaps quite as useful, it told us the exact point of the sun to which the attention was to be directed. This is a matter of very considerable importance, for in a large telescope it is only possible to see a part of the sun at once, and therefore, unless the proper part of the sun be placed in the field of view, the phenomenon may be entirely missed.

The eye-piece employed on the South equatorial may also receive a brief notice. It will, of course, be obvious that the full glare of the sun must be greatly mitigated before the eye can view it with impunity. The light from the sun falls upon a piece of transparent glass inclined at a certain angle, and the chief portion of the sun's heat, as well as a certain amount of its light, pass through the glass and are lost. A certain fraction of the light is, however, reflected from the glass, and enters the eye-piece. This light is already much reduced in intensity, but it undergoes as much further reduction as we please by an ingenious contrivance. The glass which reflects the light does so at what is called the polarising angle, and between the eye-piece and the eye is a plate of tourmaline. This plate of tourmaline can be turned round by the observer. In one position it hardly interferes with the light at all, while in the position at right angles thereto it cuts off nearly the whole of the light. By simply adjusting the position of the tourmaline, the observer has it in his power to render the image of any brightness that may be convenient, and thus the observations of the sun can be conducted with the appropriate degree of illumination.

But such appliances seemed on this occasion to be a mere mockery. The tourmaline was all ready, but up to one o'clock not a trace of the sun could be seen. Shortly after one o'clock, however, we noticed that the day was getting lighter; and, on looking to the north, from whence the wind and the snow were coming, we saw, to our inexpressible delight, that the clouds were breaking. At length the sky towards the south began to improve, and at last, as the critical moment approached, we could detect the spot where the sun was becoming visible through the clouds. But Mr. Hind's predicted moment arrived and passed, and still the sun had not broken through the clouds, though every moment the certainty that it would do so became more apparent. The external contact was therefore missed. We tried to console ourselves by the reflection that this was not, after all, a very important phase, and hoped that the internal contact would be more successful.

At length the struggling sun pierced the clouds, and I saw

the round, sharp disc of the sun in the finder, and eagerly glanced at the point on which attention was concentrated. Some minutes had now elapsed since Mr. Hind's predicted moment of first contact, and, to my delight, I saw the small dark bite out of the sun showing that the transit of Venus had commenced, and that the planet was then one-third on the sun. But the most critical moment had not yet arrived. By the expression "first internal contact" we are to understand the moment when the planet is just *exactly on* the sun. This first contact was timed to occur twenty-one minutes later than the external contact already referred to. But again the clouds disappointed our hope of seeing the internal contact. While steadily looking at the exquisitely beautiful sight of the gradual advance of the planet, I became aware that there were other objects besides Venus between me and the sun. These objects were snowflakes, which again began to fall rapidly. They were, I must admit, most singularly beautiful. The telescopic effect of a snowstorm with the sun as a background I had never before seen. It was a most remarkable sight, and reminded me of the golden rain which is sometimes seen during pyrotechnic displays; but I would gladly have dispensed with the spectacle, for it necessarily followed that the sun and Venus again disappeared from view. The clouds gathered, the snowstorm descended as heavily as ever, and we hardly dared to hope that we should see anything more. 1 h. 57 min. came and passed, the first internal contact was over, and Venus had fully entered on the sun. We had only obtained a brief view, and we had been able to make no measures or other observations that could be of service. Still, to have seen even a part of a transit of Venus is an event to remember for a lifetime, and we felt more delight than can be easily expressed at even this slight gleam of success.

But better things were in store. My assistant came over to report to me that he had also been successful in seeing Venus in the same phase as I had. We both resumed our posts, and at half-past two the clouds began to disperse, and the prospect of seeing the sun began to improve. It was now no question of the observations of contact. Venus by this time was well on the sun, and we therefore prepared to make observations with the micrometer attached to the eye-piece. The clouds at length dispersed, and at this time Venus had so completely entered on the sun that the distance from the edge of the planet to the edge of the sun was about twice the diameter of the planet. We measured the distance of the inner edge of Venus from the nearest limb of the sun. These observations were repeated as frequently as possible, but it should be added that they were only made with very considerable difficulty. The sun was now very low, and the edges of the sun and of Venus were by no means of that steady character which is suitable for micrometrical measurement. The edge of the sun was "boiling," and Venus, though it no doubt was sometimes circular, was very often distorted to such a degree as to make the measures very uncertain.

We succeeded in obtaining sixteen measures altogether; but the sun was now getting low, the clouds began again to interfere, and we saw that the pursuit of the transit must be left to the thousands of astronomers in happier climes who had been eagerly awaiting it. But before the phenomena had ceased I spared a few minutes from the somewhat mechanical work at the micrometer to take a view of the transit in the more picturesque form which the large field of the finder presented. Truly it was a most exquisite and memorable sight. The sun was already beginning to put on the ruddy hues of sunset, and there, far in on its face, was the sharp, round, black disc of Venus. It was then easy to sympathise with the supreme joy of Horrocks when, in 1639, he for the first time witnessed this spectacle. The intrinsic beauty of the phenomenon, its rarity, the fulfilment of the prediction, the noble problem which the transit of Venus enables us to solve, are all present to our thoughts when we look at this pleasing picture, the like of which will not occur again until the flowers are blooming in the June of A.D. 2004.

The occasion of a transit of Venus also affords an opportunity of studying the physical nature of the planet, and we may here briefly indicate the results that have been obtained. In the first place, a transit will throw some light on the question as to whether Venus is accompanied by a satellite. If Venus were attended by a small body in close proximity, it would be conceivable that under ordinary circumstances the brilliancy of the planet would obliterate the feeble beam of rays from the minute companion, and thus the satellite would remain undiscovered. It was therefore a matter of great interest to scrutinise the vicinity of the planet while in the act of transit over the sun. If a satellite existed—and the existence of a satellite has often been suspected—then it would be capable of detection against the brilliant background of the sun. Special attention was directed to this point during the recent transits, but no satellite of Venus was to be found. It seems, therefore, to be very unlikely that Venus can be attended by any satellite of appreciable dimensions.

The observations directed to the investigation of the atmosphere surrounding Venus have been more successful. If the planet were devoid of an atmosphere, then it would be totally invisible just before commencing to enter on the sun, and would relapse into total invisibility as soon as it had left the sun. The observations made during the transits are not in conformity with such suppositions. Special attention has been directed to this point during the recent transits. The result has been very remarkable, and has proved in the most conclusive manner the existence of an atmosphere around Venus. As the planet gradually moved off the sun the circular edge of the planet extending out into the darkness was seen to be bounded by a circular arc of light, and Dr. Copeland, who observed this transit under exceptionally favourable circumstances, was actually able to follow the planet until it had passed entirely away from the sun, at which time the globe, though itself invisible, was distinctly marked by the girdle of light by which it was surrounded. This luminous circle is inexplicable save by the supposition that the globe of Venus is surrounded by an atmospheric shell in the same way as the earth.

It may be asked, what is the advantage of devoting so much time and labour to a celestial phenomenon like the transit of Venus which has so little bearing on practical affairs? What does it matter whether the sun be 95,000,000 miles off, or whether it be only 93,000,000, or any other distance? We must admit at once that the inquiry has but a slender bearing on matters of practical utility. No doubt a fanciful person might contend, that to compute our Nautical Almanacs with perfect accuracy we require to know the distance of the sun with accuracy. Our mighty commerce depends on skilful navigation, and one factor in successful navigation is the reliability of the Nautical Almanac. The increased perfection of the almanac must therefore bear some relation to increased perfection in navigation. Now, as good authorities tell us that in running for a harbour on a tempestuous night, or in other critical emergencies, even a yard of sea-room is often of great consequence, so it may conceivably happen that to the infinitesimal influence of the transit of Venus on the Nautical Almanac is due the safety of a gallant vessel.

But the time, the labour, and the money expended in observing the transit of Venus are really to be defended on quite different grounds. We see in it a fruitful source of information. It tells us the distance of the sun, which is the foundation of all the great measurements of the universe. It gratifies the intellectual curiosity of man by a view of the true dimensions of the majestic solar system, in which the earth is seen to play a dignified, though still subordinate, part; and it leads us to a conception of the stupendous scale on which the great universe is constructed.

It is not possible for us, with a due regard to the limits of this volume, to linger any longer over the consideration of the transit of Venus. When we begin to study the details of the observations, we are immediately confronted with a multitude of technical and intricate matters. On the occasion of a transit, it has first to be decided where the observations are to be made—in itself a question that has led to no little discussion. Then the instruments that are to be used, and the description of observations to be made, have to be investigated with considerable complexity. The observers must be specially trained for the work, for even Methuselah himself could hardly have lived long enough to have had much *practice* in the observations of transits of Venus. To compensate for the inevitable want of experience, the observers had to be prepared by a special course of instruction, in which a fictitious transit was observed. Then, too, the interpretation of the observations involves many thorny and many controverted questions. To pursue all these matters so as to render them intelligible would lead us into great detail, and therefore we do not make the attempt. This course is the more advisable when it is remembered that the transit of Venus is only *one* of the methods of finding the sun's distance—a celebrated one, no doubt, but not perhaps the most reliable. It seems not unlikely that the final determination of the sun's distance will be obtained in quite a different manner. This will be explained in Chapter XI., and hence we feel the less reluctance in passing away from the consideration of the transit of Venus as a method of celestial surveying.

We must now close our description of this lovely planet; but before doing so, let us add—or in some cases repeat—a few statistical facts as to the size and the dimensions of the planet and its orbit.

The diameter of Venus is about 7,660 miles, and the planet shows no measurable departure from the globular form, though we can hardly doubt that its polar diameter must really be somewhat shorter than the equatorial diameter. This diameter is only about 258 miles less than that of the earth. The mass of Venus is about three-quarters of the mass of the earth; or if, as is more usual, we compare the mass of Venus with the sun, it is to be represented by the fraction 1 divided by 425,000. It is to be observed that the mass of Venus is not quite so great in comparison with its bulk as might have been expected. The density of Venus is about 0·850 of that of the earth. Venus would weigh 4·81 times as much as a globe of water of equal size. The gravitation at the surface of Venus will, to a slight extent, be less than the gravitation at the surface of the earth. A body here falls sixteen feet in a second; a body let fall at the surface of Venus would fall about three feet less. The time of rotation of Venus is an element about which there is still considerable uncertainty. It is supposed to occupy about twenty-three hours and twenty-one minutes.

The orbit of Venus is remarkable for the close approach which it makes to a circle. The greatest distance of Venus from the sun does not exceed by 1 per cent. the least distance. Its mean distance from the sun is about 67,000,000 miles, and the movement in the orbit has a mean velocity of nearly 22 miles per second, the entire journey being accomplished in 224·70 days.

CHAPTER IX.

THE EARTH.

The Earth is a great Globe—How the Size of the Earth is Measured—The Base Line—The Latitude found by the Elevation of the Pole—A Degree of the Meridian—The Earth not a Sphere—The Pendulum Experiment—Is the Motion of the Earth slow or fast?—Coincidence of the Axis of Rotation and the Axis of Figure—The Existence of Heat in the Earth—The Earth once in a Soft Condition—Effects of Centrifugal Force—Comparison with the Sun and Jupiter—The Protuberance at the Equator—The Weighing of the Earth—Comparison between the Weight of the Earth and an equal Globe of Water—Comparison of the Earth with a Leaden Globe—The Pendulum—Use of the Pendulum in Measuring the Intensity of Gravitation—The Principle of Isochronism—Shape of the Earth Measured by the Pendulum.

THAT the earth is a round body is a truth immediately suggested to us by the most simple astronomical considerations. The sun is round, the moon is round, telescopes show that the planets are round. No doubt comets are not round, but then a comet is not a solid body at all. We can see right through a comet, and its weight is too small for our measures to appreciate. If, then, all the solid bodies we can see are round globes, is it not likely that the earth is a globe also? But we have far more direct information than mere surmise.

There is no better way of actually seeing that the surface of the sea is curved than by watching on a clear day a distant ship. When the ship is a long way off and is still receding, its hull will gradually disappear, while the masts will remain visible. On a fine summer's day we can often see with, or indeed without an opera-glass, the top of the funnel of a steamer appearing above the sea, while the body of the steamer is below. If the sea were perfectly flat, there is nothing to obscure the body of the vessel, and it therefore would be visible as long as the funnel remains visible; but if the sea be really curved the protuberant part intercepts the view of the hull, while leaving the funnel visible.

We thus learn how the sea is curved at every part, and thus it is natural to suppose that the earth is a globe. When we make more careful measurements we find that the globe is not perfectly round. It is flattened to some extent at each of the poles. This may be easily illustrated by an indiarubber globe, which can be compressed on two opposite sides so as to bulge out at the centre. The earth is similarly flattened at the poles, and bulged out at the equator. The divergence of the earth from the truly globular form is, however, not very great, and would hardly be noticed without very careful measurements.

The determination of the size of the earth involves operations of no little delicacy. Very much skill and very much labour have been devoted to the work, and the dimensions of the earth are known with a high degree of accuracy, though perhaps not with all the precision that we may ultimately hope to attain. The scientific importance of an accurate measurement of the earth can hardly be over-estimated. The radius of the earth is itself the unit in which astronomical magnitudes are generally expressed. For example, when observations are made with the view of finding the distance of the moon, the observations, when discussed and reduced, tell us that the distance of the moon is equal to fifty-nine times the equatorial radius of the earth. If we want to find the distance of the moon in miles, we require to know the number of miles in the earth's radius.

A level part of the earth's surface having been chosen, a line a few miles long is measured. This is called the base line, and as all the subsequent measures depend on the base line, it is indispensable that this measurement be made with scrupulous accuracy. To measure a line four or five miles long with such precision as to exclude any errors greater than a few inches, demands the most minute precautions. We do not now enter upon a description of the operations that are necessary. It is a most laborious piece of work, and many ponderous volumes have been devoted to the discussion of the results. But when a few base lines have been measured in different places on the earth's surface, the measuring rods are to be laid aside, and the subsequent task of the survey of the earth is to be done by the measurement of angles and by trigonometrical calculations based thereon. Starting from a base line a few miles long, distances of greater length are measured, until at length stretches of 100 miles long, or even more, can be accomplished. It is thus possible to measure a long line running due north and south.

So far the work has been merely that of the terrestrial surveyor. The distance thus measured is then handed over to the astronomer to deduce from it the dimensions of the earth. The astronomer fixes his observatory at the northern end of the long line, and proceeds to determine his latitude by observation. There are various ways by which the latitude is to be found. They will be found fully described in works on practical astronomy. We shall here only indicate in a very brief manner the principle on which such observations are to be made.

Everyone ought to be familiar with the Pole Star, which, though by no means the most brilliant, is probably the most important star in the whole heavens. In these latitudes we are accustomed to find the Pole Star at a considerable elevation, and there, night after night, we see it, always in the same place in the northern sky. But suppose we start on a voyage to the southern hemisphere: as we approach the equator we find, night after night, the Pole Star is closer to the horizon, till at the equator the Pole Star will be on the horizon; while if we cross the line, we find on entering the southern hemisphere that the Pole Star has become invisible.

On the other hand, a traveller leaving England for Norway will find that the Pole Star is every night higher up in the heavens than he has been accustomed to see it. If he extend his journey farther north, the Pole Star will gradually rise higher and higher, until at length, when approaching the pole of the earth, the Pole Star is high up over his head. We are thus led to perceive that the higher our latitude, the higher, in general, is the elevation of the Pole Star. But we cannot use precise language until we replace the Pole Star by the pole of the heavens itself. The pole of the heavens is near the Pole Star, and the Pole Star revolves around the pole of the heavens, as all the other stars do, once every day; the circle described by the Pole Star is however so small that, unless we pay special attention to it, the motion is not perceived. The true pole is not a visible point, but it is capable of being accurately defined, and it enables us to state with the utmost precision the relation between the pole and the latitude. The statement is, that the elevation of the pole above the horizon is equal to the latitude of the place.

The astronomer stationed at one end of the long line measures the elevation of the pole above the horizon. This is an operation of some delicacy. In the first place, as the pole is invisible, he has to measure, instead of it, the height of the Pole Star when that height is greatest, and repeat the operation twelve hours later, when the height of the Pole Star is least; the mean between the two gives the height of the pole, but this has to be corrected in various ways which it is not necessary for us here to discuss. Suffice it to say that by such operations the latitude of one end of the line is determined. The astronomer then, with all his equipment of instruments, moves to the other end of the line. He there repeats the operations, and he finds that the pole has now a different elevation, corresponding to the different latitude. The difference of the two elevations thus gives him an accurate measure of the number of degrees and fractional parts of a degree between the latitudes of the two stations. This can now be compared with the actual distance in miles between the two stations, which has been ascertained by the trigonometrical survey. A simple calculation will then give the number of miles and fractional parts of a mile corresponding to one degree of latitude—or, as it is more usually expressed, the length of a degree of the meridian.

This operation is repeated in different parts of the earth—in the northern hemisphere and in the southern, in high latitudes and in low latitudes. If the sea-level over the entire earth were a perfect globe, an important consequence would follow—the length of a degree of the meridian would be the same everywhere. It would be the same in Peru as in Sweden, the same in India as in

England. But the lengths of the degrees are not all the same, and hence we learn that our earth is not really a sphere. The measured lengths of the degrees enable us to see in what way the shape of the earth departs from a perfect sphere. Near the pole the length of a degree is longer than near the equator. This shows that the earth is flattened at the poles and protuberant at the equator, and it provides us with the means of actually calculating the length of the polar and the equatorial axes.

The polar axis of the earth may be defined to be the shortest diameter of the earth. This axis intersects the surface at the north and south poles. Around this axis the earth performs one rotation every sidereal day. The sidereal day is a little shorter than the ordinary day, being only 23 hours, 56 minutes, and 4 seconds. The rotation is performed just as if a rigid axis passed through the centre of the earth; or, to use the old and homely illustration, the earth rotates just as a ball of worsted may be made to rotate around a knitting-needle thrust through its centre. The rotation of the earth upon its axis admits of being demonstrated in a very remarkable manner by actual experiment. If the pole of the earth were an attainable point, then the experiment would have a degree of simplicity that can only be partially realised at other stations. For the purpose of describing the principle of this experiment we may suppose that the observer is actually situated at the pole, and that over the point on the earth where the polar axis intersects the surface a lofty dome has been erected. From the summit of the dome a long wire descends nearly to the ground, and to this a ponderous weight is attached. The weight is to be drawn aside and then released. Slowly it will swing from side to side, but the plane in which it moves will remain invariable. If the building were also at rest, then the position of the plane of oscillation would remain in the same position relatively to the surrounding walls. But the building rotates with the earth, and will turn completely round in the sidereal day. There will accordingly be a change in the place of the plane of oscillation relatively to the building, which will appear to rotate at such a rate that it would complete a revolution in the sidereal day. If the building be situated at lower latitude the apparent change in the plane of oscillation of the pendulum is not so rapid, yet the movement of the plane is still sufficient to be perfectly observed, and the measured amount of displacement of the plane has been found to agree with the amount calculated on the supposition that the earth rotates. This is known as Foucault's celebrated pendulum experiment.

Regarding the earth as a spinning body, it might be thought that the rotation, being only once in four minutes less than a day of twenty-four hours, was a very slow movement. A wheel of ordinary dimensions would be turning very slowly at this rate; only, indeed, about half the speed of the hour-hand of a clock. When, however, the size of the earth is taken into consideration, our conception of the speed will be somewhat modified. The radius of the earth is so great that a point on the equator will have to dash along at the rate of a thousand miles per hour in order to complete its circuit in the allotted time.

It is a circumstance worthy of the very closest attention, that the axis about which the earth rotates is identical with the shortest diameter of the earth as found by actual surveying. This is a coincidence which would be utterly inconceivable if the shape of the earth was not in some way found to be connected with the fact that the earth is rotating. What connection can then be traced? Let us inquire into the subject, and we shall find that the shape of the earth is a consequence of its rotation.

The earth at the present time is subject, at various localities, to occasional volcanic outbreaks. The phenomena of volcanoes, the allied phenomena of earthquakes, the well-known fact that the heat increases the deeper we descend into the earth, the existence of hot springs, the geysers found in Iceland and elsewhere, all testify to the fact that heat exists in the interior of the earth. Whether that heat be, as some suppose, universal in the interior of the earth, or whether it be merely local at the several places where its manifestations are felt, is a matter which need not now concern us. All that is necessary for our present purpose is the admission that heat is present to some extent. This internal heat, be it much or little, has obviously a different origin to the heat which we know on the surface. The heat we enjoy is derived from the sun. The internal heat cannot have been derived from the sun; its intensity is far too great, and there are other insuperable difficulties attending the supposition that it has come from the sun. Where then has this heat come from? This is a question which at present we can hardly answer—nor, indeed, does it much concern our argument that we should answer it. The fact that the heat is there being admitted, all that we require is to apply one or two of the well-known laws of heat to the interpretation of the facts. We have first to consider the well-known law by which heat tends to diffuse itself and spread away from its original source. The heat, deep-seated in the interior of the earth, tends to penetrate through the superincumbent rocks, and slowly reaches the surface. It is true that the rocks and materials with which our earth is covered are not good conductors of heat; most of them are, indeed, extremely bad conductors of heat; but good or bad, they nevertheless are conductors, and through them the heat must creep to the surface. It cannot be urged against this conclusion that we do not feel this heat. A few feet of brickwork will so keep in the heat of a mighty blast furnace that but little will escape through the bricks; but *some* heat does escape, and the bricks have never been made, and so far as we know, never could be made, which would absolutely intercept all the heat. If a few feet of brickwork can thus nearly mask the heat of a furnace, cannot some scores of miles of rock nearly mask the heat in the depths of the earth, even though that heat were seven times hotter than the mightiest furnace that ever existed? The heat would escape slowly, and perhaps imperceptibly, but unless all our knowledge of nature is a delusion, no rocks, however thick, can prevent, in the course of time, the leakage of the heat to the surface. When this heat arrives at the surface of the earth it must, in virtue of another law of heat, gradually radiate away and be lost.

It would, perhaps, lead us too far to discuss some of the objections which may be raised against what we have here stated. It is often said that the heat in the interior of the earth is being produced by chemical combination or by mechanical process, and thus that the heat may be constantly renewed as fast or even faster than it escapes. This, however, is more a difference in form than in substance. If heat be produced in the way just supposed (and there can be no doubt that there may be such an origin for some of the heat in the interior of the globe) there must be a certain expenditure of chemical or mechanical energies that produces a certain exhaustion. For every unit of heat which escapes, there will either be a loss of an unit of heat from the globe or, what comes nearly to the same thing, a loss of an unit of heat-making power from the chemical or the mechanical energies. The substantial result is the same, the heat of the earth must be decreasing. It should, of course, be observed that a great part of the escape of heat from the earth is of an obvious character, and not dependent upon the slow processes of conduction. Each outburst of a volcano discharges a stupendous quantity of heat which disappears very speedily from the earth; while in many places, as in the hot springs, there is a perennial discharge of heat which must attain enormous proportions.

The earth is thus losing heat; but the earth never gains any heat to replace the losses of heat or of heat-making power. The consequence is obvious; the interior of the earth must be growing colder. No doubt this is an extremely slow process; the life of an individual, the life of a nation, perhaps the life of the human race itself has not been long enough to witness any pronounced change in the store of terrestrial heat. But the law is inevitable, and though the decline in heat may be slow, yet it is continuous, and in the lapse of ages must necessarily produce great and important effects.

It is not our present purpose to attempt to make any forecast as to the future effects on the earth which may arise from this process. We wish at present rather to look back into past time and see to what results we are inevitably led; and here we may at once dismiss as inappreciable such intervals of time as we are familiar with in ordinary life or even in ordinary history. As the earth is daily losing internal heat, or the equivalent of heat, the earth must have had more heat yesterday than it had to-day, more last year than this year, more twenty years ago than ten years ago. The effect has not been appreciable in historic time; but when we rise from hundreds of years to thousands of years, from thousands of years to hundreds of thousands of years, and from hundreds of thousands of years to millions of years, the effect is not only appreciable but even of startling magnitude. There must have been a time when the earth contained much more heat than at present. There must have been a time when the surface of the earth was sensibly hot from this source. We cannot pretend to say how many thousands or millions of years ago this epoch was; but of this we may be sure, that earlier still the earth was hotter still, until at length we find the temperature increase to a red heat, from a red heat we look back to a still earlier age when the earth was white hot, back again till we find the surface of our now solid earth was actually molten. We need not push the retrospect any further at present, and still less is it necessary for us to attempt to assign the probable origin of that heat. This, it will be observed, is not required in our argument. We find heat now, and we know that heat is being lost every day. From this the conclusion that we have already drawn seems inevitable, and thus we are conducted back to some remote epoch in the abyss of time past when our solid earth was a globe molten and soft throughout.

A dewdrop on the petal of a flower is nearly globular; but it is not quite a globe, because the gravitation presses it against the flower and somewhat distorts the shape. A falling drop of rain is a globe; a drop of oil suspended in a liquid with which it does not mix, forms a globe. Passing from small things to great things, let us endeavour to conceive a stupendous globe of molten matter. Let that globe be as large as the earth, and let its materials be so soft as to obey the forces of attraction exerted by each part of the globe on all the other parts. There can be no doubt as to the effect of these attractions, they would tend to smooth down any irregularities on the surface just in the same way as the surface of the ocean is smooth when freed from the disturbing influences of the wind. We might therefore expect that our molten globe, isolated from all external interference, would assume the form of a sphere.

But now suppose that this great sphere, which we have hitherto assumed to be at rest, is made to rotate round an axis passing through its centre. We need not suppose that this axis is a material object, nor are we concerned with any supposition as to how the velocity of rotation was caused. We can, however, easily see what the consequence of the rotation would be. The sphere would become deformed, the centrifugal force would make the molten body bulge out at the equator and flatten down at the poles. The greater the velocity of rotation the greater would be the bulging. To each velocity of rotation a certain degree of bulging would be appropriate. The molten earth then bulged out to an extent which was dependent upon the fact that it turned round once a day. Now suppose that the earth, while still rotating, commences to pass from the liquid to the solid state. The form which the earth would assume on consolidation would, no doubt, be very irregular on the surface; it would be irregular in consequence of the upheavals and the outbursts incident to the transformation of so mighty a mass of matter; but irregular though it be, we can be sure that, on the whole, the form of the earth's surface would coincide with the shape which it had assumed by the movement of rotation. Hence we can explain the protuberant form of the equator of the earth, and we can appeal to that form in corroboration of the view that the earth was once in a soft or molten condition.

The argument may be supported and illustrated by comparing the shape of our earth with the shapes of some of the other celestial bodies. The sun, for instance, seems to be almost a perfect globe. No measures that we can make show that the polar diameter of the sun is shorter than the equatorial diameter. But this is what we might have expected. No doubt the sun is rotating on its axis, and as it is the rotation that causes the protuberance, why should not the rotation have deformed the sun like the earth? The probability is that a difference really does exist between the two diameters of the sun, but that the difference is too small for us to measure. It is impossible not to connect this with the *slowness* of the sun's rotation. The sun takes twenty-five days to complete a rotation, and the protuberance appropriate to so low a velocity is not appreciable.

On the other hand, when we look at one of the quickly-rotating planets, we obtain a very different result. Let us take the most striking instance, which is presented in the great planet Jupiter. Viewed in the telescope, Jupiter is seen at once not to be a globe. The difference is indeed so conspicuous, that measures are not necessary to show that the polar diameter of Jupiter is shorter than the equatorial diameter. The departure of Jupiter from the truly spherical shape is indeed much greater than the departure of the earth. It is impossible not to connect this with the much more rapid rotation of Jupiter. We shall presently have to devote a chapter to the consideration of this splendid orb. We may, however, so far anticipate what we shall then say as to state that the time of Jupiter's rotation is under ten hours, and this notwithstanding the fact that Jupiter is more than one thousand times greater than the earth. The enormously rapid rotation of Jupiter has caused him to bulge out at the equator to a most remarkable extent.

The survey of our earth, and the measurement of its dimensions having been accomplished, the next operation for the astronomer is the determination of its weight. Here, indeed, is a problem which taxes the resources of science to the very uttermost. Of the interior of the earth we know little—I might almost say we know nothing. No doubt we sink deep mines into the earth. These mines enable us to penetrate half a mile, or even a whole mile, into the depths of the interior. But this is, after all, only a most insignificant attempt to explore the interior of the earth. What is an advance of one mile in comparison with the distance to the centre of the earth? It is only about one four-thousandth part of the whole. Our knowledge of the earth merely reaches to an utterly insignificant depth below the surface, and we have not a conception of what may be the nature of our globe only a few miles below where we are standing. Seeing, then, our almost complete ignorance of the solid contents of the earth, does it not seem a hopeless task to attempt to weigh the entire globe and determine its weight? Yet that problem has been solved, and the weight of the earth is known—not, indeed, with the accuracy attained in other astronomical researches, but still with a tolerable approximation.

It is needless to enunciate the weight of the earth in our ordinary units. The enumeration of billions of tons does not convey any distinct impression. It is a far more natural course to compare the mass of the earth with that of an equal globe of water. We should be prepared to find that our earth was heavier than an an equal globe of water. The rocks which form its surface are heavier, bulk for bulk, than the oceans which repose on those rocks. The abundance of metals in the earth, the gradual increase in the density of the earth, which must arise from the enormous pressure at great depths—all these considerations will prepare us to learn that the earth is very much heavier than a globe of water of equal size. Newton supposed that the earth was somewhere between five and six times as heavy as an equal bulk of water. Nor is it hard to see that such a suggestion is plausible. The rocks and materials on the surface are usually about two or three times as heavy as water. There is certainly a vast quantity of iron in the earth. It has been supposed that down in the remote depths of the earth there is a very large proportion of iron. Now an iron earth would weigh about seven times as much as an equal globe of water. We are thus led to see that the earth's weight must be probably more than three, and probably less than seven, times an equal globe of water; and hence, in fixing the density between five and six, Newton adopted a result plausible at the moment, and since shown to be probably correct. Several methods have been proposed by which this important question can be solved with accuracy. Of all these methods we shall here only describe one, because it illustrates, in a very remarkable manner, the law of universal gravitation.

In our chapter on gravitation, it was pointed out that the force of gravitation between two masses of moderate dimensions was extremely minute, and the difficulty in weighing the earth arises from this cause. The practical application of the process is encumbered by multitudinous details, which it will be unnecessary for us to consider at present. The principle of the process is simple enough. To give definiteness to our description, let us conceive a large globe about two feet in diameter; and as it is desirable to have this globe as heavy as possible, let us suppose it to be made of lead. A small globe brought near the large globe is attracted by the force of gravitation. The amount of this attraction is extremely small, but, nevertheless, it can be measured by a refined process which renders extremely small forces sensible. The intensity of this attraction depends both on the masses of the globes and on their distance apart, as well as on the force of gravitation. We can also readily measure the attraction of the earth upon the small globe. This is in fact nothing more nor less than the weight of the small globe in the ordinary acceptation of the word. We thus can compare the attraction exerted by the leaden globe with the attraction exerted by the earth. If the centre of the earth and the centre of the leaden globe were at the same distance from the attracted body, then the intensity of their attractions would give at once the ratio of their masses by simple proportion. In this case, however, matters are not so simple; the leaden ball is only distant by a few inches from the attracted ball, while the centre of the earth's attraction is nearly 4,000 miles away at the centre of the earth. Allowance has to be made for this difference, and the attraction of the leaden sphere has to be reduced to what it would be were it removed to a distance of 4,000 miles. This can fortunately be effected by a simple calculation depending upon the great law that the intensity of gravitation varies inversely as the square of the distance. We can thus, partly by calculation and partly by experiment, compare the intensity of the attraction of the leaden sphere with the attraction of the earth. It is known that the attractions are proportional to the masses, so that thus the comparative masses of the earth and of the leaden sphere have been measured; and it has been ascertained that the earth is about half as heavy as a globe of lead of equal size would be. We may thus state finally that the mass of the earth is a little more than five times as great as the mass of a globe of water equal to it in bulk.

We have, in our chapter on Gravitation, mentioned the fact that a body let fall near the surface of the earth moves through sixteen feet in the first second. This distance varies slightly at different parts of the earth. If the earth were perfectly round, then the attraction would be the same at every part, and the body would fall through the same distance everywhere. The earth is not round, so the distance which the body falls in one second differs slightly at different places. At the pole the radius of the earth is shorter than at the equator, and accordingly the attraction of the earth at the pole is greater than at the equator. If we had the means of measuring exactly the distance a body would fall in one second at the pole and at the equator, we would have the means of ascertaining the actual shape of the earth.

It is, however, difficult to measure accurately the distance a body will fall in one second. We have, therefore, been obliged to resort to other means for measuring the force of attraction of the earth at the equator and other accessible parts of its surface. The methods adopted are founded on the pendulum, which is at the same time the simplest and one of the most useful of philosophical instruments. The ideal pendulum is a small and heavy weight suspended from a fixed point by a fine and flexible wire. If we draw the pendulum aside from its vertical position and then release it, the weight will swing to and fro.

For its journey to and fro the pendulum requires a certain amount of time, but this time does not appreciably depend on the length of the arc through which the pendulum swings. To verify this very remarkable law, suspend another pendulum beside the first, both being of the same length; draw both pendulums aside and release them: they swing together and return together. This might have been expected. But if we draw one pendulum a great deal to one side, and the other only a little, the two pendulums still swing sympathetically. This, perhaps, would not have been expected. Try it again, with even a still greater difference in the arc of vibration, and still we see the two weights occupy the same time for the swing. We can vary the experiment in another way. Let us change the weights on the pendulums, so that they are of unequal size, though both of iron. Shall we find any difference in the periods of vibration? We try again: the period is the same as before; swing them through different arcs, large or small, the period is still the same. But it may be said that this is due to both weights being of the same material. Try it again, using a leaden weight instead of one of the iron weights; the result is identical. Even with a ball of wood the period of oscillation is the same as that of the ball of iron, and this is true no matter what be the arc through which the vibration takes place.

If, however, we change the *length* of the wire by which the weight is supported, then the period will not remain unchanged. This can be very easily illustrated. Take a short pendulum with a wire only one-fourth of the length of the long pendulum; suspend the two close together, and compare the periods of vibration of the short pendulum with that of the long one, and we find that the short pendulum has a period only half that of the long one. We may state the result generally, and say that the time of vibration of a pendulum is proportional to the square root of its length. If we quadruple the length of the pendulum, we double the time of its vibration; if we increase the length of the pendulum ninefold, we increase its period of vibration threefold.

It is the gravitation of the earth which makes the pendulum swing. The greater the gravitation, the more rapidly will the pendulum oscillate. This will be easily accounted for. If the earth pulls the weight down very quickly, the time will be very short; if the power of the earth's attraction be lessened, then it cannot pull the weight down so quickly, and the period will be lengthened.

It is possible to determine the time of vibration of the pendulum with great accuracy. Let it swing for 10,000 oscillations, and measure the time that these oscillations have consumed. The arc through which the pendulum swings may not be quite constant, but, as we have seen, this does not appreciably affect the *time* of its oscillation. Suppose, then, that an error of a second is made in the determination of the time of 10,000 oscillations; this will only entail an error of the ten-thousandth part of the second in the time of a single oscillation, and will give a correspondingly accurate determination of the gravity.

Take a pendulum to the equator. Let it perform 10,000 oscillations, and determine carefully the *time* that these oscillations have required. Bring the same pendulum to another part of the earth, and repeat the experiment. We have thus a means of comparing the gravitation at the two places. There are no doubt a multitude of precautions to be observed which need not here concern us. We need not here enter into details as to the manner in which the motion of the pendulum is to be sustained, nor as to the effects of changes of temperature in the alteration of its length. It will suffice for us to see how the time of the pendulum's swing can be measured accurately, and how from that measurement the intensity of gravitation can be calculated.

The pendulum thus enables us to make, with the highest degree of accuracy, a gravitational survey of the surface of the earth. We cannot, however, infer that gravity alone affects the oscillations of the pendulum. We have seen how the earth rotates on its axis, and we have attributed to this rotation the bulging of the earth at the equator. But the centrifugal force arising from the rotation affects bodies on the surface of the earth. It has the effect of decreasing the weight of bodies, and this effect is greatest at the equator, and lessens gradually as we approach the poles. From this cause alone the attraction of the pendulum at the equator is less than elsewhere, and therefore the oscillations of the pendulum there will take a longer time than at other localities. A part of the apparent change in gravitation is thus merely due to the centrifugal force; but there is, besides this, a real change in gravitation. At a hasty glance it might be thought that, as there was a protuberance of matter at the equator, there ought to be a greater attraction at the equator than elsewhere. This is not so. The effect of the additional matter is more than compensated by the greater distance of the pendulum from the centre of the earth.

Indeed, a moment's reflection will show that the pendulum at the pole is, on the whole, nearer to the mass of the earth generally than the pendulum is at the equator. It shows, in a very marked way, how the researches in different branches of science are interwoven when we find that, by merely swinging a pendulum at different parts of the earth, we are enabled to determine its *shape* as accurately as by the elaborate measurements of the arcs of the meridian made at different parts of the earth.

We defer to a future chapter the important phenomenon of the earth's movement, which is known as "precession." In a work on astronomy, it does not come within our scope to enter into further detail on the subject of our planet. The surface of the earth, its contour and its oceans, its mountain chains and its rivers, are for the physical geographer; while its rocks and their contents, its volcanoes and its earthquakes, are to be studied by the geologists and the physicists.

CHAPTER X.

MARS.

Our nearer Neighbours in the Heavens—Surface of Mars can be Examined in the Telescope—Remarkable Orbit of Mars—Resemblance of Mars to a Star—Meaning of Opposition—The Eccentricity of the Orbit of Mars—Different Oppositions of Mars—Apparent Movements of the Planet—Effect of the Earth's Movement—Measurement of the Distance of Mars—Theoretical Investigation of the Sun's Distance—Drawings of the Planet—Is there Snow on Mars?—The Rotation of the Planet—Gravitation on Mars—Has Mars any Satellites?—Mr. Asaph Hall's great Discovery—The Revolutions of the Satellites—Deimos and Phobos—Gulliver's Travels.

The special relation in which we stand to one planet of our system has necessitated a somewhat different trèatment of that planet from the treatment appropriate to the others. We discussed Mercury and Venus as distant objects known chiefly by telescopic research, and by calculations of which astronomical observations were the foundation. Our knowledge of the earth is of a different character, and attained in a different way. Yet it was necessary for symmetry that we should discuss the earth after the planet Venus, in order to give to the earth its true position in the solar system. But now that the earth has been passed in our outward progress from the sun, we come to the planet Mars; and here again we resume, though in a somewhat modified form, the methods that were appropriate to Venus and to Mercury.

Venus and Mars have, from one point of view, quite peculiar claims on our attention. They are our nearest planetary neighbours, on either side. We may naturally expect to learn more of them than of the other planets farther off. In the case of Venus, however, this anticipation can hardly be realised, for, as we have already pointed out, its dazzling brilliancy prevents us from making a really satisfactory telescopic examination. When we turn to our other planetary neighbour, Mars, we are enabled to learn a good deal with regard to his appearance. Indeed, with the exception of the moon, we are better acquainted with the details of the surface of Mars than with those of any other celestial body.

This very beautiful planet offers many features for consideration besides those presented by its physical structure. The orbit of Mars is a very remarkable one, and it was by the observations of this orbit that the celebrated laws of Kepler were discovered. During the occasional approaches of Mars to the earth, it has been possible to measure its distance with accuracy, and thus another method of finding the sun's distance has arisen, which, to say the least, may compete in precision with that afforded by the transit of Venus; and it must also be observed that the greatest achievement in pure telescopic research which this century has witnessed was that of the discovery of the satellites of Mars.

To the unaided eye, Mars generally appears like a star of the first magnitude. It is usually to be distinguished by its ruddy colour, but the beginner in astronomy cannot rely, for the identification of Mars, on its colour only. There are several stars nearly, if not quite, as ruddy as Mars. This planet has often been mistaken for the bright star Aldebaran, the brightest star in the constellation of the Bull. It often resembles Betelgueze, a very brilliant point in the constellation of Orion. Mistakes of this kind will be impossible if the beginner has first learned the principal constellations and the principal bright stars. He will then find great interest in tracing out the positions of the planets, and in watching their ceaseless movements. The position of each planet at any time can always be ascertained from the almanac. Sometimes the planet will be too near the sun to be visible. It will rise with the sun and set with the sun, and consequently will not be above the horizon during the night. The best time for seeing one of the planets exterior to the earth will be during what is called its opposition. The opposition of Mars occurs when the earth comes directly between Mars and the sun. In this case, the distance from Mars to the earth is less than at any other time. There is also another advantage in viewing Mars during opposition. The planet is then at one side of the earth and the sun at the opposite side, so that when Mars is high in the heavens the sun is directly beneath the earth; in other words, the planet is then at its greatest elevation above the horizon at midnight. Some oppositions of Mars are,

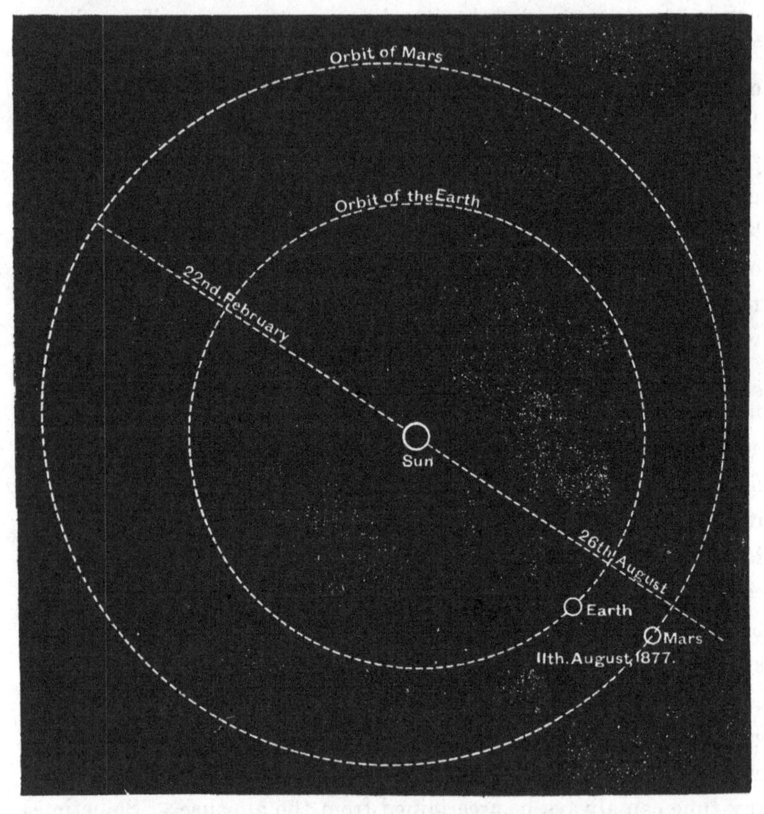

Fig. 45.—The orbits of the Earth and of Mars, showing the favourable Opposition of 1877.

however, much more favourable than others. This is distinctly shown in Fig. 45, which represents the orbit of Mars and the orbit of the Earth accurately drawn to scale. It will be seen that while the orbit of the earth is very nearly circular, the orbit of Mars has a very decided degree of eccentricity; indeed, with the exception of the orbit of Mercury, that of Mars has the greatest eccentricity of any orbit of the larger planets in our system. It will thus be observed that the value of an opposition of Mars for telescopic purposes will vary greatly. The favourable oppositions will be those which occur as near as possible to the 26th of August. The opposite extreme will be found in an opposition which occurs near the 22nd of February. In the latter case the distance between the planet and the earth is nearly twice as great as the former. The last favourable opposition occurred in the year 1877. During that opposition, Mars was a magnificent object, and received much and deserved attention. The favourable oppositions follow each other at somewhat irregular intervals; the next will occur in the year 1892, and another in the year 1909.

The apparent movements of Mars are by no means simple. We can imagine the embarrassment of the early astronomer who first undertook the task of observing or attempting to decipher these movements. The planet is seen to be a brilliant and conspicuous object. It attracts the astronomer's attention; he looks carefully, and he sees how the planet lies among the constellations with which he is familiar. A few nights later he looks at the planet again; but is it exactly in the same place? He thinks not. He notes more carefully than before the place of the planet. He sees how it is situated with regard to the stars. Again, in a few days, his observations are repeated. There is no longer a trace of doubt about the matter—Mars has decidedly changed his position. It is a planet, a wanderer. Night after night the primitive astronomer is at his post. He notes the changes of Mars. He sees that it is now moving even more rapidly than it was at first. Is it going to complete the circuit of the heavens? The astronomer determines to watch the planet, and see whether this surmise is justified. He pursues his task night after night, and at length he begins to think that the planet is not moving quite so rapidly as at first. A few nights more, and he is sure of the fact: the planet is moving more slowly. Again a few nights more, and he begins to surmise that the motion may cease; after a short time the motion does cease, and the planet seems to rest; but is it going to remain at rest for ever? Has its long journey been finished? For many nights this seems to be the case, but at length the astronomer begins to suspect that the planet must be commencing to move backwards. A few nights more, and the fact is confirmed beyond possibility of doubt, and the extraordinary discovery of the direct and the retrograde movement of Mars has been accomplished.

In the greater part of its journey around the heavens, Mars seems to move steadily from the west to the east. It moves

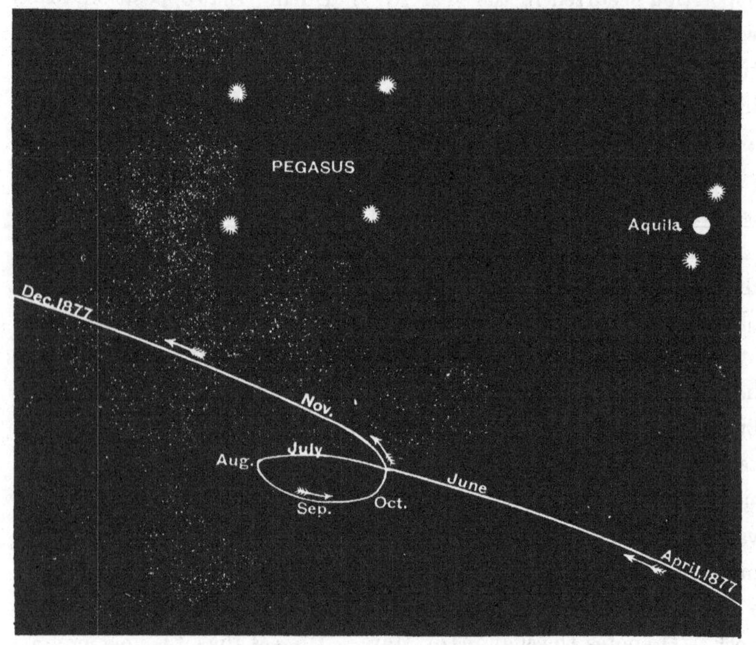

Fig. 46.—The apparent movements of Mars in 1877.

backwards, in fact, as the moon moves and as the sun moves. It is only during a comparatively small part of its path that these complex movements are accomplished which presented such an enigma to the primitive observer. We have in the adjoining picture the track of the actual journey which Mars accomplished in the opposition of 1877. The figure only shows that part of its path which presents the anomalous features; the rest of the path is pursued, not indeed with uniformity of velocity, but with uniformity of direction.

This complexity of the apparent movements of Mars seems, at first sight, fatal to the acceptance of any really simple and elementary explanation of the planetary motion. If the motion of Mars were purely elliptic, how, it may well be said, could it perform this extraordinary evolution? The explanation is to be sought in the fact that the earth on which we stand is moving. Even if Mars were at rest, the fact that our earth is moving would make Mars appear to move. The apparent movements of Mars are thus combined with the real movements. This circumstance will not embarrass the geometer. He is able to disentangle the true movement of the planet from its association with the apparent movement, and to account completely for the very complicated observed movement of Mars. Could we transfer our point of view from the ever-shifting earth to an immovable standpoint like the sun, we should then see that the true shape of the orbit of Mars was an ellipse, performed by revolving round the sun in conformity with the laws of Kepler, discovered by observations on this planet.

Under the most favourable circumstances, the planet Mars, at the time of opposition, may approach the earth to a distance not greater than about 35,000,000 miles. No doubt this seems an enormous distance, when estimated by any standard adapted for terrestrial measurements; it is, however, hardly greater than the distance of Venus when nearest, and it is much less than the distance from the earth to the sun. We have already explained how the *form* of the solar system is known from Kepler's laws, and how the absolute size of the system and of its various parts can be known when the direct measurement of any one part has been accomplished. A close approach of Mars affords a favourable opportunity for measuring his distance, and thus, in a different way, solving the same problem as that presented in the transit of Venus. We are thus a second time led to a knowledge of the distance of the sun and the distances of the planets generally, and to many other numerical facts about the solar system.

A very elaborate and successful attempt was made on the occasion of the opposition of Mars in 1877 to apply this refined method to the solution of the great problem. It cannot be said to have been the first occasion on which this method was suggested, or even practically attempted. But the observations and the discussions of 1877 were conducted with a degree of skill and a minute attention to all the necessary precautions which render them an important contribution to astronomy. Mr. David Gill, now Her Majesty's Astronomer at the Cape of Good Hope, undertook a journey to the Island of Ascension for the purpose of observing the parallax of Mars in 1877. On this occasion, Mars approached to the earth so closely as to afford an admirable opportunity for the application of the method. Mr. Gill succeeded in obtaining a most valuable series of measurements, and from them he concluded the distance of the sun with an accuracy not inferior to that attainable by the transit of Venus.

There is yet another method by which Mars can be made to give us information as to the distance of the sun. This method is one of very great beauty, and is deeply interesting from its connection with some of the loftiest mathematical enquiries. It was foreshadowed in the Dynamical theory of Newton, and was wrought to perfection by Le Verrier. It is based upon the great law of gravitation, and is intimately associated with the splendid discoveries in planetary perturbation which form so striking a chapter in modern astronomical discovery.

There is a certain relation known between two quantities which at first sight seem quite independent. These quantities are the mass of the earth and the distance of the sun. It follows, from the measurements of the intensity of gravitation on the earth's surface, and from the relative movements of the sun and of the earth, that the sun's parallax is proportional to the cube root of the mass of the earth. There is no uncertainty about this result, and the consequence is obvious. If we have the means of weighing the earth in comparison with the sun, then the distance of the sun can be immediately deduced. How, then, are we to place our great earth in the weighing scales? This is the problem which Le Verrier has shown us how to solve, and he does so by invoking the aid of the planet Mars.

If Mars in his revolution around the sun were solely swayed by the attraction of the sun, he would, in accordance with the well-known laws of planetary motion, follow for ever the same elliptic path. At the end of one century, or even of many centuries, the shape, the size, and the position of that ellipse would remain unaltered. Fortunately for our present purpose, a disturbance in the orbit of Mars is produced by the earth. Although the mass of the earth is so much less than that of the sun, yet the earth is still large enough to exercise an appreciable attraction on Mars. The ellipse described by Mars is consequently not always the same. The shape of that ellipse and its position gradually change, so that the position of the planet depends upon the mass of the earth. The place in which the planet is found can be determined by observation; the place which the planet would have had if the earth were absent can be found by calculation. The difference between the two is due to the attraction of the earth, and, when it has been measured, the mass of the earth can be ascertained. The amount of displacement increases from one century to another, but as the rate of growth is small, ancient observations are necessary to enable the measures to be made with accuracy.

A remarkable occurrence which took place more than two centuries ago enables the place of Mars to be determined with great precision at that date. On the 1st of October, 1672, three independent observers witnessed the occultation of a star in Aquarius by the planet Mars. The place of the star is known with accuracy, and hence we are provided with the means of accurately defining the point in the heavens occupied by Mars on the day in question. From this result, combined with the modern meridian observations, we learn that the displacement of Mars by the attraction of the earth has, in the lapse of two centuries, grown to about five minutes of arc (294 seconds). It has been maintained that this cannot be erroneous to the extent of more than a second, and hence it would follow that the earth's mass is determined to about one three-hundredth part of its amount. If no error were present this would give the sun's distance to about one nine-hundredth part. Notwithstanding the intrinsic beauty of this method, and the very high auspices under which it has been introduced, it will, we think, hardly be worthy of reliance in comparison with some of the other methods. We make no impeachment of the fidelity of the observations; and we feel no doubt that the displacement of the planet is mainly, if not entirely, due to the disturbing effect of the earth's attraction; but it seems quite impossible to be sure that some other cause, minute though it must be, may not also have contributed to the result. We cannot be absolutely sure that the theory is above suspicion. Interesting and beautiful though this method may be, we must rather regard it as a striking confirmation of the law of gravitation, than as affording an accurate means of measuring the sun's distance.

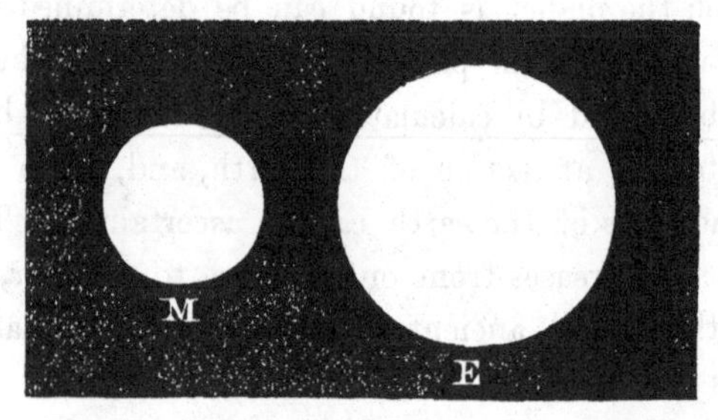

Fig. 47.—Relative sizes of Mars and the Earth.

The close approaches of Mars to the earth afford us opportunities for making a careful telescopic scrutiny of his surface. It must not be expected that the details on Mars could be delineated with the same minuteness as those on the moon. Even under the most favourable circumstances, Mars is still more than a hundred times as far as the moon, and, therefore, the features on Mars have to be at least one hundred times as large if they are to be seen as distinctly as the features on the moon. Mars is very much smaller than the earth. The diameter of the planet is 4,200 miles, but little more than half that of the earth. Fig. 47 shows the comparative sizes of the two bodies.

We here present to the reader two drawings of the telescopic appearance of the planet Mars. They were drawn by the late Mr. Charles E. Burton,* who employed in the work a

* "Transactions of the Royal Dublin Society," 1880.

telescope of moderate dimensions though of great optical perfection. It seems hard to decline the suggestion that these markings on Mars may really correspond to the divisions of land and water on that planet. There are circumstances which strongly suggest that water may be present on Mars. At the poles of Mars are large white regions, shown to some extent on Mr. Burton's drawings, which undergo periodic changes; and it has been surmised that these are due to an accumulation of ice or of snow on the polar regions of the planet. On some occasions, indeed, an "ice-cap"

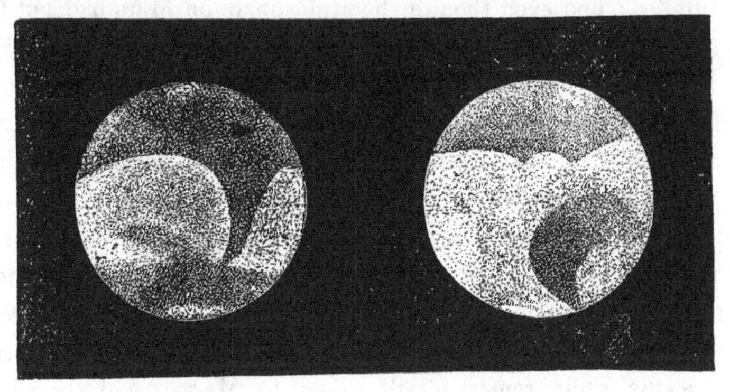

May 12, 1871. April 11, 1873.
Fig. 48.—Views of Mars by C. E. Burton.

on Mars, with its brilliancy and its sharply-defined margin, is a most striking feature in the telescopic view of the planet.

In making an examination of the planet, it is to be observed that it does not, like the moon, always present the same face towards the telescope. Mars rotates upon an axis in exactly the same manner as the earth. It is, indeed, not a little remarkable that the period required by Mars for one complete rotation is only about half an hour more than the period of rotation of the earth. The exact period is 24 hours, 37 minutes, $22\frac{3}{4}$ seconds (Proctor). It, therefore, follows that the aspect of the planet changes from hour to hour; the western side gradually sinks from view, the eastern side gradually rises into view. In twelve hours the aspect of the planet is completely changed. These changes, together with the inevitable effects of foreshortening, render it often difficult to identify the objects on the planet with those on the map. The maps, it must be confessed, fall very far short of the maps of the moon in definiteness and in certainty; yet there can hardly be a doubt that the main features of the planet are to be regarded as thoroughly established, and some astronomers have given names to all the prominent objects.

On the question as to the possibility of life on Mars, a few words may be added. If we could be certain of the existence of water on Mars, then one of the most fundamental conditions would be fulfilled; and even though the atmosphere on Mars had but few points of resemblance in composition or in density to the atmosphere of the earth, life might still be possible. Yet even if we could suppose that a man would find suitable nutriment for his body and suitable air for his respiration, it yet seems very doubtful whether he would be able to live. Owing to the small size of Mars and the smallness of its mass in comparison with the earth, the intensity of the gravitation on the surface of Mars would be very different to the gravitation on the surface of the earth. We have already alluded to the small gravitation on the moon, and in a lesser degree the same remarks will apply to Mars. A body which weighs on the earth two pounds would on the surface of Mars weigh only one pound. Nearly the same exertion which will raise a 56 lb. weight on the earth would lift two similar weights on Mars. The effect of such changes on man would be indeed remarkable. He would experience a buoyancy quite unfamiliar to his terrestrial experience. His labours would no doubt be greatly lightened under such circumstances. The fatigue of walking would be reduced to one-half—he would walk up two steps of stairs at a time with the same exertion required here for one step. If he could jump four feet high on the earth, he could with precisely the same exertion jump eight feet high on Mars. Yet, notwithstanding all this, it may well be doubted whether our organism would be adapted to a change so radical and so complete. We should imagine that the circulation of the blood and the various movements which constitute life would be greatly, and probably even fatally, deranged by any such changes. The effect on our system which would arise from a great diminution of gravity may be illustrated in the following way:—In the act of jumping from a chair to the ground, our body during the act of jumping is to a certain extent in the condition in which it would be if we were resident on a planet which possessed no power of attraction. A moment's reflection will show this to be the case. There is no doubt that a man could not sustain a ton weight in his hand while he stood on the earth, yet, while jumping down, a ton or any other weight could be held in the hand; for, as it could but fall, and as he is falling simultaneously, the exertion of holding it in the hand vanishes. During the act of falling, therefore, the weight is held in the hand with no exertion, and this is what we should find if we were standing on a planet devoid of attracting power.

The earth is attended by one moon. Jupiter is attended by four moons. Mars is a planet revolving between the orbits of the earth and of Jupiter. It is a body of the same general type as the earth and Jupiter. It is ruled by the same sun, and all three planets form part of the same system; but as the earth has one moon and Jupiter four moons, why should not Mars also have a moon? No doubt Mars is a small body, less even than the earth, and vastly less than Jupiter. We could not expect Mars to have large moons, but why should it be unlike its two neighbours, and not have any moon at all? So reasoned astronomers, but yet until recently no satellite of Mars could be found. For centuries the planet has been diligently examined with this special object, and as failure after failure came to be recorded, the conclusion seemed almost to be justified that the chain of analogical reasoning had broken down. The moonless Mars was thought to be an exception to the rule, that all the great planets outside Venus were dignified by an attendant retinue of satellites.

It seemed almost hopeless to begin again a research which had so often been tried, and had so invariably led to disappointment; yet, fortunately, the present, generation has witnessed still one more attack, conducted with consummate skill, and with the most perfect equipment. This attempt has obtained the success it so well merited, and the result has been the memorable discovery of two satellites of Mars.

This, perhaps the most important telescopic discovery of this century, was made by Professor Asaph Hall, the distinguished astronomer, at the observatory of Washington. Mr. Hall was provided with an instrument of colossal proportions and of exquisite workmanship, known as the great Washington refractor. It is the product of the celebrated workshop of Messrs. Alvan Clark and Sons, from which so many large telescopes have proceeded, and in its noble proportions far outstripped any other telescope ever devoted to the same research. The object-glass measures twenty-six inches in diameter, and is hardly less remarkable for the perfection of its definition than for its size. But even the skill of Mr. Hall, and the space-penetrating power of his telescope, would not on ordinary occasions have been able to discover the satellites of Mars. Advantage was prudently taken of that memorable opposition of Mars in 1877, when, as we have already described, the planet came unusually near the earth.

It was known that Mars could not have a large moon. Had Mars been attended by a moon which was one-hundredth part of the bulk of our moon, it must long ago have been discovered. Mr. Hall, therefore, knew that if there were any satellites to Mars, they must be extremely small, and he braced himself for a severe and diligent search. The circumstances were all favourable. Not only was Mars as near as it could well be to the earth; not only was the great telescope at Washington the most powerful refractor then in existence; but the situation of Washington is such, that Mars was seen from the observatory at a high elevation in a pure air. It was while the British Association were meeting at Plymouth in 1877, that a telegram flashed across the Atlantic. Brilliant success had rewarded Mr. Hall's efforts. He had hoped to discover one satellite. The discovery of even one would have made the whole scientific world ring; but fortune smiled on Mr. Hall. He discovered first one satellite, and then he discovered a second; and in connection with these satellites, he further discovered a unique fact in the solar system.

The outer of the satellites revolves around the planet in the period of 30 hours, 17 min., 54 secs.; it is the inner satellite which has commanded the attention and curiosity of every astronomer in the world. Mars turns round on his axis in a martial day, which is very nearly the same length as our day of twenty-four hours. The inner satellite of Mars moves round in 7 hours, 39 min., 14 secs. The inner satellite, in fact, revolves three times round Mars in the same time that Mars can turn round once. This circumstance is unparalleled in the solar system; indeed, as far as we know, it is unparalleled in the universe. In the case of our own planet, the earth rotates twenty-seven times for one revolution of the moon. To some extent the same may be said of Jupiter and of Saturn, while in the great system of the sun himself and the planets, the sun rotates on his axis several times for each revolution of the planets. There is no other known case where the satellite revolves around the primary more quickly than the primary rotates on its axis. The anomalous movement of the satellite of Mars has, however, been accounted for. In a subsequent chapter we shall again allude to this, as it is connected with an important department of modern astronomy.

It will be interesting to endeavour to obtain some notion of the probable size of these two satellites of Mars. The average light of our full moon has been found, by actual measurement, to be 24,000 times that of Vega, a bright star of the first magnitude. Were our moon removed to the distance of Mars when the planet is nearest to us, due allowance being made for the enfeeblement of the solar light, the moon would still appear to us more than half as bright (0·5887) as a first magnitude star. But Deimos was estimated to be no brighter than a star of the twelfth magnitude, of which it takes about 25,120 to give as much light as Vega. From this it is evident that if Deimos is made of materials similar to our moon, its face must have an area of about one 15,000th part of the moon's, or it must be about 18 miles in diameter. Phobos is brighter by about half a magnitude, and would require a diameter of 22⅓ miles of moon material in order to reflect the light we actually receive. But Phobos is so close to

Mars that, notwithstanding its smallness, it will seem, when seen overhead from the surface of the planet, to be about two-thirds the size of our moon. The more distant and smaller satellite will appear but about a quarter of the size of its rival.

As the satellites revolve in paths vertically above the equator of their primary, the one less than 4,000 miles and the other only some 14,500 miles above the surface, it follows that they can never be visible from the poles of Mars; indeed, to see Phobos, the observer's planetary latitude must not be above 68¾°, or the satellite will be hidden by the body of Mars, just as we, in the British Islands, would be unable to see an object revolving round the earth a few hundred miles above the equator. For the same reason Phobos will only be rather more than one-third of each revolution above the horizon of a spectator on the planet.

Before passing from the attractive subject of the satellites, we may just mention two points of a literary character. Mr. Hall consulted his classical friends as to the names which should be conferred on the two satellites. Homer was referred to, and a passage in the "Iliad" suggested the names Deimos and Phobos. These personages were the attendants of Mars, and the lines in which they occur have been thus construed by my friend Professor Tyrrell:—

> "Mars spake, and called Dismay and Rout
> To yoke his steeds, and he did on his harness sheen."

A curious circumstance with respect to the satellites of Mars will be familiar to those who are acquainted with "Gulliver's Travels." The astronomers on board the flying Island of Laputa had, according to Gulliver, keen vision and good telescopes. Gulliver announced that they had found two satellites to Mars, one of which revolves round Mars in ten hours. The author has thus not only made a correct guess about the number of the satellites, but he actually stated the periodic time with considerable accuracy! We do not know what can have suggested the latter guess. A few years ago any astronomer reading the voyage to Laputa would have said this was absurd. There might be two satellites to Mars no doubt; but to say that one of them goes round in ten hours would be to assert what was "impossible!" Yet the truth has been even stranger than fiction.

And now we must bring to a close our account of this beautiful and interesting planet. There are many additional features of it over which we are tempted to linger, but there are so many other bodies which claim our attention in the solar system, so many other bodies which exceed Mars in size and intrinsic importance, that we are obliged to desist. Our next step will not, however, at once conduct us to the giant planets. We find outside Mars a host of planets, small indeed, but of the greatest interest; and with these we shall find abundant occupation for the following chapter.

CHAPTER XI.

THE MINOR PLANETS.

The Lesser Members of our System—Bode's Law—The Vacant Region in the Planetary System—The Research—The Discovery of Piazzi—Was the small Body a Planet ?—The Planet becomes Invisible—Gauss undertakes the Search by Mathematics—The Planet Recovered—Further Discoveries—Number of Minor Planets now known—The Region to be Searched—The Construction of the Chart for the Search for Small Planets—How a Minor Planet is Discovered —Physical Nature of the Minor Planets—Small Gravitation on the Minor Planets—The Berlin Computations—How the Minor Planets tell us the Distance of the Sun—Accuracy of the Observations—How they may be Multiplied—Victoria and Sappho—The most perfect Method.

In our chapters on the Sun and Moon, on the Earth and Venus, and on Mercury and Mars, we have been usually discussing the features and the movements of globes of vast dimensions. The least of all these bodies is the moon, but even our moon is a ball 2,000 miles from one side to the other. In approaching the subject of the minor planets, we must be prepared to find planets of dimensions quite inconsiderable in comparison with the great globes of our system. No doubt these minor planets are all of them some few miles, and some of them a great many miles, in diameter. Were they close to the earth they would be conspicuous, and even splendid objects; but as they are so distant they do not, even in our greatest telescopes, become very remarkable, while to the unaided eye they are totally invisible.

In a diagram of the orbits of the various planets, it is shown that a wide space exists between the orbit of Mars and the orbit of Jupiter. It was often surmised that this wide region must be tenanted by some other planet. The presumption became much stronger when a remarkable law was discovered which exhibited, with considerable accuracy, the relative distances of the great planets of our system. Take the series of numbers 0, 3, 6, 12, 24,

48, 96, whereof each number (except the second) is double of the number which precedes it. If we now add four to each, we have the series 4, 7, 10, 16, 28, 52, 100. With the exception of the fifth of these numbers (28) they are all sensibly proportional to the distances of the various planets from the sun. In fact, the distances are as follows:—Mercury, 3·9; Venus, 7·2; Earth, 10; Mars, 15·2; Jupiter, 52·9; Saturn, 95·4. Although we have no physical reason to offer why this law—known as Bode's law—should be true, yet the fact that it is so nearly true in the case of all the known planets, tempts us to ask whether there may not also be a planet revolving around the sun at the distance represented by 28.

So strongly was this felt that a large number of astronomers decided, at the end of the last century, to make a united effort to search for the unknown planet. It seemed certain that the planet could not be large, otherwise it would have been found long ago. If it should exist, then means must be found for discriminating between the planet and the hosts of stars strewn along its path.

The search for the small planet was soon rewarded by a success which has rendered the evening of the first day in this century memorable in astronomy. It was in the pure skies of Palermo that the observatory was situated, where the memorable discovery of the first minor planet was accomplished by Piazzi. This laborious and accomplished astronomer had organised a most ingenious system of exploring the heavens, which was eminently calculated to discriminate a planet from among the stars. On a certain night he would select a series of stars to the number of fifty, more or less according to circumstances. With his meridian circle he determined the places of all the fifty stars. The following night, or at all events as soon as convenient, he re-observed the whole fifty stars with the same instrument and in the same manner, and the whole operation was repeated on two, or perhaps more, nights subsequently. When the observations were compared together he was in possession of some four or more places of each one of the stars on four nights, and the whole series was complete. He was persevering enough to carry on these observations for very many groups of stars, and at length he was rewarded by a success which amply compensated him for all his toil. It was on the 1st of January, 1801, that he commenced for the one hundredth and fifty-ninth time to observe a group of stars. Fifty stars this night were viewed in his telescope, and their places were carefully recorded. Of these objects the first twelve were undoubtedly stars, and so to all appearance was the thirteenth, a star of the eighth magnitude in the constellation of Taurus. There was nothing to distinguish the telescopic appearance of this object from all the stars which preceded or followed it. The following night Piazzi, according to his custom, re-observed the whole fifty stars, and he did the same again on the 3rd of January, and once again on the 4th. He then, as usual, brought together the four places he had found for each of the stars on his list. When this was done it was at once seen that the thirteenth object on the list was quite a different body from the remainder and from all the other stars which he had ever observed before. The four places of this mysterious object were all different; in other words, the object was in movement, and was therefore a planet.

A few days' observation sufficed to show how this little object, afterwards called Ceres, revolved around the sun, and how it circulated in that vacant path intermediate between the path of Mars and the path of Jupiter. Great, indeed, was the interest aroused by this discovery, and great, indeed, is the influence which it has exercised on the progress of astronomy. The majestic planets of our system were now to admit a much more humble object to a share of the benefits dispensed by the sun.

After Piazzi had obtained a few observations of the object, and after he had fully identified its character as a planet, the season for observing this part of the heavens had passed away, and the new planet of course ceased to be visible. In a few months, no doubt, the same part of the sky would be above the horizon after dark, and the stars would of course be seen as before. The planet, however, was moving, and would continue to move, and by the time the next season had arrived it would have passed off into some distant region, and would be again confounded with the stars which it so closely resembled. How, then, was the planet to be pursued through its period of invisibility and identified when it again came within reach of observation?

This difficulty attracted the attention of astronomers, and they sought for some method by which the place of the planet could be recovered so as to prevent Piazzi's discovery from falling into oblivion. A young German mathematician, whose name was Gauss, opened his distinguished career by his successful attempt to solve this problem. A planet, as we have shown, describes an ellipse around the sun, and the sun lies at a focus of that ellipse. It can be demonstrated that when three positions of a planet are known, then the ellipse in which that planet moves is completely determined. Piazzi had seen the planet on at least three occasions, and he had on each occasion measured the place which the planet occupied. This information was available to Gauss, and the problem which he had to solve may be thus stated. Knowing the place of the planet on three nights, it is required, without any further observations, to tell what the place of the planet will be on a special occasion some months in the future. Mathematical calculations, based on the laws of Kepler, will enable this problem to be solved, and Gauss succeeded in solving it. He showed that though the telescope of the astronomer was unable to detect the planet during its long season of invisibility, yet the pen of the mathematician could follow it with unfailing certainty. When, therefore, the progress of the seasons permitted the observations to be renewed, the search was recommenced. The telescope was directed to the point which Gauss's calculations indicated, and there was the little Ceres. Ever since its rediscovery, the planet has been so completely bound in the toils of mathematical reasoning, that its place every night of the year can be indicated with a fidelity approaching to that attainable with the moon or with the great planets of our system.

The discovery of one minor planet was quickly followed by similar discoveries, so that within seven years Pallas, Juno, and Vesta were added to the Solar system. The orbits of all these bodies lie in the region between the orbit of Mars and of Jupiter, and for many years it seems to have been thought that our planetary system was now complete. Forty years later the career of discovery was again commenced. Planet after planet was added to the list; gradually the discoveries became a stream of increasing volume, until in 1884 the total number of the known minor

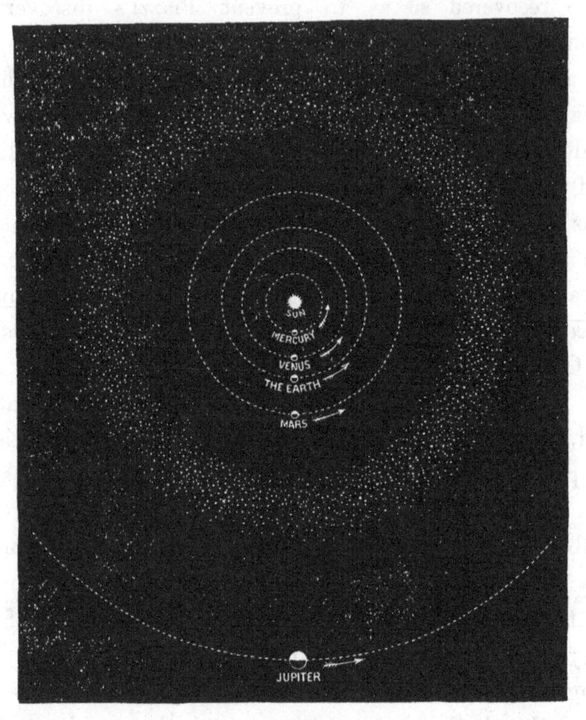

Fig. 49.—The Zone of Minor Planets between Mars and Jupiter.

planets exceeded 240. Their distribution in the solar system is somewhat as represented in Fig. 49.

These numerous discoveries are not now accomplished exactly in the same way as the original discovery of Piazzi. By the improvement of astronomical telescopes, and by the devotion with which certain astronomers have applied themselves to this interesting research, a special method of observing has been created with the distinct purpose of searching for these little objects.

It is known that the paths in which all the great planets move through the heavens coincide very nearly with the path which the sun appears to follow among the stars, and which is known as the ecliptic. It is natural to assume that the small planets also move in the same great highway, which leads them through all the signs of the zodiac in succession. Some of the small planets, no doubt, deviate rather widely from the track of the sun, but the great majority are still tolerably near it. This consideration at once simplifies the search for the new planets. A certain zone extending around the heavens is to be examined, but there is little advantage in pushing the examination into other parts of the sky.

The next step is to construct a map containing all the stars in this region. This is a task of very great labour; the stars visible in the large telescopes are so numerous that many tens of thousands, perhaps we should say hundreds of thousands, are included in the region even so narrowly limited. The fact is that many of the minor planets now discovered are objects of extreme minuteness; they can only be seen with very powerful telescopes, and for their detection it is necessary to use charts on which even the faintest stars are depicted. Many astronomers have concurred in the labour of producing these charts. We cannot here mention them all. We will therefore only allude to the last series which have appeared—they have been prepared by Professor Peters with exquisite care, and as, in his hands, they have already served to enable him to discover very many new planets, so now he has placed them in the hands of astronomers at large who may be capable of carrying on this very fascinating inquiry.

The astronomer about to seek for a new planet directs his telescope towards that part of the sun's path which is on the meridian at midnight; there, if anywhere, lies the chance of success, because that is the region in which a planet is nearer to the earth than at any other part of its course. Point by point he compares his chart with the heavens, and usually he finds the stars in the heavens and the stars in the chart to correspond; but sometimes it will happen that a point in the heavens is missing from the chart. His attention is at once arrested; he follows the object with care, and if it moves it is a planet. Still he cannot be sure that he has really made a discovery; he has found a planet no doubt, but it may be one of the host of planets already known. To clear up this point he must under take a further, and sometimes a very laborious, inquiry; he must trace out all the known planets and see whether it is possible for one of them to have been in the position on the night in question. If he can satisfy himself that no known planet could have been there, he is then entitled to announce to his brother astronomers the discovery of a new planet. It seems certain that all the more important of the minor planets have been long since discovered. The recent additions to the list are generally extremely minute objects, beyond the powers of small telescopes.

Of the physical nature of these bodies and of the character of their surface we are entirely ignorant. It may be, for anything we can tell, that these little bodies are miniature globes like our earth, diversified by continents and by oceans. If there be life on such bodies, whose diameter is often not very many miles, that life must be something totally different to the life with which we are familiar. Setting aside every other difficulty from the possible absence of water and from the great improbability of an atmosphere of a density and a composition suitable for respiration, the simple fact of gravitation alone would present the most profound contrast to the condition of life on the earth.

Let us attempt to illustrate this point, and suppose that we take the case of a minor planet eight miles in diameter, or in round numbers one-thousandth part of the diameter of the earth. If we further suppose that the materials of the planet are the same as the materials of the earth, it is easy to prove that the gravity on the surface of the planet will be only one-thousandth part of the gravity of the earth. It follows that the weight of an object on the earth would be reduced to the thousandth part if that object were transferred to the planet. This would not be disclosed of course by an ordinary weighing scales, where the weights are placed in one pan and the body to be weighed in the other. Tested in this way a body would of course weigh precisely the same anywhere; for if the gravitation of the body is altered, so is also in equal proportion the gravitation of the weights which balance it. But weighed with a spring balance the change would be at once evident, and the effort with which a weight could be raised would be reduced to one-thousandth part. A load of one thousand pounds could be lifted from the surface of the planet by the same effort which would lift one pound on the earth; the effects which this would produce are very remarkable. Let us instance one of them, which involves a point of no little interest in some other branches of astronomy.

A stone thrown up into the air soon falls to the ground. A rifle bullet fired straight up into the air will ascend higher and higher, until at length its motion ceases, it begins to turn, and falls down to the ground. Let us for the moment suppose that we had a rifle of infinite strength and gunpowder of unlimited power. As we increase the charge we find that the bullet will ascend higher and higher, and each time it will take a longer period before it returns to the ground. The return of the bullet to the ground is, of course, due to the attraction of the earth. This attraction acts on the bullet at all times, and gradually lessens the velocity, ultimately overcomes the upward motion, and then brings the bullet back. It must be remembered that the efficiency of the attraction decreases when the height is increased. Consequently when the bullet has a prodigiously great initial velocity, in consequence of which it ascends to an enormous height, the return of the bullet is retarded by a twofold cause. In the first place the distance through which it has to be recalled is greatly increased, and in the second place the efficiency of gravitation in effecting its recall has decreased. The greater the velocity, the feebler must be the capacity of gravitation for bringing back the body, and thus at length the velocity can be increased to that point at which the gravitation, constantly declining as the body ascends, is never able to entirely neutralise the velocity, and hence we have the remarkable case of a body projected away never to return.

It is possible to exhibit this reasoning in a numerical form, and it can be shown that a velocity of six miles a second vertically upwards, would suffice to convey a body entirely away from the gravitation of the earth. This velocity is far beyond the utmost limits of our artillery. It is, indeed, from twenty to thirty times as swift as a cannon shot; and even if we could produce it, the resistance of the air would present an insuperable difficulty. Such reflections, however, do not affect the conclusion that there is for each planet a certain specific velocity appropriate to that planet, and depending solely upon the size and the mass of the planet.

If we assume for the sake of simplicity that the materials of the planets are similar, then the law which expresses the critical velocity for each planet can be stated in an extremely simple form. It is, in fact, simply proportional to the diameter of the planet. For a minor planet whose diameter was one-thousandth part of that of the earth, the critical velocity would be the thousandth part of six miles a second—that is about thirty feet per second. This is a low velocity compared with ordinary standards. A child can here toss a ball up fifteen or sixteen feet high, yet to carry a ball up this height it must be projected with a velocity of thirty feet per second. A child, standing upon a planet eight miles in diameter, throws his ball vertically upwards; up and up the ball will soar to an amazing elevation. If the original velocity were less than thirty feet per second the ball will at length gradually cease moving, will begin to turn, will fall with gradually accelerating velocity, until at length it returns with the same velocity with which it had been projected. But if the original velocity had been as much or more than thirty feet per second, then the ball would soar up and up never to return. In a future chapter it will be necessary for us to refer again to this subject.

The atmosphere surrounding a small planet must, under ordinary circumstances, be one of very great rarity. Even supposing that the quantity of air over each square foot on the small planet was as much as that over each square foot on the earth, yet the difference in gravitation would cause the widest difference in the density at the surface. It is the weight of the superincumbent air which presses on the lower strata and gives them their density. If the gravitation were reduced, then the pressure would be lessened and the air would expand. We should therefore expect to find that the atmosphere surrounding a small planet is of extreme rarity, though possibly of enormous volume.

A great increase in the number of minor planets has rewarded the zeal of those astronomers who have devoted their labours to this subject. Their success has entailed a vast amount of labour on the computers of the "Berlin Year-book." That useful work in this respect occupies a position which has not been taken by our own "Nautical Almanac," nor by the similar publications of other countries. A skilful band of computers make it their duty to provide for the "Berlin Year-book" detailed information as to the movements of the minor planets. As soon as a few observations of a planet have been obtained the little object passes into the secure grasp of the computer; he is able to predict the career of that planet for years to come, and such predictions for all the known minor planets are to be found in the annual volumes of the work referred to. The growth of discovery has been so rapid, that the necessary labour for the preparation of such predictions is now enormous. It must be confessed that many of the minor planets are very faint and otherwise devoid of interest, so that astronomers are sometimes tempted to concur with the suggestion that a proportion of the astronomical labour now devoted to the computation of the paths of these bodies might be more profitably applied. For this it would only be necessary to cast adrift all the less interesting members of the host and allow them to pursue their paths unwatched by the telescope, or by the still more ceaseless tables of the mathematical computer.

The sun, which controls the mighty planets of our system, does not disdain to guide, with equal care, the tiny globes which form the minor planets. Each revolves in an elliptic orbit, and at certain times some of them approach near enough to the earth to have their distances measured. The observations can be made with very great precision; they can be multiplied to any extent that may be desired. Some of these little bodies have consequently a great astronomical future, inasmuch as they seem destined to determine more accurately than Venus or than Mars the true distance from the earth to the sun. The smallest of these planets will not answer; they can only be seen in powerful telescopes, and they do not admit of being measured with the necessary accuracy. It is also obvious that the planets to be chosen must come as near the earth as possible. They should have orbits possessing a high degree of eccentricity, so that when in perihelion and in opposition simultaneously, the approach may be made advantageously. Under favourable circumstances, some of the minor planets will approach the earth to a distance little more than three-quarters of the distance of the sun. These various conditions limit the number of planets available for this purpose to about a dozen, of which one or two will usually be suitably placed each year.

For the determination of the sun's distance, this method by the minor planets offers unquestionable advantages. The planet itself is a minute star-like point, and the measures are made from it to the other stars which are seen near it in the telescope. A few words will, perhaps, be necessary at this place as to the nature of the measurements referred to. When we speak of the measures from the planet to the star we do not refer to what would be perhaps the most ordinary acceptation of the expression. We do *not* mean the actual measurement of the number of miles in a straight line between the planet and the star. This element, even if attainable, could only be the result of a protracted series of observations of a nature which will be explained later on when we come to speak of the distances of the stars. The measures now referred to are of a very much more simple type; they merely refer to the apparent distance of the objects expressed in angular measure. This angular measurement is of a wholly different character to the linear measurement, and the two methods may, indeed, lead to results that would at first seem paradoxical. Let us give, as an illustration, the case of the group of stars forming the Pleiades, and those which form the Great Bear. The latter is a large group, the former is a small one. But why do we think the words large and small rightly applied here. Each pair of stars of the Great Bear makes a large angle with the eye. Each pair of stars in the Pleiades makes a small angle, and it is these angles which are the direct objects of astronomical measurement. We speak of the distance of the stars, meaning thereby the angle which is bounded by the two lines from the eye to the two stars. This is what our instruments are able to measure, and it is to be observed that there is no reference to linear magnitude at all. Indeed, if we allude to linear magnitudes, it is quite possible, for anything we can tell, that the Pleiades are really a much larger group than the Great Bear, and that the apparent superiority of the latter is merely due to its being closer to us. The most accurate of these angular measures are obtained when two stars, or two star-like points, are so close together as to enable them to be included in one field of view of the telescope. There are many different forms of apparatus which enable the astronomer to give to such measures a precision unattainable in the measurement of objects less definitely marked, or at a greater apparent distance. The measures of the distance of the small star-like planet from a star are characterised by great accuracy, and this is one source of the accuracy of the minor planet method of finding the sun's distance.

But there is another and, perhaps, a weightier argument in its favour, the full importance of which has lately begun to be realized. The real strength of the minor planet method rests hardly so much on the individual accuracy of the observations, as on the fact that from the nature of the method a considerable number of observations can be concentrated on the result. It will, of course, be understood that when we speak of the accuracy of an observation, it is not to be presumed that the observation is entirely free from error. Errors always exist, and though the errors be small, yet if the total quantity to be measured is small, the error may amount to an appreciable fraction of the total. The one way by which the effect of errors of observation can be subdued, is by taking the mean of a large number of observations. This is the real source of the value of the minor planet method. We have not to wait for the occurrence of rare events like the transit of Venus. Each year will witness the approach of some one or more minor planets sufficiently close to the earth to render the observations possible. The varied circumstances attending each planet, and the great variety of the observations which may be made upon it, will all tend further and further to eliminate errors.

As the planet pursues its course through the sky, and as the sky is everywhere studded over with countless myriads of minute stars, it is evident that the planet, itself so like a star, will always have some stars in its immediate neighbourhood. As the movements of the planet are well known, it is possible to foretell where it will be on each night that it is to be observed. It is thus possible to pre-arrange with observers in widely-different parts of the earth as to the observations to be made on each particular night.

An attempt has only recently been made, on the suggestion of Mr. Gill, to carry out this method on a scale commensurate with its importance. The planets Victoria and Sappho happened, in the year 1882, to approach so close to the earth that arrangements were made for simultaneous measurements in both the northern and the southern hemispheres. A scheme was completely drawn up many months before the observations were to commence. Each observer who participated in the work was thus advised beforehand of the stars which were to be employed each night. Viewed from any part of the earth, from the Cape of Good Hope or from Great Britain, the positions of the stars are absolutely unchanged. Their distance is so stupendous that a change of place on the earth displaces them to no appreciable extent. But the case is different with a minor planet. It is hardly one-millionth part of the distance of the stars, and the displacement of the planet when viewed from the Cape and when viewed from Europe, is a measurable quantity.

The displacement of the planet is to be elicited by comparison between the observations in the northern hemisphere with those in the southern hemisphere. The observations in both hemispheres must be as nearly simultaneous as possible, due allowance being made for the motion of the planet in whatever interval there may be. Although every precaution is taken to eliminate the errors of each observation, yet the fact remains that we compare the measures made by observers in the northern hemisphere, with those made by different observers, and of course different instruments, in the southern hemisphere. In this respect we are at no greater disadvantage than in observing the transit of Venus, yet it is possible to obviate even this difficulty, and thus to give the minor planet method a great advantage over its rival. The difficulty would be overcome if we could conceive that an observer and his observatory, after making a set of observations on a fine night in the northern hemisphere, were by means of the lamp of Aladdin, or some similar agency, to be transferred, instruments and all, to the southern hemisphere, and there to repeat the observations. An equivalent transformation can be effected without any miraculous agency, and in it we have undoubtedly the most perfect mode of measuring the sun's distance with which we are acquainted. This method has already been applied with success by Mr. Gill in the case of Juno, but there are other planets more favourably situated.

Take, for instance, one of those minor planets, which sometimes approach to within 70,000,000 miles of the earth. When the opposition is drawing near, a skilled observer is to be placed at some suitable station near the equator. The instrument he is to use should be that marvellous piece of mechanical and optical skill known as the heliometer.* It can be used to measure stars at a greater range than is obtainable with the filar micrometer. The measurements are to be made in the evening as soon as the planet has risen high enough to enable it to be seen distinctly. The observer and the observatory is then to be transferred to the other side of the earth. How is this to be done? Say, rather, how we could prevent it being done. Is not the earth rotating on its axis, so that in the course of a few hours the observator on the equator is carried bodily round for thousands of miles? As the morning approaches the observations are to be repeated. The planet is found to have changed its place very considerably with regard to the stars. This is partly due to its own motion, but it is also largely due to the parallactic displacement arising from the rotation of the earth, which may amount to as much as twenty

* The heliometer is a telescope with its object-glass cut in half along a diameter. One or both of these halves can be moved transversely by a screw. Each half gives a complete image of the object. The measures are effected by observing how many turns of the screw convey the image of the star formed by one half of the object-glass to coincidence with the image of the planet formed by the other.

seconds. The measures on a single night with the heliometer should not have a mean error greater than one-fifth of a second, and we might reasonably expect that observations could be secured on about twenty-five nights during the opposition. Four such groups might be expected to give the sun's distance without any uncertainty greater than the thousandth part of the total amount. The chief difficulty of the process arises from the movement of the planet during the interval which divides the evening from the morning observations. This, it must be admitted, is a drawback to the method. It can, however, be avoided by diligent and repeated measurements of the place of the planet with respect to the stars among which it passes. Let us hope that long ere the next transit of Venus approaches, the problem of the sun's distance will have been satisfactorily solved by the minor planets.

CHAPTER XII.

JUPITER.

The great Size of Jupiter—Comparison of his Diameter with that of the Earth—Dimensions of the Planet and his Orbit—His Rotation—Comparison of his Weight and Bulk with that of the Earth—Relative Lightness of Jupiter—How explained—Jupiter still probably in a Heated Condition—The Belts on Jupiter—Spots on his Surface—Time of Rotation of different Spots various—Storms on Jupiter—Jupiter not Incandescent—The Satellites—Their Discovery—Telescopic Appearance—Their Orbits—The Eclipses and Occultations—A Satellite in Transit—The Velocity of Light Discovered—How is this Velocity to be Measured experimentally?—Determination of the Sun's Distance by the Eclipses of Jupiter's Satellites—Jupiter's Satellites demonstrating the Copernican System.

In our exploration of the beautiful series of bodies which form the solar system, we have gradually proceeded step by step outwards from the sun. In the pursuit of this method we have now come to the splendid planet Jupiter, which wends its majestic way in a path immediately outside those orbits of the minor planets which we have just been considering. Great, indeed, is the contrast between these tiny globes and the stupendous globe of Jupiter. Had we adopted a somewhat different method of treatment—had we, for instance, discussed the various bodies of our planetary system in the order of their magnitude—then the minor planets would have been the last to be considered, while the leader of the host would be Jupiter. To this position Jupiter is entitled without an approach to rivalry. The next greatest on the list, the beautiful and interesting Saturn, comes a long distance behind. Another great descent in the scale of magnitude has to be made before we reach Uranus and Neptune, while still another step downwards must be made before we reach that lesser group of planets which includes our earth. So conspicuously does Jupiter tower over the rest, that even if Saturn were to be augmented by all the other globes of our system rolled into one, the united mass would still not equal the great globe of Jupiter.

The adjoining picture shows the relative dimensions of Jupiter and the earth, and it conveys to the eye a more vivid impression of the enormous bulk of Jupiter than we can readily obtain by merely considering the numerical statements by which his bulk is to be accurately estimated. As, however, it will be necessary to place

Fig. 50.—The relative dimensions of Jupiter and the Earth.

the numerical facts before our readers, we do so at the outset of this chapter.

Jupiter revolves in an elliptic orbit around the sun in the focus, at a mean distance of 482,000,000 miles. The path of Jupiter is thus about 5·2 times as great in diameter as the path pursued by the earth. The shape of Jupiter's orbit departs very appreciably from a circle, the greatest distance from the sun being 5·45, while the least distance is about 4·95, the earth's distance from the sun being taken as unity. Under the most favourable circumstances for seeing Jupiter at opposition, it must still be about four times as far from the earth as the earth is from the sun. Jupiter will also illustrate the law, that the more distant a planet is, the slower is the velocity with which the orbital motion of the planet is accomplished. While the earth passes over eighteen miles each second, Jupiter only accomplishes eight miles. Thus for a two-fold reason the time occupied by an exterior planet in completing a revolution is greater than the period of the earth. Not only has the outer planet to pursue a longer course than the earth, but the speed is less; it thus happens that Jupiter requires 4,332·6 days, or about fifty days less than twelve years, to make a circuit of the heavens.

The mean diameter of the great planet is about 85,000 miles. We say the *mean* diameter, because there is a conspicuous difference in the case of Jupiter between his equatorial and his polar diameter. We have already seen that there is a similar difference in the case of the earth, where we find the polar diameter to be shorter than the equatorial; but the disproportion of these two diameters is very much larger in Jupiter than in the earth. The equatorial diameter of Jupiter is 87,500 miles, while the polar diameter is not more than 82,500 miles. The ellipticity of Jupiter thus produced, is sufficiently marked to be obvious without any refined measures. Around this short diameter the planet spins with what must be considered an enormous velocity when we reflect on the size of the globe. Each rotation is completed in about 9 hrs. 55½ min. We may naturally contrast this with the much slower rotation of our earth in twenty-four hours. The difference becomes much more striking if we consider the relative speeds at which points on the equator of the earth and on the equator of Jupiter actually move. As the diameter of Jupiter is nearly eleven times that of the earth, it will follow that the actual speed of the equator on Jupiter must be about twenty-seven times as great as that on the earth. It is, no doubt, to this high velocity of rotation that we must ascribe the extraordinary ellipticity of Jupiter; for the rapid rotation causes a great degree of centrifugal force, and this bulges out the pliant materials of which he seems to be formed.

Jupiter is, so far as we can see, not a solid body. This is a very important circumstance; and, therefore, it will be necessary to discuss the matter at some little length, as we here perceive a wide contrast between this great planet and the other planets which have previously occupied our attention. From the measurements already given it is easy to calculate the bulk or the volume of Jupiter. It will be found that Jupiter is about 1,200 times as large as the earth; in other words, it would take 1,200 globes, each as large as our earth, all rolled into one, to form a single globe as large as Jupiter. If, therefore, the materials of which Jupiter is composed were of a nature analogous to the materials of the earth, we might expect that the weight of Jupiter would exceed the weight of the earth in something like the proportion of their volumes. This is the matter now proposed to be brought to trial. Here we may at once be met with the query, as to how we are to find the weight of Jupiter. It is not an easy matter even to weigh the earth on which we stand. How, then, can we weigh a mighty planet vastly larger than the earth, and distant from the earth by some hundreds of millions of miles? Truly, this is a noble problem. Yet the intellectual resources of man are able to achieve this mighty feat of celestial engineering. They are not, perhaps, actually able to make the ponderous weighing scales in which the great planet is to be cast, but they are able to divert to this purpose certain natural phenomena which yield the information that is required. All such investigations repose on universal gravitation as their foundation. The mass of Jupiter attracts other masses in the solar system. The efficiency of that attraction is more particularly shown on the bodies which are near Jupiter. In virtue of this attraction certain movements are impressed on those bodies. We can observe these movements in our telescopes, we can measure their amount, and from such measurements we can calculate the mass of the body by which the movements have been produced. This is the sole method which we possess for the investigation of the masses of the planets; and, though it may be difficult in its application—not only from the observations which are required, but from the intricacy and the profundity of the calculations to which those observations must be submitted—yet in the case of Jupiter, at least, there is no uncertainty about the result. The task is, indeed, peculiarly simplified in the case of Jupiter by reason of the beautiful system of four moons by which he is attended. These little moons move under the guidance of Jupiter, and their movements are not otherwise interfered with in a way calculated to prevent their use for our present purpose. It is from the observations of the satellites of Jupiter that we are enabled to measure his attractive power, and thence to conclude the mass of the mighty planet.

To those not specially conversant with the principles of mechanics, it may seem difficult to realise the accuracy of which such a method is capable. Yet there can be no doubt that his moons really teach us the mass of Jupiter, and do not leave a margin of inaccuracy so great as one hundredth part of the total amount. If other confirmation be needed, then other confirmation is forthcoming. A minor planet occasionally draws near to the orbit of Jupiter and experiences his attraction; the planet is forced to swerve from its path, and the amount of its deviation can be measured. From that measurement the mass of Jupiter can be computed by a calculation, of which it would be impossible here even to give an outline. The mass of Jupiter, as found from the minor planet, agrees with the mass obtained in a totally different manner from the satellites. Nor have we here exhausted the resources of astronomy in its bearing on this question. We can even discard the whole planetary system entirely, and we can invite the occasional aid of a comet which, flashing through the orbits of the planets, experiences large and sometimes enormous disturbances. For the present it suffices to remark, that on one or two occasions it has happened that venturous comets have been near enough to Jupiter to feel the effect of his ponderous body, and then to proclaim by their altered movements the magnitude of the mass which has affected them. The satellites of Jupiter, the minor planets, and the comets, all tell us the mass of Jupiter; and, as they all give us the same result (at least within extremely narrow limits), we cannot hesitate to conclude that the mass of the greatest planet of our system has been determined with accuracy.

The results of these measures must now be stated. They show, of course, that Jupiter is vastly inferior to the sun—that, in fact, it would take about 1,048 Jupiters, all rolled into one, to form a globe equal in *weight* to the sun. They also show us that it would take 310 globes as heavy as our earth to counterbalance the we ght of Jupiter.

No doubt this shows Jupiter to be a body of the most majestic mass; but the real point for consideration is not that Jupiter is 310 times as heavy as the earth, but that he is not a great deal more. Have we not stated that Jupiter is 1,200 times as *large* as the earth? How then comes it that he is only 310 times as *heavy?* This points at once to some fundamental contrast between the constitution of Jupiter and of the earth. How are we to explain this difference? We can conceive of two explanations. In the first place, it might be supposed that Jupiter is constituted of materials partly or wholly unknown to us on the earth. There is, however, an alternative supposition at once more philosophical and more consistent with the evidence. It is true that we know little or nothing of what the elementary substances on Jupiter may be, but one of the great discoveries of modern astronomy had taught us something of the elementary bodies present in other bodies of the universe, and has fully demonstrated that to a large extent they are identical with the elementary bodies on the earth. We shall have occasion to dwell on this very important subject in a future chapter, and accordingly we now merely refer to it for the purpose of justifying the assumption, that the materials of Jupiter are not entirely different from those on the earth. And if Jupiter be composed of bodies resembling those on the earth, there is one way, and only one, in which we can account for the disparity between his size and his mass. Perhaps the best way of stating the argument will be found in a retrospective glance at the probable past history of the earth itself, for it seems not impossible that the present condition of Jupiter was itself foreshadowed by the condition of our earth countless ages ago.

In a previous chapter we had occasion to point out how the earth seemed to be cooling from an earlier and highly-heated condition. The further we look back the hotter seems to have been the earth, and if we project our glance back to an epoch suffi ciently remote, we see that it must have been so hot that life on its surface would be impossible. Back still earlier, we see the heat to have been such that water could not rest on the earth; and hence it seems likely that at some incredibly remote epoch all the oceans now reposing on the surface of the earth, and perhaps a considerable portion of its more solid materials, must have been actually in vapour. Such a transformation of the globe would not alter its *mass*, for the materials weigh the same whatever be their condition as to temperature, but it would alter to the most remarkable extent the *size* of our globe. If all our oceans were transformed into vapour, our atmosphere, charged with mighty clouds, would have a mass some hundreds of times greater than that which it has at present, and the size of the globe would be correspondingly swollen. Viewed from a distant planet, the cloud-laden atmosphere would seem to indicate the size of our globe, and its density would accordingly be concluded to be very much less than it is at present.

From these considerations it will be manifest that the discrepancy between the size and the weight of Jupiter, as contrasted with our earth, would be completely removed if we supposed that Jupiter was at the present day a highly-heated body in the condition of our earth countless ages ago. Every circumstance of the planet tends to justify this reasoning. We have assigned the smallness of the moon as a reason why the moon has cooled sufficiently to make its volcanoes silent and still. In the same way the smallness of the earth, as compared with Jupiter, accounts for the fact that Jupiter still retains a large part of its original heat, while the smaller earth has dissipated most of its store. This argument is illustrated and strengthened when we introduce other planets into the comparison. As a general rule we find that the smaller planets, like the earth and Mars, have a high density, indicative of a low temperature, while the giant planets, like Jupiter and Saturn, have a low density, suggesting that they still retain a large part of their original heat. We say "original heat" for the want, perhaps, of a more correct expression; it will, however, indicate that we do not in the least refer to the solar heat, of which, indeed, the great outer planets receive much less than those nearer the sun. Where the original heat may have come from is a matter still confined to the province of speculation.

A complete justification of these views with regard to Jupiter is to be found when we make a minute telescopic scrutiny of its surface; and it fortunately happens that the size of the planet is so great that, even at a distance of more millions of miles than there are days in the year, we can still trace out on its surface some of the characteristic features.

Plate XI. gives a series of four different views of the planet Jupiter. It has been copied from the very admirable drawings of this object made by Mr. L. Trouvelot at the Astronomical Observatory of Harvard College, United States. The first picture shows the appearance of the planet on February 2nd, 1872, through a powerful refracting telescope. We at once notice in this drawing that the outline of Jupiter is distinctly elliptical. The surface of the planet usually shows the remarkable series of belts here represented; these belts are nearly parallel to each other, and to the equator of the planet.

When Jupiter is observed for an hour or two, the appearance of the belts undergoes some changes. These are partly due to the regular rotation of the planet on its axis, which, in a period of less than five hours, will completely carry away the hemisphere we first saw and replace it by the hemisphere originally at the other side. But besides the changes thus arising, the belts and other features on the planet are also very variable. Sometimes new belts or markings appear, and old ones disappear; in fact, a thorough examination of Jupiter will demonstrate the very remarkable fact that there are no permanent marks whatever upon the planet. We are here immediately reminded of the contrast between Jupiter and Mars; on the smaller planet the markings are almost entirely permanent, and it has been possible to construct maps of the surface with tolerably accurate detail; a map of Jupiter is, however, an impossibility—the drawing of the planet which we make to-night will be different to the drawing of the same hemisphere made a few

PLATE XI.

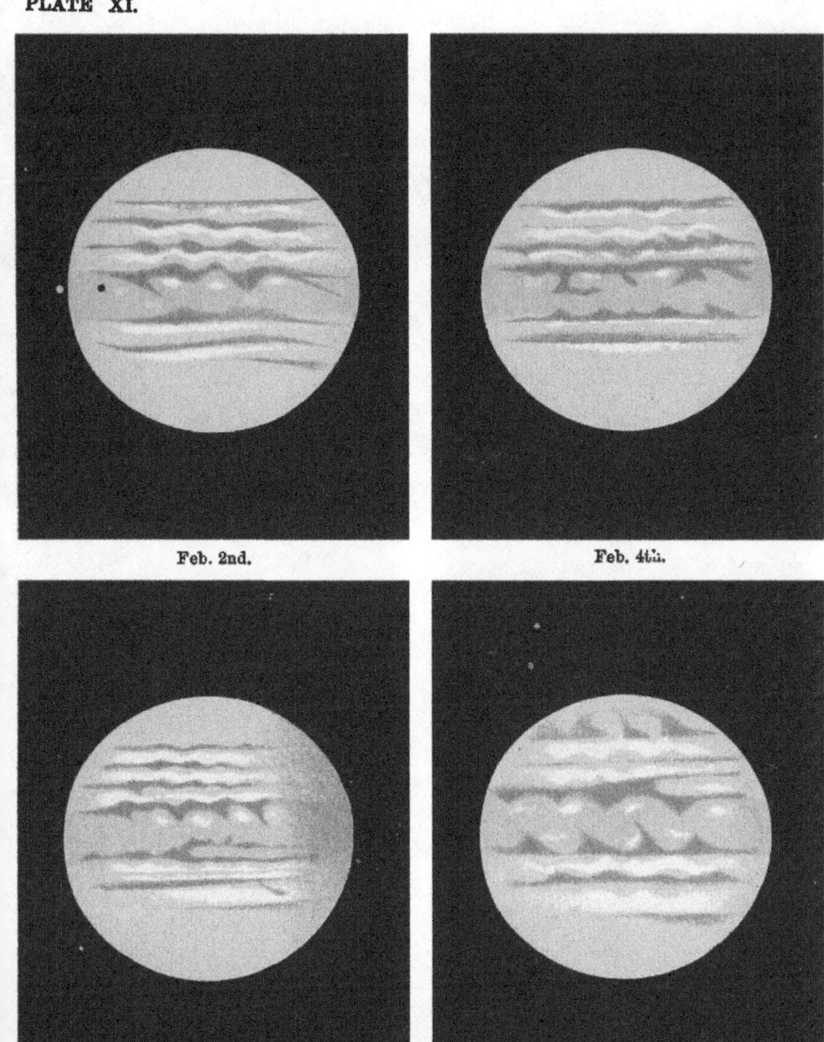

Feb. 2nd. Feb. 4th.

Feb. 12th. Feb. 28th.

THE PLANET JUPITER,

1872.

weeks hence. In this respect there is an analogy between the appearance of Jupiter and the appearance of the sun.

It should, however, be noticed that spots occasionally appear on the planet, which seem of a rather more permanent character than the belts. We may especially mention the object known as the great Red Spot, which has been a very remarkable feature upon Jupiter since the year 1878, and is probably shown on some of the drawings made several years earlier.

The conclusion is irresistibly forced upon us, that when we look at the surface of Jupiter we are not looking at any solid body. The want of permanence in the features of the planet would be intelligible, if the surface we see be merely an atmosphere laden with clouds of such density that our vision can never penetrate to the interior of the planet. The belts especially support this view; we are at once reminded of the equatorial zones on our own earth, and it is not at all unlikely that an observer sufficiently remote from the earth to obtain a just view of its appearance would see upon its surface more or less perfect cloud-belts suggestive of those on Jupiter. A view of our earth would be, as it were, intermediate between a view of Jupiter and of Mars. In the latter case the appearance of the permanent features of the planet is but to a trifling extent ever obscured by clouds floating over the surface. Our earth would always be partly, and often perhaps very largely, covered with cloud, while Jupiter seems at all times completely enveloped.

From another kind of observation we are also taught the important truth that Jupiter is not, superficially at least, a solid body. The period of the rotation of Jupiter around its axis is derived from the observations of certain spots, which present sufficient definiteness and sufficient permanence to be suitable for the purpose. Suppose one of these spots to lie at the centre of the planet's disc; its position is carefully measured, and the time is noted. As the hours pass on, the spot moves to the edge of the disc, then round the other side of the planet, and back again to the visible disc. When the spot regains the position originally occupied the time is again taken, and the interval which has elapsed is called the period of rotation of the planet. If Jupiter were a solid, and if all these features were fixed upon its surface, then it is perfectly clear that the time of rotation as found by any one spot would coincide precisely with the time yielded by any other spot; but this is not found to be the case. In fact, it would be nearer the truth to say that each spot gives us its own special period. Nor are the differences very minute. It has been found that the time in which the red spot is carried round is five minutes longer than that required by certain white spots near the equator, while certain small black spots have been found to accomplish the journey in even two minutes less. It may, therefore, be regarded as certain that the globe of Jupiter, so far as we can see it, is not a solid body. It consists, on the exterior at all events, of clouds and vaporous masses, which seem to be agitated by storms of the utmost intensity, if we may judge from the ceaseless changes of the planet's surface.

It is well known that the tempests by which the atmosphere surrounding the earth is convulsed, are all ultimately to be attributed to the heat of the sun. It is the heat of the sun which, striking on the vast continental masses, warms the air in contact therewith. This heated air becomes lighter and rises, while air to supply its place must flow in along the surface. The current of air so produced forms a breeze or a wind; while, under exceptional circumstances, we have the phenomena of cyclones and of hurricanes, all ultimately due to the sun's heat. Need we add that the rains, which so often accompany the storms, have also arisen from the sun's heat, which has distilled from the wide expanse of ocean the moisture by which the earth is refreshed?

The storms on Jupiter seem to be vastly greater than those on the earth. Yet the intensity of the sun's heat on Jupiter is only a mere fraction, less indeed than the twenty-fifth part, of the sun's heat on the earth. It is incredible that the motive power of the appalling tempests on the great planet can be entirely, or even largely, due to the feeble influence of solar heat. We are, therefore, led to seek for some other source for these disturbances. That source appears obvious when we admit that Jupiter still retains a large proportion of its primitive internal heat. Just as the sun himself is distracted by the most violent tempests in consequence of his internal heat, so, in a lesser degree, do we observe the same phenomena in Jupiter. It may also be noticed that the spots on the sun are usually, in more or less regular zones, parallel to the equator of the sun, and in this respect, not dissimilar to the belts on Jupiter.

It being admitted that Jupiter still retains some of its internal heat, the question remains as to how much. It is, of course, obvious that the heat of Jupiter is incomparably less than the heat of the sun. The brilliancy of Jupiter, which makes it an object of such splendour in our midnight sky, is only derived from the same source which illuminates the earth, the moon, or the other planets. Jupiter, in fact, shines by reflected sunlight, and not in virtue of any brilliancy of his own. A beautiful proof of this truth is familiar to every user of a telescope. The little satellites of Jupiter sometimes come between him and the sun, and they cast a shadow on Jupiter. That shadow is black, or, at all events, it seems black, relatively to the brilliant surrounding surface of the planet, whence it must be obvious that Jupiter is indebted only to the sun for its brilliancy. The satellites supply another interesting proof of this truth. A satellite sometimes enters into the shadow of Jupiter, and lo! the satellite vanishes. It vanishes because Jupiter has cut off the supply of sunlight which previously rendered the satellite visible. But Jupiter is not himself able to offer the satellite any light in place of the sunlight which he has intercepted.*

Enough, however, has been demonstrated to enable us to pronounce on the question as to whether Jupiter can be a body inhabited by living beings, as we understand the term. Obviously it cannot. The internal heat and the fearful tempests seem to preclude the possibility of organic life, even were there not other arguments against it. It may, however, be contended, with perhaps some plausibility, that Jupiter has in the distant future the prospect of a

* It is only right to add that some observers believe that, under exceptional circumstances, points of Jupiter have shown some slight degree of intrinsic light.

glorious career as the residence of organic life. The time will assuredly come when the internal heat must subside, when the clouds will gradually condense into oceans. On the surface it may then be that dry land will appear, and thus Jupiter may be rendered habitable.

From this sketch of the planet itself we now turn to the interesting and beautiful system of four satellites with which Jupiter is attended. We have, indeed, already found it necessary to allude more than once to these little bodies, but not to such an extent as to interfere with the more formal treatment which they are now to receive.

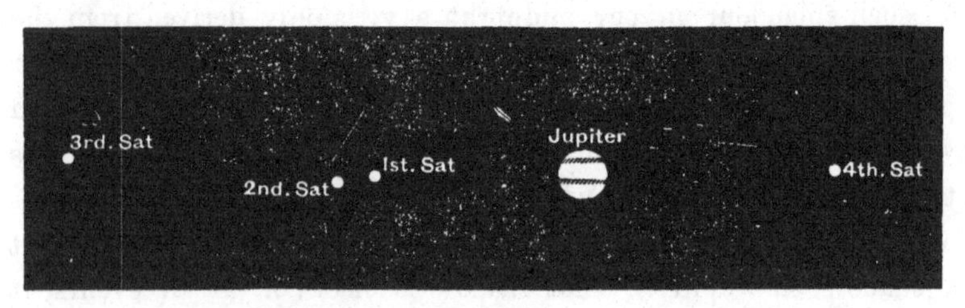

Fig. 51.—Jupiter and his Four Satellites as seen in a telescope of low power.

The discovery of these satellites may be regarded as marking an important epoch in the history of astronomy. They are objects situated in a remarkable manner on the border line which divides the objects visible to the unaided eye from those which require telescopic aid. It has been stated frequently that these objects have been seen with the unaided eye; but without entering into any controversy on the matter, it is sufficient to recite the well-known fact that, notwithstanding Jupiter had been familiar to the unaided eyes for countless centuries, and in the clearest skies, yet no one ever discovered the satellites until Galileo turned his newly-invented telescope upon them. This telescope was no doubt a very feeble instrument, but the feeblest instrument is quite adequate to show objects so close to the limit of visibility.

The view of the planet and its elaborate system of four satellites, as shown in a telescope of moderate power, is represented in Fig. 51. We here see the globe of the planet, and nearly in a line with its centre lie four small objects, three on side and one on the other. These little objects resemble stars, but they can be distinguished from stars by their ceaseless movements around the planet which they never fail to accompany during his entire circuit of the heavens. There is no more pleasing spectacle for the student of the heavens than to follow with his telescope the movements of this beautiful system.

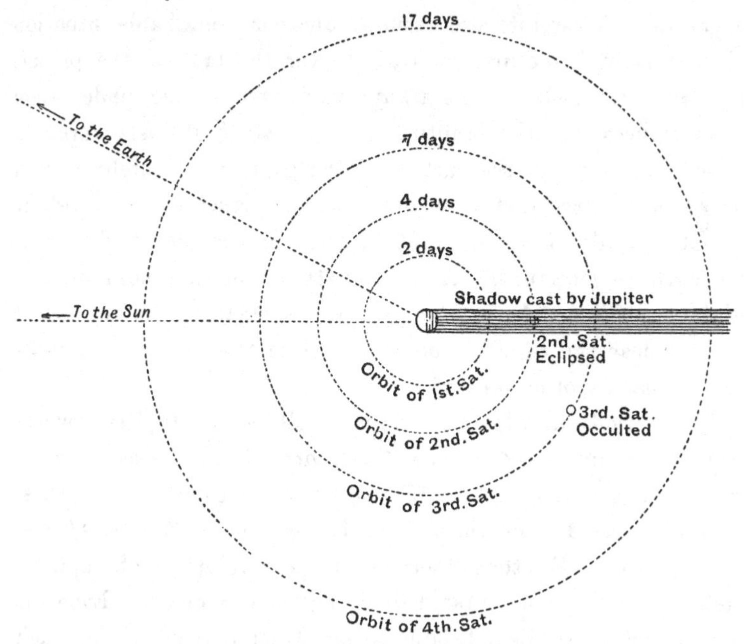

52.—Disappearances of Jupiter's Satellites.

In Fig. 52 we have represented some of the various phenomena which the satellites present. The long black shadow is that produced by the interposition of Jupiter in the path of the sun's rays. The second satellite is immersed in this shadow, and consequently eclipsed. The eclipse of a satellite must not be attributed to the intervention of the body of Jupiter between the satellite and the earth. This occurrence is called an occultation, and the third satellite is shown in this condition. The second and the third satellites are thus alike invisible, but the cause of the invisibility is quite different in the two cases. The eclipse is much the more striking phenomenon of the two, because the satellite, at the moment it plunges into the shadow, is still at some apparent distance from the edge of the planet, and is thus seen clearly up to the moment of the eclipse. In an occultation the satellite in passing behind the planet is, at the time of disappearance, close to the planet's bright edge, and the extinction of the light of the satellite cannot be observed with the same distinctness as the occurrence of an eclipse. A satellite also assumes another remarkable situation when it is in the course of transit over the face of the planet. The satellite itself is not always very easy to see under such circumstances, but the beautiful shadow which it casts forms a sharp black spot on the surface of the planet; the satellite may, indeed, sometimes cast a shadow on the planet even though it be not actually at the moment in front of the planet, so far as the earth is concerned. A case of this kind has been already exhibited in Plate XI., in which we see the small disc of the second satellite near the edge of the planet while the shadow is a well-marked black spot on its surface.

The periods in which the four satellites of Jupiter revolve around him are respectively, 1 day 18 hrs. 27 min. 34 secs. for the first; 3 days 13 hrs. 13 min. 42 secs. for the second; 7 days 3 hrs. 42 min. 33 secs. for the third; and 16 days 16 hrs. 32 min. 11 secs. for the fourth. We thus observe that the revolutions of Jupiter's satellites are much more rapid than that of our moon. Even the slowest and most distant satellite on Jupiter requires for each revolution less than two-thirds of an ordinary lunar month. The innermost satellite, revolving as it does in less than two days, presents a striking series of ceaseless and rapid changes, and during every revolution it becomes eclipsed. The distance from the centre of Jupiter to the orbit of the innermost satellite is about a quarter of a million miles, while the radius of the outermost is a little more than a million miles. The second of the satellites proceeding outwards from the planet is almost the same size as our moon; the other three satellites are larger; the third is the greatest of all, and has a diameter considerably greater than that of the moon.

Among the many interesting astronomical investigations to

which the observations of Jupiter's satellites has given rise, no doubt the most celebrated is that made by Roemer, which led to the discovery of the velocity of light.

The eclipses of Jupiter's satellites had been observed for many years, and the times of their occurrence had been recorded. At length it was perceived that a certain order reigned among the eclipses of the satellites, as among all other astronomical phenomena. When once the laws, according to which the eclipses recurred, had been perceived, the usual consequence followed. It became possible to predict the time at which the eclipses would occur in future. Those predictions were made, and it was found that the predictions were approximately verified. Further improvements in the calculations were made, and it was sought to predict the time with still greater accuracy. But when it came to naming the actual minute at which the eclipse should occur, the predictions were not always successful. Sometimes the eclipse was five or ten minutes too soon. Sometimes it was five or ten minutes too late. Discrepancies of this kind always demand attention. It is, indeed, by such discrepancies that discoveries are often made, and one of the most interesting discoveries is that now before us.

The irregularity in the recurrence of the eclipses was at length perceived to observe certain rules. It was noticed that when the earth was near to Jupiter the eclipse generally occurred before the predicted time; while when the earth happened to be at the side of its orbit away from Jupiter, the eclipse occurred after the predicted time. Once this was proved, the great discovery was quickly made. When the satellite enters the shadow its light gradually decreases until it disappears. It is the last ray of light from the eclipsed satellite that gives the time of the eclipse; but that ray of light has to travel from the satellite to the earth and enter our telescope before we can note the occurrence. It used to be thought that light travelled instantaneously, so that the moment the eclipse occurred was assumed to be the moment when the eclipse was seen in the telescope. This was now perceived to be incorrect. It was found that light took time to travel. When the earth was near Jupiter the light had cnly a short journey, the intelligence of the eclipse arrived quickly, and the eclipse happened sooner than the calculations indicated. When the earth occupied a position far from Jupiter the light had a longer journey, and took more than the average time, so that the eclipse was later than the prediction. This simple explanation removed the difficulty attending the predictions of the eclipses of the satellites. But the discovery had a significance far more momentous. We learned from it that light had a measurable velocity, which, according to recent researches, amounts to 185,000 miles per second.

One of the most celebrated methods of measuring the distance of the sun is derived from a combination of experiments on the velocity of light with astronomical measurements. This is a method of considerable refinement and interest, and although it does not satisfy all the necessary conditions of a perfectly satisfactory method, yet it is impossible to avoid some reference to it here. Notwithstanding that the velocity of light is so stupendous, it has been found possible to measure that velocity by actual trial. This is one of the most delicate experimental researches that has ever been undertaken. If it be difficult to measure the speed of a rifle bullet, what shall we say of the speed of a ray of light, which is nearly a million times as great? How shall we devise an apparatus subtle enough to measure the velocity which would girdle the earth at the equator no less than seven times in a single second of time? All ordinary contrivances for the measurement of velocity are here futile; we have to devise an instrument of a wholly different character.

In the attempt to measure the speed of a moving body we first mark out a certain distance, and then measure the time which the body requires to traverse that distance. We determine the velocity of a railway train by the time it takes to pass from one mile post to the next. We measure the speed of a rifle bullet by an ingenious contrivance really founded on the same principle. The greater the velocity, the more desirable is it that the distance traversed during the trial shall be as large as possible. In dealing with the measurement of the velocity of light, we therefore choose for our measured distance the greatest length that may be convenient. It is, however, necessary that the two ends of the line shall be visible from each other. A hill a mile or two away will form a suitable site for the distant station, and the distance of the selected point on the hill from the observer must be carefully measured. The problem is now easily stated. A ray of light is to be sent from the observer to the distant station, and the time occupied by that ray in the journey is to be measured. We may suppose that the observer, by a suitable contrivance, has arranged a lantern from which a thin ray of light issues. Let us assume that this ray of light travels all the way to the distant station, and there falls upon the surface of a reflecting mirror. Instantly the ray of light will be diverted by reflection into a new direction depending upon the inclination of the mirror. By suitable adjustment the latter can be so placed that the light shall fall perpendicularly upon it, in which case the ray will of course simply return along the direction in which it came. Let the mirror be fixed in this position throughout the course of the experiments. It follows that a ray of light starting from the lantern will be returned to the lantern after it has made the journey to the distant station and back again. Imagine, then, a little shutter placed in front of the lantern. We open the shutter, the ray streams forth to the remote reflector, and back again through the opening. But now, after having allowed the ray to pass through the shutter, suppose we try and close it before the ray has had time to get back again. What fingers could be nimble enough to do this? Even if the distant station were ten miles away, so that the light had a journey of ten miles in going to the mirror and ten miles in coming back, yet the whole course would be accomplished in about the nine thousandth part of a second—a period so short that even were it a thousand times as long it would hardly enable manual dexterity to close the aperture. Yet a shutter can be constructed which shall be sufficiently delicate for the purpose.

The principle of this beautiful method will be sufficiently obvious from the diagram on the next page (Fig. 53). The figure exhibits a rough view of the lantern and the observer, and in front of both is a large wheel with projecting teeth. Each tooth as it passes round eclipses the beam of light emerging from the lantern, and also the eye, which is of course directed to the mirror at the distant station. In the position of the wheel here shown the ray from the lantern will pass to the mirror and back so as to be visible to the eye, but if the wheel be rotating it may so happen that the beam after leaving the lantern will not have time to return before the next tooth of the wheel comes in front of the eye and screens it. If the wheel be urged still faster the next tooth may have

Fig. 53.--Mode of Measuring the Velocity of Light.

passed the eye so that the ray again becomes visible. The speed at which the wheel is rotating can be measured. We can thus determine the time taken by one of the teeth to pass in front of the eye; we have thus a measure of the time occupied by the ray of light in the double journey, and hence we have a measurement of the velocity of light.

It thus appears that we can tell the velocity of light either by the observations of Jupiter's satellites or by experimental inquiry. If we take the latter method, then we are entitled to deduce remarkable astronomical consequences. We can, in fact, employ this method for solving that great problem so often referred to—the distance from the earth to the sun—though it cannot compete in accuracy with some of the other methods.

The dimensions of the solar system are so considerable that a sunbeam requires an appreciable interval of time to span the abyss which separates the earth from the sun. Eight minutes is approximately the duration of the journey, so that at any moment we see the sun as it appeared eight minutes earlier to an observer in its immediate neighbourhood. In fact, if the sun were to be suddenly blotted out it would still be seen shining brilliantly for eight minutes after it had really disappeared. We can determine this period of time from the eclipses of Jupiter's satellites.

So long as the satellite is shining it radiates a stream of light across the vast space between Jupiter and the earth. When the eclipse has commenced the little orb is no longer luminous, but there is, nevertheless, a long slender stream of light on its way, and until all this has poured into our telescopes we still see the satellite shining as before. If we could calculate the moment when the eclipse really took place, and if we could observe the moment at which the eclipse is seen, the difference between the two gives the time which the light occupies on the journey. This can be found with some accuracy; and, as we already know the velocity of light, we can ascertain the distance of Jupiter from the earth; and hence deduce the scale of the solar system. It must, however, be remarked that at both extremities of the process there are characteristic sources of uncertainty. The occurrence of the eclipse is not an instantaneous phenomenon. The satellite is large enough to require an appreciable time in crossing the boundary which defines the shadow, so that the observation of an eclipse is not sufficiently precise to form the basis of an important and accurate measurement.* Still greater difficulties accompany the attempt to define the true moment of the occurrence of the eclipse as it would be seen by an observer in the vicinity of the satellite. For this we would require a far more perfect theory of the movements of Jupiter's satellites than is at present attainable. This method of finding the sun's distance holds out no prospect of a

* Professor Pickering, of Cambridge, Mass., has, however, effected the important improvement of measuring the decline of light of the satellite undergoing eclipse by the photometer. Much additional precision may be anticipated in the results of such observations.

result accurate to the one-thousandth part of its amount, and we may discard it, inasmuch as the other methods available seem to admit of much higher accuracy.

The satellites of Jupiter have special attractions for the mathematician, who finds in them a most striking instance of the universality of the law of gravitation. The satellites are, of course, mainly controlled in their movements by the attraction of the great planet; but they also attract each other, and two curious consequences are the result.

The mean motion of the first satellite in each day about the centre of Jupiter is 203°·4890. That of the second is 101°·3748, and that of the third is 50°·3177. These quantities are so related that the following law will be found to be observed:

The mean motion of the first satellite added to twice the mean motion of the third is exactly equal to three times the mean motion of the second.

There is another law, which is of an analogous character, and is thus expressed (the mean longitude being the angle between a fixed line and the radius to the mean place of the satellite): If to the mean longitude of the first satellite we add twice the mean longitude of the third, and subtract three times the mean longitude of the second, the difference is always 180°.

It was from observation that these laws were first discovered to be true. Laplace, however, showed that if the satellites revolved nearly in this way, then their mutual perturbations, in accordance with the law of gravitation, would preserve them in this relative position for ever.

It follows from the second law that when two of the satellites are in line on one side of Jupiter the other satellite must be in the same line on the opposite side of the planet. The fourth satellite does not enter into consideration in these laws.

We shall conclude with the remark, that the discovery of Jupiter's satellites was really the great confirmation of the Copernican theory. Copernicus had asked the world to believe that our sun was a great body, and that the earth and all the other planets were small bodies revolving around the great one. This doctrine, so repugnant to the theories previously held, and to the immediate evidence of our senses, could only be established by a refined course of reasoning. The discovery of Jupiter's satellites was very opportune. Here we had an exquisite ocular demonstration of a system, though, of course, on a much smaller scale, precisely identical with that which Copernicus had proposed. The astronomer who had watched Jupiter's moons circling around their primary, who had noticed their eclipses and all the interesting phenomena attendant on them, saw before his eyes, in a manner wholly unmistakable, that the great planet controlled these small bodies, and forced them to revolve around him, and thus exhibited a miniature of the great solar system itself. "As in the case of the spots on the sun, Galileo's announcement of this discovery was received with incredulity by those philosophers of the day, who believed that everything in nature was described in the writings of Aristotle. One eminent astronomer, Clavius, said that to see the satellites one must have a telescope which would produce them; but he changed his mind as soon as he saw them himself. Another philosopher, more prudent, refused to put his eye to the telescope lest he should see them and be convinced. He died shortly afterwards. 'I hope,' said the caustic Galileo, 'that he saw them while on his way to heaven.'" *

* "Newcomb's Popular Astronomy," p. 336.

CHAPTER XIII.

SATURN.

The Position of Saturn in the System—Saturn one of the Three most Interesting Objects in the Heavens—Compared with Jupiter—Saturn to the Unaided Eye—Statistics relating to the Planet—Density of Saturn—Lighter than Water—The Researches of Galileo—What he found in Saturn—A Mysterious Object—The Discoveries made by Huyghens half a Century later—How the Existence of the Ring was Demonstrated—Invisibility of the Rings every Fifteen Years—The Rotation of the Planet—The Celebrated Cypher—The Explanation—Drawing of Saturn—The Dark Line—W. Herschel's Researches—Is the Division in the Ring really a Separation?—Possibility of Deciding the Question—The Ring in a Critical Position—Are there other Divisions in the Ring?—The Third Ring—Has it appeared but recently?—Physical Nature of Saturn's Rings—Can they be Solid?—Can they even be Slender Rings?—A Fluid—Probable Nature of the Rings—A Multitude of Small Satellites—Analogy of the Rings of Saturn to the Group of Minor Planets—Problems Suggested by Saturn—The Group of Satellites to Saturn—The Discoveries of Additional Satellites—The Orbit of Saturn not the Frontier of our System.

At a profound distance in space, which sometimes approaches to nearly a thousand millions of miles, the planet Saturn performs its mighty revolution around the sun in a period of over a quarter of a century. This gigantic orbit formed the boundary to the planetary system, so far as it was known to the ancients.

Although Saturn is not so great a body as Jupiter, yet it vastly exceeds the earth in bulk and in mass, and is, indeed, much greater than any one of the planets, Jupiter alone excepted. But while Saturn must yield the palm to Jupiter so far as mere dimensions are concerned, yet it will be generally admitted that even Jupiter, with all the retinue by which he is attended, cannot compete in beauty with the marvellous system of Saturn. To the present writer it has always seemed that Saturn is one of the three most interesting celestial objects visible to observers in northern latitudes. The two others will occupy our attention in future chapters. They are the great nebula in Orion, and the star cluster in Hercules.

So far as the globe of Saturn is concerned, we do not meet with any features which give to the planet any exceptional interest. The globe is less than that of Jupiter, and as the latter is also much nearer to us, the apparent size of Saturn is in a two-fold way much smaller than that of Jupiter. It should also be noticed that, owing to the greater distance of Saturn from the sun, its intrinsic brilliancy is less than that of Jupiter. There are, no doubt, certain marks and bands often to be seen on Saturn, but they are not nearly so striking nor so characteristic as the ever-variable belts upon Jupiter. The telescopic appearance of the globe of Saturn must also be ranked as greatly inferior in interest to that of Mars. The delicacy of detail which we can see on Mars when favourably placed has no parallel whatever in the dim and distant Saturn. Nor has Saturn, regarded again merely as a globe, anything like the interest of Venus. The great splendour of Venus is altogether out of comparison to that of Saturn, while the brilliant crescent of Venus is infinitely more pleasing than any telescopic view of the globe of Saturn. Yet even while we admit all this to the fullest extent, it does not invalidate the claim of Saturn to be one of the most supremely beautiful and interesting objects in the heavens. This interest is not due to his globe; it is due to that marvellous system of rings by which Saturn is surrounded—a system wonderful from every point of view, and, so far as our knowledge goes, without a parallel in the wide extent of the universe.

To the unaided eye Saturn usually appears like a star of the first magnitude. Its light alone would hardly be sufficient to discriminate it from many of the brighter fixed stars. Yet the ancients were acquainted with Saturn, and they knew it as a planet. It was included with the other four great planets—Mercury, Venus, Mars, and Jupiter—in the group of wanderers, which were bound to no fixed points of the sky like the stars. On account of the great distance of Saturn, its movements are much slower than those of the other planets known to the ancients. Twenty-nine years and a half are required for this distant object to complete its circuit of the heavens; and, though this movement is slow compared with the incessant changes of Venus, yet it is rapid enough to attract the attention of any careful observer. In a single year Saturn moves through a distance of about twelve degrees, a quantity sufficiently large to be conspicuous to casual observation. Even in a month the planet traverses an arc of the sky which can be detected by any one who will take the trouble to mark the place of the planet with regard to the stars in its vicinity. Those who are privileged to use accurate astronomical instruments, can readily detect the motion of Saturn in a few hours.

The average distance from the sun to Saturn is about 884 millions of miles. The path of Saturn, as of every other planet, is really an ellipse with the sun in the focus. In the case of Saturn the shape of this ellipse is very appreciably different from a purely circular path. Around this path Saturn moves with an average velocity of 5·96 miles per second.

The mean diameter of the globe of Saturn is about 71,000 miles. Its equatorial diameter is about 74,000 miles, and its polar diameter 68,000 miles—the ratio of these numbers being approximately that of 10 to 9. It is thus obvious that Saturn departs from the truly spherical shape to a very marked extent. The protuberance at its equator must, no doubt, be attributed to the high velocity with which the planet is rotating. The velocity of rotation of Saturn is more than double as fast as that of the earth, though it is not quite so fast as that of Jupiter. Saturn makes one complete rotation in 10 hrs. 14 min. 23·8 secs.

We have already commented upon the low density of the materials of Jupiter, and we have endeavoured to account for that low density by the supposition that internal heat is still present in Jupiter to a considerable extent. We can apply the same reasoning with still greater emphasis to Saturn. The lightness of this planet is such as to be wholly incompatible with the supposition that its Globe is constituted of solid materials at all comparable with those of which the crust of our earth is composed. The satellites, which surround Saturn and form a system only less interesting than the renowned rings themselves, enable us to weigh the planet in comparison with the sun, and hence to deduce its actual mass relatively to the earth. The result is not a little remarkable. It appears that the density of the earth is eight times as great as that of Saturn. In fact, the density of the latter is less than that of water itself, so that a mighty globe of water, equal in bulk to Saturn, would actually weigh more. If we could conceive a vast ocean into which a globe equal to Saturn in size and weight were cast, the great globe would not sink like our earth or like any of the other planets; it would float buoyantly at the surface with one-fourth of its bulk out of the water. We thus learn with high probability that what our telescopes show upon Saturn is not a solid surface, but merely a vast envelope of clouds surrounding a highly heated interior. It is impossible to resist the suggestion that this planet, like Jupiter, has still retained its heat because its mass is so large. We must, however, allude to a circumstance which perhaps may seem somewhat inconsistent with the view here taken. We have found that Jupiter and Saturn are, both of them, much less dense than the earth. When we compare the two planets together, it appears that Saturn is much less dense than Jupiter. In fact, every cubic mile of Jupiter weighs nearly twice as much as each cubic mile of Saturn. This would seem to point to the conclusion that Saturn is the more heated of the two bodies. Yet, as Jupiter is the larger, it might more reasonably have been expected to be the hotter. We do not attempt to reconcile this discrepancy; in fact, in our ignorance as to the material constitution of these bodies, it would be idle to discuss the question.

Even if we allow for the lightness of Saturn, as compared bulk for bulk with the earth, yet the volume of Saturn is so enormous that the planet weighs more than eighty times as much as the earth. The adjoining view represents the relative sizes of Saturn and the earth (Fig. 54).

As the unaided eye discloses none of those marvels by which Saturn is surrounded, the interest which attaches to this planet may be said to commence from the time when it began to be observed with the telescope. The history must be briefly alluded to, for it was only step by step that the real nature of this complicated object was understood. When Galileo completed his little refracting telescope, which, though it only magnified thirty times, was yet an enormous addition to the powers of unaided vision, he made with it his memorable review of the heavens. He saw the spots on the sun and the mountains on the moon; he noticed the crescent of Venus and the satellites of Jupiter. Stimulated and encouraged by such brilliant discoveries, he naturally sought to examine the other planets, and accordingly directed his telescope to Saturn. Here, again, Galileo at once made a discovery. He saw that Saturn presented a visible form like the other planets,

Fig. 54.—Relative Sizes of Saturn and the Earth.

but that it differed from any other telescopic object, inasmuch as it appeared to him to be composed of three bodies which always touched each other and always maintained the same relative positions. These three bodies were in a line—the central one was the largest, and the two others were east and west of it. There was nothing he had hitherto seen in the heavens which filled his mind with such astonishment, and which seemed so wholly inexplicable.

In his endeavours to study this object thoroughly, Galileo continued his observations during the year 1610, and, to his amazement, he saw the two lesser bodies gradually become smaller and smaller until, in the course of the two following years, they had entirely vanished, and the planet simply appeared with a round disc like

Jupiter. Here, again, was a new source of anxiety to Galileo. He had at that day to contend against the advocates of the ancient system of astronomy, who derided his discoveries and refused to accept his theories. He had announced his observation of the composite nature of Saturn; he had now to tell of the gradual decline and the ultimate extinction of these two auxiliary globes, and he naturally feared that his opponents would seize the opportunity of pronouncing that the whole of his observations were illusory.* "What," he remarks, "is to be said concerning so strange a metamorphosis? Are the two lesser stars consumed after the manner of the solar spots? Have they vanished and suddenly fled? Has Saturn, perhaps, devoured his own children? Or were the appearances indeed illusion or fraud, with which the glasses have so long deceived me, as well as many others to whom I have shown them? Now, perhaps, is the time come to revive the well-nigh withered hopes of those who, guided by more profound contemplations, have discovered the fallacy of the new observations, and demonstrated the utter impossibility of their existence. I do not know what to say in a case so surprising, so unlooked for, and so novel. The shortness of the time, the unexpected nature of the event, the weakness of my understanding, and the fear of being mistaken, have greatly confounded me."

But Galileo was not mistaken. The objects were really there when he first began to observe, they really did decline, and they really disappeared; but this disappearance was only for a time—they again came into view. They were then subjected to ceaseless examination, until gradually their nature became unfolded. With increased telescopic power it was found that the two bodies which Galileo had described as globes, or spheres, near Saturn, were not really spherical—they were rather two luminous crescents with the concavity of each turned towards the great central globe. It was also perceived that these objects underwent a remarkable series of periodic changes. At the beginning of such a series the planet was found with a truly circular disc. The appendages first appeared

* *See* Grant, "History of Physical Astronomy," page 255.

as two arms extending directly outwards on each side of the planet; then these arms gradually opened into two crescents, resembling handles to the globe, and attained their maximum width after about seven or eight years; then they began to contract, until after the lapse of about the same time they vanished again.

The true nature of these objects was at length discovered by Huyghens in 1655, nearly half a century after Galileo had first detected their appearance. He discovered the shadow thrown by the ring upon the globe, and his explanation of the phenomena was obtained in a very philosophical manner. He noticed that the earth, the sun, and the moon, rotated upon their axes, and he therefore regarded it as a general law that each one of the bodies in the system rotates about an axis. No doubt at that early period no observations had been made which actually showed that Saturn was also rotating; but it would be highly, nay indeed infinitely, improbable that any planet should be devoid of rotation. All the analogies of the system pointed to the conclusion that the velocity of rotation would be considerable. One satellite of Saturn was already known to revolve in a period of sixteen days, being little more than half our month. Huyghens assumed—and he was entitled to the assumption—that Saturn in all probability rotated rapidly on its axis. It was also to be observed that if these remarkable appendages were attached by an actual bodily connection to the planet they must rotate with Saturn. If, however, the appendages were not actually attached it would still be necessary that they should rotate if the analogy of Saturn to other objects in the system were to be in any degree preserved. We see satellites near Jupiter which revolve around him. We see, nearer home, how the moon revolves around the earth. We see how all the planetary system revolves around the sun. All these considerations were present to Huyghens when he came to the conclusion that, whether the curious appendages were actually attached to the planet or were physically free from it, they must still be in rotation.

Provided with such reasonings it soon became easy to conjecture the true nature of the Saturnian system. We have seen how once every fifteen years the appendages declined to invisibility and then gradually reappeared in the form, at first, of rectilinear arms projecting outwards from the planet. The progressive development is a slow one, and for weeks and months, night after night, the same appearance is presented with but little change. But all this time both Saturn and the mysterious objects around him are rotating. Whatever these may be, they present the same appearance to the eye, notwithstanding their ceaseless motion of rotation. Now what must be the shape of an object which satisfies the conditions here implied? It will obviously not suffice to regard the projections as two spokes diverging from the planet. They would change from visibility to invisibility in every rotation, and thus there would be ceaseless alterations of the appearance instead of that slow and gradual change which requires fifteen years for a complete period. There are, indeed, other considerations which preclude the possibility of the objects being anything of this character, for they are always of the same length as compared with the diameter of the planet. A little reflection will show that one supposition—and indeed only one—will meet all the facts of the case. If there were a thin symmetrical ring rotating in its own plane around the equator of Saturn, then the persistence of the object from night to night would be accounted for. This at once removes the greater part of the difficulty. For the rest, it was only necessary to suppose that the ring was so thin that when turned actually edgewise to the earth it became invisible, and then as the illuminated side of the plane became turned more and more towards the earth the appendages to the planet gradually increased. The handle-shaped appearance which the planet periodically assumed demonstrated that the ring was not attached to the globe. At length Huyghens found that he had the clue to the great enigma which had perplexed astronomers for the last fifty years. He saw that the ring was an object of the most astonishing interest, unique at that time, as it is, indeed, unique still. He felt, however, that he had hardly demonstrated the matter with all the certainty which it merited, and which he thought that by further attention he could secure. Yet he was loth to hazard the loss of his discovery by an undue postponement of its announcement, lest some other astronomer might intervene. How, then, was he to secure his priority if the discovery should turn out correct, and at the same time be enabled to perfect it at his leisure? He adopted the course not unusual at the time, of making his first announcement in cipher, and accordingly, on March 5th, 1656, he published a tract, which contained the following proposition:—

aaaaaaa ccccc d eeeee g. h
iiiiiii llll mm nnnnnnnnn
oooo pp q rr s ttttt uuuuu

Perhaps some of those curious persons whose successors now devote so much labour to double acrostics, may have pondered on this renowned cryptograph, and even attempted to decipher it. But even if such attempts were made we do not learn that they were successful. A few years of further study was thus secured to Huyghens. He tested his theory in every way that he could devise, and he found it verified in every detail. He therefore thought that it was needless for him any longer to conceal from the world his great discovery, and accordingly in the year 1659 —about three years after the appearance of his cryptograph— he announced the interpretation of it. By restoring the letters to their original arrangement the discovery was enunciated in the following words:—"*Annulo cingitur, tenui, plano, nusquam cohærente, ad eclipticam inclinato,*" which may be translated into the statement:—"The planet is surrounded by a slender flat ring everywhere distinct from its surface, and inclined to the ecliptic."

Huyghens was not content with merely demonstrating how fully this assumption explained all the observed phenomena. He submitted it to the further and most delicate test which can be applied to any astronomical theory. He attempted by its aid to make a prediction the fulfilment of which would necessarily give his theory the stamp of truth. From his calculations he saw that the planet would appear circular, about July or August in 1671. This anticipation was practically verified, for the ring was seen to vanish in May of that year. No doubt, with our modern calculations founded on long-continued and accurate observation, we are now enabled to make forecasts as to the appearance or the disappearance of Saturn's ring with far greater accuracy; but, remembering the early stage in the history of the planet at which the prediction of Huyghens was made, we must regard its fulfilment as quite sufficient, and as confirming in a satisfactory manner the theory of Saturn and his ring.

The ring of Saturn having thus been thoroughly established as an astronomical certainty, each generation of astronomers has laboured, and successfully laboured, to find out more and more of its marvellous features. In the frontispiece we have a view of the planet as seen at the Harvard College Observatory, U.S.A., between July 28th and October 20th, 1872. It has been drawn by the skilful astronomer and artist– Mr. L. Trouvelot—and gives a faithful and beautiful representation of this unique object. To realise from the plate the stupendous size of the whole system, it may be remembered that the volume of Saturn is 700 times that of the earth.

The next great discovery in the Saturnian system after those of Huyghens, showed that the ring surrounding the planet was marked by a dark concentric line, which divided the ring into two parts—the outer being narrower than the inner. This line was first seen by J. D. Cassini, when Saturn emerged from the rays of the sun in 1675. That this black line is not merely a black mark on the ring, but that it is actually a separation, was rendered very probable by the researches of Maraldi in 1715, followed many years later by those of Sir William Herschel, who, with that thoroughness which was a marked characteristic of the man, made a most minute and scrupulous examination of Saturn. Night after night he followed it for hours with his unrivalled instruments, and considerably added to our knowledge of the planet and his system. He devoted very particular attention to the examination of the line dividing the ring. He saw that the colour of this line was not to be distinguished from the colour of the space intermediate between the globe and the ring. He observed it for ten years on the northern face of the ring, and during that time it continued to present the same breadth and colour and sharpness of outline. He was then fortunate enough to observe the southern side of the ring. There again could the black line be seen, corresponding both in appearance and in position with the dark line as seen on the northern side. Hardly a doubt could remain as to the fact that Saturn was girdled by two concentric rings equally thin, and the outer edge of one closely approaching to the inner edge of the other. At the same time it is right to add, that the only absolutely indisputable proof of the division between the rings has not yet been yielded by the telescope. The appearances noted by Herschel would be consistent with the view that the black line was merely a part of the ring extending through its thickness, and composed of materials very much less capable of reflecting light than the rest of the ring. It is still a matter of doubt how far it is ever possible actually to see through the dark line. There is apparently only one satisfactory method of accomplishing this. It would only occur under rare circumstances, and it does not seem that the opportunity has as yet arisen. Suppose that in the course of its motion through the heavens the path of Saturn happened to cross directly between the earth and a fixed star. The telescopic appearance of a star is merely a point of light much smaller than the globe and rings of Saturn. If the ring passed in front of the star and the black line on the ring came over the star, we would, if the black line were really an opening, see the star shining through the narrow aperture. Up to the present, we believe, there has been no opportunity of submitting the question of the duplicity of the ring to this crucial test. Let us hope that as Saturn is now in a position well suited for examination, and as there are now so many telescopes in use adequate to deal with the subject, there may, ere long, be observations made which will decide the question. It can hardly be expected that a very small star would be suitable. No doubt the smallness of the star would render the observations more delicate and precise if the star were visible; but we must remember that the star will be thrown into contrast with the bright rings of Saturn on each margin, so that unless the star were of considerable magnitude it would hardly answer. It has, however, been recently observed that the globe of the planet can be, in some degree, discerned through the dark line;

this is practically a demonstration of the fact that the line is at all events partly transparent.

The outer ring is also divided into two by a line much fainter than that just described. It requires a good telescope and a good night, combined with a good position of the planet, to render this line a well marked object. It is most easily seen at the extremities of the ring most remote from the planet. To the present writer, who has examined the planet with the twelve inch refractor of the South equatorial at Dunsink Observatory, this outer line appears as broad as the well-known line; but it is unquestionably fainter and has a more shaded appearance. It certainly does not suggest the appearance of being actually an opening in the ring. It rather seems as if the ring were at this place thinner, and possessed less substance without being actually divided.

On these points it may be expected that much additional information will be acquired when next the ring places itself in such a position that its plane, if produced, would pass between the earth and the sun. Such occasions are but rare, and even when they do occur, it may happen that the planet will not be well placed for observation. The next really good opportunity will not occur till 1907. In this case the sunlight illuminates one side of the ring, while it is the other side of the ring that is presented towards the earth. Powerful telescopes are necessary to deal with the planet under such circumstances; but it may be reasonably hoped that the questions relating to the division of the ring, as well as to many other matters, will then receive some further elucidation.

Occasionally, other divisions of the ring, both inner and outer, have been recorded. It may, at all events, be stated that no such divisions can be regarded as permanent features. Yet their existence has been so frequently enunciated by skilful observers that it is impossible to doubt that they have been sometimes seen.

It was about 200 years after Huyghens had first offered the true theory of Saturn, that another very interesting and important discovery was effected. It had, up to the year 1850, been always supposed that the two rings, divided by the well-known black line, comprised the entire ring system surrounding the planet. In the year just mentioned, Professor Bond, the distinguished astronomer of Cambridge, Mass., startled the whole astronomical world by the announcement of his discovery of a third ring surrounding Saturn. As so often happens in such cases, the same object was discovered independently by the English astronomer Dawes. This third ring lay just inside the inner of the two well-known rings, and extended to within about half the distance towards the body of the planet. This new ring is therefore of considerable size, and occupies a conspicuous position: how then came it to pass that it eluded not only the telescopes of Galileo and Huyghens, but many more perfect instruments subsequently constructed? How, in fact, did this third ring succeed in escaping the penetration of the great Sir William Herschel, who devoted so much attention to observing Saturn and his system? The third ring seems to be of a totally different character to the two others, in so far as they present a comparatively substantial appearance. We shall, indeed, presently show that they are in all probability not solid —not even liquid bodies—but still when compared with the third ring the older ones are of quite a substantial character. They can receive and exhibit the deeply marked shadow of Saturn, and they can throw a deep and black shadow upon Saturn themselves; but the third ring seems of a much more spiritual texture. It has not the brilliancy of the others, it is rather of a dusky semi-transparent appearance, and the expression "crape ring," by which it is often designated, is by no means inappropriate. It is the faintness of this crape ring which led to its being so frequently overlooked by the earlier observers of Saturn.

It has often been noticed that when an astronomical discovery has been made with a good telescope, it afterwards becomes possible for the same object to be observed with instruments of much inferior power. No doubt, when the observer knows what to look for, he will often be able to see what would not otherwise have attracted his attention. It may be regarded as an illustration of this principle, that the crape ring of Saturn has become an object familiar to those who are accustomed to work with good telescopes; but it may, nevertheless, be doubted whether the ease and distinctness with which the crape ring is now seen can be entirely accounted for by this supposition. Indeed, it seems hardly possible to resist the supposition that the crape ring has, from some cause or other, gradually become more and more visible. The supposed recent appearance of the crape ring is one of those arguments now made use of, to prove that in all probability the rings of Saturn are at this moment undergoing gradual transformation; but observations of Hadley would seem to show that the crape ring was perhaps seen by him in 1720, or even previously by J. J. Cassini.

The various features of the rings are well shown in the beautiful drawing of Trouvelot. We here see the inner and the outer ring, and the line of division between them. We see in the outer ring the faint traces of the line by which it is divided, and inside the inner ring we have a view of the curious and semi-transparent crape ring. The black shadow of the planet is cast upon the ring, thus proving that the ring, no less than the body of the planet, shines only in virtue of the sunlight which falls upon it. This shadow presents some anomalous features, and its curious irregularity may be, to some extent, an optical illusion. The drawing contains no trace of those other and finer lines which are more or less problematical, but it is a faithful representation of the planet, under good seeing conditions, and viewed in a telescope of considerable power.

There can be no doubt that any attempt to depict the rings of Saturn can only represent the salient features of that marvellous system. We are situated at such a great distance that all objects not of colossal dimensions are invisible. We have, indeed, only an outline, which makes us wish to be able to fill in the details. We long, for instance, to see the actual texture of the rings, and to learn of what materials they are made; we wish to comprehend the strange and filmy crape ring, so unlike any other object known to us in the heavens. There is no doubt that much may even yet be learned under all the disadvantageous circumstances of our position; there is still room for the labour of whole generations of astronomers provided with splendid instruments. We want accurate drawings of Saturn under every conceivable aspect in which it may be presented. We want incessantly repeated measurements, of the most fastidious accuracy. These measures are to tell us the sizes and the shapes of the rings; they are to measure with fidelity the position of the dark lines and the boundaries of the rings. These measures are to be protracted for generations and for centuries; then and then only can terrestrial astronomers learn whether this elaborate system has really the attributes of permanence; or whether it may be undergoing changes of marvellous rapidity, when the gigantic nature of the objects involved is considered.

We have been accustomed to find that the law of universal gravitation pervades every part of our system, and to look to gravitation as the source of the explanation of many phenomena otherwise inexplicable. We have good reasons for knowing that in this marvellous Saturnian system, the law of gravitation is paramount. There are satellites revolving around Saturn as well as a ring; these satellites move, as other satellites do, in conformity with the laws of Kepler; and from them we learn that Saturn is really an attracting body, and, therefore, any theory as to the nature of Saturn's ring must be formed subject to the condition that it shall be attracted by the gigantic planet situated in its interior.

To a hasty glance nothing might seem easier than to reconcile the phenomena of the ring with the attraction of the planet. We might suppose that the ring stands at rest symmetrically around the planet. At its centre the planet pulls in the ring equally on all sides, so that there is no tendency in it to move in one way rather than another; and, therefore, it will stay at rest. This will not do. A ring composed of materials almost infinitely rigid might possibly, under such circumstances, be for a moment at rest; but it could not remain permanently at rest any more than can a needle balanced vertically on its point. In each case the equilibrium is unstable. If the slightest cause of disturbance arise, the equilibrium is destroyed, and the ring would inevitably fall in upon the planet. Such causes of derangement are incessantly present, so that unstable equilibrium cannot be an appropriate explanation of the phenomenon.

Even if this difficulty could be removed, there is still another, which would be quite insuperable if the ring be composed of any materials with which we are acquainted. Let us ponder for a moment on the matter, as it will lead up naturally to that explanation of the rings of Saturn which is now most generally accepted.

Imagine that you stood on the planet Saturn, near his equator; over your head stretches the ring which sinks down to the horizon in the east and in the west. The half-ring above your horizon would then resemble a mighty arch, whose span was about a hundred thousand miles. Every particle of this arch is drawn towards Saturn by gravitation, and if the arch continue to exist, it must do so in pursuance of the ordinary mechanical laws which regulate the railway arches with which we are familiar. The subsistence of these arches depends upon the resistance of the stones forming them to a crushing force. Each stone of an arch is subjected to a vast pressure, but stone is a material capable of resisting such pressure, and the arch remains. The wider the span of the arch, the greater is the pressure to which each stone is exposed. At length a span is reached which corresponds to a pressure as great as the stones can safely bear, and accordingly we thus find the limiting span over which a single arch of masonry can be constructed. Apply these principles to the stupendous arch formed by the ring of Saturn. It can be shown that the pressure on the materials of the arch capable of spanning an abyss of such awful magnitude would be something so enormous that no materials we know of would be capable of bearing it. Were the ring formed of the toughest steel that was ever made, the pressure would be so great that the metal would be squeezed like a liquid, and the mighty structure would collapse and fall down on the surface of the planet. It is not credible that any materials could exist capable of sustaining a stress so stupendous. The law of gravitation accordingly bids us to search for a method by which the intensity of this stress can be mitigated.

One method is at hand, and is obviously suggested by analogous phenomena everywhere in our system. We have spoken of the ring as if it were at rest; let us now suppose it to be animated by a motion of rotation in its plane around Saturn as a centre. Instantly we have a force developed directly antagonistic to the gravitation of Saturn. This force is the so-called centrifugal force. If, then, we imagine the ring to rotate, the centrifugal force at all points acts in an opposite direction to the attractive force, and hence the enormous stress on the ring can be abated, and one difficulty can be overcome.

The theory of gravitation alone would thus tell us that the ring must rotate, but we are fortunately not left merely to such speculative inferences. That the ring does rotate, William Herschel proved in 1789. In July of that year it happened that the edge of the ring was directly turned towards the earth. In a small telescope the thin edge of the ring ceases, under such circumstances, to be visible; but in Herschel's great reflecting telescope, twenty feet long, the edge still remained as a broken line of light. He was able to distinguish certain spots of light on this line, and he perceived that they gradually moved; he saw that they travelled from one end of the fine line to the other; he ascertained that this was really due to the rotation of the ring; and he determined the period of rotation from these observations to be about 10 hours, 32 minutes, and 15 seconds.

Such is the period of rotation of the outer ring, and we can also attribute to the inner ring a rotation which will partly relieve it from the stress the arch would otherwise have to sustain. But we cannot admit that the difficulty has been fully removed. Suppose that the outer ring revolve at such a rate as shall be appropriate to neutralise the gravitation on its outer edge, the centrifugal force will be less at the interior of the ring, while the gravitation will be greater; and hence vast stresses will be set up in the interior parts of the outer ring. Or suppose the ring to rotate at such a rate as would be adequate to neutralise the gravitation at its inner margin; then the centrifugal force at the outer parts will largely exceed the gravitation, and there will be a tendency to disruption of the ring outwards. To obviate this tendency we may suppose the outer parts of each ring to rotate more slowly than the inner parts. This naturally requires that the parts of the ring shall be mobile relatively to one another, and thus we are conducted to the suggestion that perhaps the rings are really composed of matter in a fluid state. The suggestion is a plausible one, and superficially it will, no doubt, account for the phenomena; each part of each ring can then move with an appropriate velocity, and the rings would thus exhibit a number of concentric circular currents with different velocities. The mathematician can push this inquiry a little farther, and he can study how this fluid would behave under such circumstances. His symbols can pursue the subject into the intricacies which cannot be described in general language. The mathematician finds that waves would originate in the supposed fluid, and that as these waves would lead to disruption of the rings, the fluid theory must be abandoned.

But we can still make one or two more suppositions. What if it be really true that the ring consist of an incredibly large number of concentric rings, each animated precisely with the velocity which would be suitable to the production of a centrifugal force just adequate to neutralise the attraction? No doubt this meets many of the difficulties: it is also suggested by those observations which have shown the presence of several dark lines on the ring. Here again dynamical considerations must be invoked for the reply. Such a system of solid rings revolving in the way that would be required is not compatible with the laws of dynamics. We are, therefore, compelled to make one last attempt, and still further to subdivide the ring. It may seem rather startling to abandon entirely the supposition that the ring is in any sense a continuous body, but there remains no alternative. Look at it how we will, we seem to be conducted to the conclusion that the ring is really an enormous shoal of extremely minute bodies; each of these little bodies pursues an orbit of its own around the planet, and is, in fact, merely a satellite. These bodies are so numerous and so close together that they seem to us to be continuous, and they may be very minute—perhaps not larger than the globules of water found in an ordinary cloud over the surface of the earth, which, even at a short distance, seems like a continuous body.

As Saturn's ring is itself unique, we cannot find elsewhere any very pertinent illustration of a structure so remarkable as that now claimed for the ring. Yet the solar system does show some analogous phenomena. There is, for instance, one on a very grand scale surrounding the sun himself. We allude to the multitude of minor planets, all confined within a certain region of the system. Imagine these planets to be vastly increased in number, and the orbits of the irregular ones flattened down and otherwise adjusted, and we should have a ring surrounding the sun, thus producing an arrangement not dissimilar from that now attributed to Saturn. On another and indeed a truly majestic scale, the heavens present us with an imposing spectacle of an enormous myriad of bodies forming a ring. We shall in a future chapter refer to that most extensive object, the Milky Way, which is a vast zone of stars girdling the entire heavens. No one supposes that the individual points forming the Milky Way are small objects. It would be more reasonable to suppose that they were splendid suns, perhaps in many cases rivalling, or it may be excelling, our own sun in brilliancy. But even admitting this, the objects forming the Milky Way are extremely small when compared with the dimensions of the vast zone in which they are placed; and to this extent at least we may find in them an analogue on the most stupendous scale to the phenomenon of Saturn's ring.

It is tempting to linger still longer over this beautiful system, to speculate on the appearance which the ring would present to an inhabitant of Saturn, to conjecture whether it is to be regarded as a permanent feature of our system in the same way as we attribute permanence to our moon or to the satellites of Jupiter. Looked at from every point of view, the question is full of interest, and it provides occupation abundant for the labours of every type of astronomer. If he be furnished with a good telescope, then has he ample duties to fulfil in the task of surveying, of sketching, and of measuring. If he be one of those useful astronomers who devote their energies not to actual telescopic work, but to forming calculations based on the observations of others, then the beautiful system of Saturn provides copious material. He has to foretell the different phases of the ring, to announce to astronomers when each feature can be best seen, and at what hour each element can be best determined. He has also to predict the times of the movements of Saturn's satellites, and the other phenomena of a system more elaborate than that of Jupiter; or, lastly, if the astronomer be one of that class—perhaps, from some points of view, the highest class of all—who employ the most profound researches of the human intellect to unravel the dynamical problems of astronomy, he, too, finds in Saturn problems which task to the utmost, even if they do not utterly transcend, the loftiest flights of analysis. He finds in Saturn's ring an object so utterly unlike anything else, that new weapons of analysis have to be forged for the encounter. He finds in the system so many extraordinary features, and such delicacy of adjustment, that he is constrained to admit that if he did not actually see Saturn's rings before him, he would not have thought that such a system was possible. The mathematician's labours on this wondrous system are at present only in their infancy. Not alone are the researches of so abstruse a character as to demand the highest mathematical genius, but even yet the materials for the inquiry have not been accumulated. In a discussion of this character, observation must precede calculation. The scanty observations hitherto obtained, however they may illustrate the beauty of the system, are still utterly insufficient to form the basis of that great mathematical theory of Saturn which must eventually be written.

But Saturn possesses an interest for a far more numerous class of persons than those who are specially devoted to astronomy. It is of interest, it must be of interest, to every cultivated person who has the slightest love for nature. A lover of the picturesque cannot behold Saturn in a telescope without feelings of the liveliest emotion; while, if his reading and reflection have previously rendered him aware of the colossal magnitude of the object at which he is looking, he will be constrained to admit that no more remarkable spectacle is presented in the whole realm of natural scenery.

We have pondered so long over the fascinations of Saturn's ring, that we can only give a very brief account of that system of satellites by which the planet is attended. We have already had occasion to allude more than once to these bodies; it only remains now to enumerate a few further particulars.

It was on the 25th of March, 1655, that the first satellite of Saturn was detected by Huyghens, to whose penetration we owe the discovery of the true form of the ring. On the evening of the day referred to, Huyghens was examining Saturn with a telescope constructed with his own hands, when he observed a small starlike object near the planet. The next night he repeated his observations, and it was found that the star was accompanying the planet in its progress through the heavens. This showed that the little object was really a satellite to Saturn, revolving around him in a period of 15 days, 22 hours, 41 minutes. Such was the commencement of that numerous series of discoveries of satellites which accompany Saturn. One by one they were detected, so that at the present time no fewer than eight are known to attend the great planet through his wanderings. The subequent discoveries were, however, in no case made by Huyghens, for he abandoned the search for any further satellites on grounds which sound strange to modern ears. It appears that from some principle of symmetry, Huyghens thought that it would accord with the fitness of things that the number of satellites, or secondary planets, should be equal in number to the primary planets themselves. The primary planets, including the earth, numbered six; and Huyghens' discovery now brought the total number of satellites to be also six. The earth had one, Jupiter had four, Saturn had one, and the system was complete.

Nature, however, knows no such arithmetical doctrines as those which Huyghens attributed to her. Had he been less influenced by such prejudices, he might, perhaps, have anticipated the labours of Cassini, who, by discovering other satellites of Saturn, demonstrated the absurdity of the doctrine of numerical equality between planets and satellites. As further discoveries were made, the number of satellites was at first raised above the number of planets; but in recent times, when the swarm of minor planets came to be discovered, the number of planets speedily reached and speedily passed the number of satellites.

It was in the year 1671, about sixteen years after the discovery of the first satellite of Saturn, that a second was discovered by Cassini. In the following year he discovered another; and twelve years later, in 1684, still two more; thus making a total of five satellites to this planet.

The complexity of the Saturnian system had now no rival in the heavens. Saturn had five satellites, and Jupiter had but four, while at least one of the satellites of Saturn, named Titan, was larger than any satellite of Jupiter. The discoveries of Cassini had been made with telescopes of most portentous dimensions. The length of the instrument, or rather the distance at which the object-glass was placed, was one hundred feet or more from the eye of the observer. It seemed hardly possible to push telescopic research farther with instruments of this cumbrous type. At length, however, the great reformation in the construction of astronomical telescopes began to dawn. In the hands of Herschel, it was found possible to construct reflecting telescopes of manageable dimensions, vastly more powerful and vastly more accurate than the long-focussed lenses of Cassini. A great instrument of this kind forty feet long, just completed by Herschel, was directed to Saturn on the 28th of August, 1789. Never before had the wondrous planet been submitted to a scrutiny so minute. Herschel was familiar with the labours of his predecessors. He had often looked at Saturn and his five moons in inferior telescopes; now again he saw the five moons and a star-like object so near the plane of the ring that he conjectured this to be a sixth satellite. A speedy method of testing this conjecture was at hand. Saturn was then moving rapidly over the heavens. If this new object were in truth a satellite, then it must be carried on by Saturn. Herschel watched with anxiety to see whether this would be the case. A short time sufficed to answer the question: in two hours and a half the planet had moved to a distance quite appreciable, and had carried with him not only the five satellites already known, but also this sixth object. Had this been a star it would have been left behind; it was not left behind, and hence it, too, was a satellite. Thus, after the long lapse of a century, the telescopic discovery of satellites to Saturn recommenced. Herschel, as was his wont, observed this object with unremitting ardour, and discovered that it was much nearer to Saturn than any of the previously-known satellites. In accordance with the general law, that the nearer the satellite the shorter the period of revolution, Herschel found that this little moon completed a revolution in about 1 day, 8 hours, 53 minutes. The same great telescope used with the same unrivalled skill, soon led Herschel to a still more interesting discovery. An object so small as only to appear like a very minute point in the great forty-foot reflector was also detected by Herschel, and was by him proved to be a satellite, so close to the planet that it completed a revolution in the very brief period of 22 hours and 37 minutes. This is an extremely delicate object, only to be seen in the best telescopes in the brief intervals when it is not entirely screened from view by the ring.

Again another long interval elapsed, and for almost fifty years the Saturnian system was regarded as consisting of the series of rings and of the seven satellites. The next discovery has a singular historical interest. It was made simultaneously by two observers—Professor Bond, of Cambridge, Mass., and Mr. Lassell, of Liverpool—for on the 19th September, 1848, each of these astronomer verified that a small point seen by each on previous nights was really a satellite. This object is, however, at a considerable distance from the planet, and requires 21 days, 7 hours, 28 minutes for each revolution.

Such is at present the system of Saturn as we know it. It seems certain that if there be any more satellites than the ample number of eight already discovered, they must be objects much more minute than those already detected. Accomplished observers, provided with telescopes more powerful than that of Bond, and in some cases more powerful than that of Lassell, have examined and re-examined this object. They have made out many further details of the Saturnian system, but no further satellite has been added.

A law has been observed by Professor Kirkwood, which connects together the movements of four of the satellites of Saturn. This law is fulfilled in such a manner as leads to the supposition that it arises from the mutual attraction of the satellites. We have already described a similar law relative to three of the satellites of Jupiter. The problem relating to Saturn, involving as it does no fewer than four satellites, is one of no ordinary complexity. It involves the theory of Perturbations to a greater degree than that to which mathematicians are accustomed in their investigation of the more ordinary features of our system. To express this law it is necessary to have recourse to the daily movements of the satellites; these are respectively—

SATELLITE.	DAILY MOVEMENT.
I.	382°·2.
II.	262°·74.
III.	190°·7.
IV.	131°·4.

The law states that if to five times the movement of the first satellite we add that of the third and four times that of the fourth, the whole will equal ten times the movement of the second satellite. The calculation stands thus:—

5 times I.	equals	1911°·0		
III.	,,	190°·7		II. ... 262°·74
4 times IV.	,,	525°·6		10
		2627°·3	equals	2627°·4 nearly.

Nothing can be simpler than the verification of this law; but the task of showing the physical reason why it should be fulfilled has not yet been accomplished.

Saturn was the most distant planet known to the ancients. It revolves in an orbit far outside the other ancient planets, and until the discovery of Uranus a century ago, the orbit of Saturn might well be regarded as the frontier of the solar system. Saturn was indeed an object worthy to occupy a position so distinguished. But we now know that the mighty orbit of Saturn does not extend to the frontiers of the solar system; a splendid discovery, leading to one still more splendid, has vastly extended the boundary, by revealing two mighty planets, revolving in dim telescopic distance, far outside the path of Saturn. These objects have not the beauty of Saturn; they are much less brilliantly attended; they are, indeed, in no sense effective telescopic pictures. Yet these outer planets awaken an interest of a most special and remarkable kind. The discovery of each is a classical event in the history of astronomy, and the opinion has been maintained, and perhaps not without good reason, that the discovery of Neptune, the more remote of the two, is the greatest achievement in astronomy made since the time of Newton. To the development of these discoveries our next chapters must be devoted.

CHAPTER XIV.

URANUS.

Contrast between Uranus and the other great Planets—William Herschel—His Birth and Parentage—Herschel's Arrival in England—His love of Learning—Commencement of his Astronomical Studies—The Construction of Telescopes—Reflecting Telescopes—Construction of Mirrors—The Professor of Music becomes an Astronomer—The Methodical Research—The 13th March, 1781—The Discovery of Uranus—Delicacy of Observation—Was the Object a Comet?—The Significance of this Discovery—The Fame of Herschel—George III. and the Bath Musician—The King's Astronomer at Windsor—Caroline Herschel—The Planet Uranus—Numerical Data with reference thereto—The Four Satellites of Uranus—Their Circular Orbits—Early Observations of Uranus—Flamsteed's Observations—Lemonnier saw Uranus—Utility of their Measurements—The Elliptic Path—The great Problem thus Suggested.

To the present writer it has always seemed that the history of Uranus, and of the circumstances attending its discovery, form one of the most pleasing and interesting episodes in the whole history of science. We here occupy an entirely new position in the study of the solar system. All the other great planets were familiarly known to the ancients, however erroneous might be the ideas entertained in connection with them. They were conspicuous objects, and by their movements could hardly fail to attract the attention of those whose pursuits led them to observe the stars. But now we come to a great planet, whose very existence was utterly unknown to the ancients; and hence, in approaching the subject, we have first to describe the actual discovery of this object, and then to consider what we can learn as to its physical nature.

We have, in preceding pages, had occasion to mention the revered name of William Herschel in connection with various branches of astronomy; but we have hitherto designedly postponed any more explicit reference to this extraordinary man until we have arrived at the present stage of our work. The story of Uranus, in its earlier stages at all events, is the story of the early career of William Herschel. It would be alike impossible and undesirable to attempt to separate them.

William Herschel, the illustrious astronomer, was born at Hanover in 1738. His father was an accomplished man, pursuing, in a somewhat humble manner, the calling of a Professor of Music. He had a family of ten children, of whom William was the fourth; and it may be noted that all the family of whom any record has been preserved inherited their father's musical talents, and became accomplished performers. Pleasing sketches have been given of this interesting family, of the unusual aptitude of William, of the long discussions on music and on philosophy, and of the little sister Caroline, destined in later years for an illustrious career. William soon learned all that his master could teach him in the ordinary branches of knowledge, and by the age of fourteen he was already a competent performer on the oboe and the viol. He was engaged in the Court orchestra at Hanover, and was also a member of the band of the Hanoverian Guards. Troublous times were soon to break up Herschel's family. The French invaded Hanover, the Hanoverian Guards were overthrown in the battle of Hastenbeck, and young William Herschel had some unpleasant experience of actual warfare. He was not wounded, but after passing the night in a ditch, he decided that he would make a change in his profession. His method of doing so is one which his biographers can scarcely be expected to defend; for, to speak plainly, he deserted, and succeeded in making his escape to England. It is stated on unquestionable authority that on Herschel's first visit to King George III. more than twenty years afterwards, his pardon was handed to him by the King himself, written out in due form.

At the age of nineteen, the young musician began to seek his fortunes by his profession in England. He met at first with very considerable hardship, but industry and skill conquered all difficulties, and by the time he was twenty-six years of age, he was thoroughly settled in England, and doing well in his profession. In the year 1766, we find Herschel occupying a position of some distinction in the musical world: he had become the organist of the Octagon Chapel at Bath, and his time was fully employed in giving lessons to his numerous pupils, and with his preparation for concerts and for oratorios.

Notwithstanding his busy professional life, Herschel still retained that insatiable thirst for knowledge which he had when a boy. Every moment he could snatch from his musical engagements was eagerly devoted to study. In his desire to perfect his knowledge of the more abstruse parts of the theory of music, he had occasion to learn mathematics; from mathematics the transition to optics was a natural one; and once he had commenced to study optics, he was of course brought to a knowledge of the telescope, and thence to astronomy itself.

His beginnings were made on a very modest scale. It was through a small and imperfect telescope that the great astronomer obtained his first view of the celestial glories. No doubt he had often before looked at the heavens on a clear night, and admired the thousands of stars with which it was adorned; but now, when he was able to increase his powers of vision even to a slight extent, he obtained a view which fascinated him. The stars he saw before, he now saw far more distinctly; but more than this, he found that myriads of others previously invisible were now revealed to him. Glorious, indeed, is this spectacle to any one who possesses a spark of enthusiasm for natural beauty. To Herschel this view immediately changed the whole current of his life. His success as a Professor of Music, his oratorios and his pupils, were speedily to be forgotten, and the rest of his life was to be devoted to the absorbing pursuit of one of the noblest of the sciences.

Herschel could not remain contented with the small and imperfect instrument which first interested him. Throughout his career he determined to see everything for himself in the best manner which his utmost powers could command. He at once determined to have a better instrument, and he wrote to a celebrated optician in London with a view of purchasing one. But the price which the optician demanded seemed more than Herschel thought he could or ought to give. Instantly his determination was taken. A good telescope he must have, and as he could not buy one, he resolved to make one. It was alike fortunate, both for Herschel and for science, that circumstances impelled him to this determination. Yet, at first sight, how unpromising was the enterprise! That a music teacher, busily employed day and night, should, without previous training, expect to succeed in a task where the highest mechanical and optical skill was required, seemed indeed unlikely. But enthusiasm and genius know no insuperable difficulties. From conducting a brilliant concert in Bath, when that city was at the height of its fame, Herschel would rush home, and without even delaying to take off his lace ruffles, he would plunge into his manual labours of grinding specula and polishing lenses. No alchemist of old ever was more deeply absorbed in a project for turning lead into gold than Herschel in his determination to have a telescope. He transformed his home into a laboratory; of his drawing-room he made a carpenter's shop. Turning lathes were the furniture of his best bedroom. A telescope he must have, and as he progressed he determined, not only that he should have a good telescope, but a very good one; and as success cheered his efforts, he ultimately succeeded in constructing the greatest telescope that the world had up to that time ever seen. Though it is as an astronomer that we are concerned with Herschel, yet we must observe that, even as a telescope maker, great fame and no small degree of commercial success flowed in upon him. When the world began to ring with the glorious discoveries of Herschel, and when it was known that the Bath musician had made these discoveries with telescopes which were the work of his own hands, a demand sprang up for instruments of his construction. It is stated that he made upwards of eighty large telescopes, as well as many others of smaller size. Several of these instruments were purchased by foreign princes and potentates.* We have never heard

* Extract from "Three Cities of Russia," by C. Piazzi Smyth, Vol. ii., p. 164. "In the year 1796. It then chanced that George III., of Great Britain, was pleased to send as a present to the Empress Catharine of Russia a ten-foot reflecting telescope constructed by Sir William Herschel. Her Majesty immediately desired to try its powers, and Roumovsky was sent for from the Academy to repair to Tsarskoe-Selo where the court was at the time residing. The telescope was accordingly unpacked, and for eight long consecutive evenings the Empress employed herself ardently in observing the moon, planets, and stars; and more than this, in inquiring

that any of these illustrious personages became assiduous or celebrated astronomers, but, at all events, they seem to have paid Herschel handsomely for his skill, so that by the sale of large telescopes he was able to realise what may be regarded as a fortune in the moderate horizon of the man of science.

The instruments made by Herschel were entirely of that class which we have already described in an earlier chapter under the title of Reflecting Telescopes. These instruments were known long before Herschel's time. It was, however, left for him to discover the means by which they could be constructed of large size, while he also wrought the specula with unprecedented care. The success of a telescope of this kind mainly depends upon the accuracy with which the speculum receives that particular shape which is necessary if every ray of light is to travel to its precise destination. The methods of Herschel are not very fully known; he never published them with the same degree of detail as was subsequently given to the world by Lord Rosse and by Lassell, when they described their experience in following up the path which Herschel had indicated. It seems, however, that Herschel relied, to a very large extent, upon manual skill rather than, as his successor, did, on mechanical contrivance. It is narrated how, during the critical operation of polishing one of the great mirrors, Herschel has sat with his hands on the mirror for very many hours consecutively. On such occasions he was constantly tended by the ever-faithful sister Caroline, who sat by his side to place the necessary food between his lips, and to beguile the time by reading the "Arabian Nights." What could be more appropriate reading for such an occasion than the story of Aladdin? The wonders which Aladdin saw when he polished his lamp may well be compared with what Herschel saw when he had polished his mirror.

Up to the present point of our narrative, we have seen

into the state of astronomy in her dominions. Then it was that Roumovsky set before the Imperial view the Academy's idea of removing their observatory, detailing the necessity for, and the advantages of, such a proceeding. Graciously did the 'Semiramis of the North,' the 'Polar Star,' enter into all these particulars, and warmly approve of the project; but death closed her career within a few weeks after, and prevented her execution of the design."

Herschel as a laborious musician, with considerable renown in his profession, not only in Bath, but throughout the west of England. His telescope making was merely the occupation of his spare moments, and was quite unknown to most of those who knew and respected his musical attainments. It was in 1774 that Herschel first enjoyed a view of the heavens through an instrument built with his own hands. It was but a small one in comparison with those which he afterwards fashioned, but at once he experienced the advantage of being his own instrument maker. Night after night he was able to make one improvement after another; at one time he was enlarging the size of his mirrors; at another he was reconstructing the mounting, and trying to remedy defects in the eye-pieces. With unwearying perseverance he aimed at the highest excellence, and with each successive advance he found that he was able to see more and more. His enthusiasm attracted to him a few friends who were, like himself, ardently attached to science. The mode in which he first made the acquaintance of Sir William Watson, who afterwards became his warmest friend, was characteristic of both. Herschel was observing the mountains in the moon, and as the hours passed on, he had occasion to bring his telescope into the street in front of his house to enable him to continue his work. Sir William Watson happened to pass by, and was arrested by the somewhat unusual spectacle of an astronomer in the public street, at the dead of night, using a large and somewhat quaint-looking instrument. Having a love for astronomy, he stopped, and when Herschel took his eye from the telescope, asked if he might be allowed to have a view of the moon. The request was readily granted. Probably Herschel at that time found but few in the gay city who cared for such matters; he was quickly drawn to Sir W. Watson, who at once reciprocated the feeling, and thus began a friendship which bore important fruit in Herschel's subsequent career.

At length the year 1781 approached, which was to witness his great achievement. Herschel had now some six or seven years' practical experience in astronomy, and he had completed a telescope of exquisite optical perfection, though greatly inferior in point of size to those which he afterwards erected. With this telescope Herschel commenced a methodical piece of observation. He formed the scheme of systematically examining all the stars which were above a certain degree of brightness. It does not quite appear what object Herschel proposed to himself when he undertook this labour, but, in any case, he could hardly have anticipated the extraordinary success with which the work was to be crowned. In the course of this review, the telescope was directed to a star; that star was examined; then another was brought into the field of view, and it too was examined. Every star under such circumstances merely shows itself as a point of light; the point may be brilliant or not, according as the star is bright or not; the point will also, of course, show the colour of the star, but it cannot show any recognisable size or shape. The greater, in fact, the perfection of the telescope, the smaller is the telescopic image of a star.

How many stars Herschel looked at in this review, we are not told; but at all events, on the ever-memorable night of the 13th of March, 1781, he was pursuing his self-allotted task among the stars in the constellation Gemini. Doubtless one star after another was admitted to view, and was allowed to pass away. At length, however, a star was placed in the field which differed from every other star. It was not a mere point of light; it had a minute, but still a perfectly recognisable, disc. We say the disc was perfectly recognisable, but we should be careful to add that it was so to the delicate eye of Herschel alone. Other astronomers had often seen this object before. It had actually been most deliberately and carefully measured no fewer than seventeen times on former occasions before the Bath musician, with his home-made telescope, looked at it. Herschel perceived that this object was not what previous astronomers had supposed it to be; it was, indeed, not a star at all. Even after the discovery was made, and when well-trained observers with good instruments looked again under the direction of Herschel at this object, one after another has borne testimony to the extraordinary delicacy of Herschel's perception, which enabled him almost at the first glance to discriminate between this object and a star.

If not a star, what, then, could it be? The first step to enable this question to be answered was to observe the body for some time. This Herschel did. He looked at it one night after another, and soon he discovered another most fundamental difference between this object and an ordinary star. The stars are, of course, characterised by their fixity, but this object was not fixed; night after night the place it occupied changed with respect to the stars. No longer could there be any doubt that this body was a member of the solar system, and that an interesting discovery had been made; many months however elapsed before Herschel knew the real merit of his achievement. It does not seem at first to have occurred to him that he had made the superb discovery of another mighty planet revolving outside Saturn; he thought at first that it could only be a comet. No doubt this object looks very different from the ordinary conception of a comet, which is decorated with a tail. It was not, however, so entirely different from some forms of telescopic comet as to make the suggestion of its being a body of this kind unlikely; and, in fact, the discovery was at first announced to be that of a new comet. Time was necessary before the true character of the object could be ascertained. It must be followed for a considerable distance along its path, and measures of its position at different epochs must be effected, before it is practicable for the mathematician to calculate the path which the body pursues; once, however, attention was devoted to the subject, many astronomers aided in making the necessary observations. These were placed in the hands of mathematicians, and the result was proclaimed that this body was not a comet, but that, like all the planets, it revolved in a nearly circular path around the sun, and that the path lay millions of miles outside the path of Saturn, which had so long been regarded as the boundary of the solar system.

It is hardly possible to over-estimate the significance of this splendid discovery. The five planets had been known from all antiquity; they were all, at suitable seasons, brilliantly conspicuous to the unaided eye. But Herschel showed that, far outside the outermost of these planets, there was another splendid planet, larger than Mercury or Mars, larger—far larger—than Venus and the earth, and only surpassed in bulk by Jupiter and by Saturn. This superb new planet was plunged into space to such a depth that, notwithstanding its noble proportions, it seemed merely a tiny star, being only on rare occasions within reach of the unaided eye. Herschel showed that this great planet required a period of eighty-seven years to complete its majestic path, and that the diameter of that path was 3,600,000,000 miles.

Although the history of astronomy is the record of brilliant discoveries—of the discoveries of Copernicus, and of Kepler—of the telescopic achievements of Galileo, and the splendid theory of Newton—of the refined discovery of the aberration of light—of many other imperishable triumphs of intellect—yet this achievement of the organist at the Octagon Chapel occupies a totally different position from any other. There never before had been any historic record of the discovery of one of the bodies of the particular system to which the earth belongs. The older planets were no doubt discovered by some one, but we can say little more about these discoveries than we can about the discovery of the sun or of the moon; all are alike prehistoric. Here, then, was the first recorded instance of the discovery of a planet which, like our earth, revolves around the sun, and, like our earth, may conceivably be an inhabited globe. So unique an achievement instantly arrested the attention of the whole scientific world. The music-master at Bath, hitherto unheard of as an astronomer, was speedily placed in the very foremost rank of those entitled to the name. On all sides the greatest interest was manifested about the unknown philosopher. The name of Herschel, then unfamiliar to English ears, appeared in every journal which pretended to the slightest importance, and a curious list has been preserved of the number of blunders which were made in spelling the name. The different scientific societies hastened to convey their congratulations on an occasion so memorable.

Tidings of the discovery made by the Hanoverian musician reached the ears of George III., and he sent for Herschel to come to the Court, that the King might learn what his achievement actually was from the discoverer's own lips. Herschel brought with him one of his telescopes, and he provided himself with a chart of the solar system, with which to explain precisely wherein the significance of the discovery lay. The King was greatly interested in Herschel's narrative, and not less in Herschel himself. The telescopes were erected at Windsor, and, under the astronomer's guidance, the King was shown Saturn and other celebrated objects. It is also told how the ladies of the Court the next day asked Herschel to show them the objects which had so pleased the King. The telescope was duly erected in a window of one of the Queen's apartments, but when evening arrived the sky was found to be overcast with clouds, and no stars could be seen. This was an experience with which Herschel, like every other astronomer, is unhappily only too familiar. But it is not every astronomer who would have had the readiness of Herschel to escape gracefully from the position. He showed to his lady pupils the construction of the telescope; he explained the mirror, and how he had fashioned it and polished it; and then, seeing the clouds were inexorable, he proposed that, as he could not show them the real Saturn, permission should be given to him to exhibit an artificial one. The permission granted, Herschel turned the telescope away from the sky, and pointed it towards the wall of a distant garden. On looking into the telescope there was Saturn, his globe and his system of rings, so faithfully shown that, says Herschel, even a skilful astronomer might have been deceived. The fact was that during the course of the day Herschel saw that the sky would probably be overcast in the evening, and he had provided for the emergency by cutting a hole in a piece of cardboard, the shape of Saturn, which was then placed against the distant garden wall, and illuminated by a lamp at the back.

This visit to Windsor was productive of consequences momentous to Herschel, momentous to science. He had made so favourable an impression, that the King proposed to create for him the special appointment of King's Astronomer at Windsor. The King was to provide the means for erecting the great telescopes, and he allocated to Herschel a salary of £200 a year, the figures being based, it must be admitted, on a somewhat moderate estimate of the requirements of an astronomer's household. Herschel mentioned these particulars to no one save to his constant and generous friend, Sir W. Watson, who exclaimed, "Never bought monarch honour so cheap." To all his other inquiring friends, Herschel merely said that the King had provided for him. In accepting this post, Herschel took no doubt a most serious step. He at once sacrificed entirely his musical career, now, from many sources, a lucrative one; but his determination was speedily taken. The splendid earnest that he had already given of his devotion to astronomy was, he knew, only the commencement of a series of most memorable labours. He had indeed long been feeling that it was his bounden duty to follow that path in life which his genius indicated. He was no longer a young man. He had attained middle age, and the years had become especially precious to one who knew that he had still a mighty life-work to accomplish. He at one stroke freed himself from all distractions; his pupils and concerts, his whole connection at Bath, were immediately renounced; he accepted the King's offer with alacrity, and after one or two changes settled permanently at Slough, near Windsor.

With him went his sister Caroline, also a recipient of the King's bounty, as assistant to her brother. It hardly comes within our scope to narrate the wonderful career of this brother and sister. It would lead us too far to attempt any description of the herculean labours by which mighty telescopes were erected, and of the indomitable perseverance with which the observations were conducted. How, from dusk to dawn, the great astronomer stood at the telescope throughout the winter's night; how he narrated the wonders that he saw; how the faithful sister, sitting by, transcribed those observations till the ink sometimes froze in her pen; how she devoted her days to making the necessary calculations; how for nearly forty years this wonderful brother and sister worked together; and how many and marvellous were the discoveries which were made! All these matters cannot here be more fully described. In the course of this work we shall still have occasion to refer to Herschel's discoveries in particular branches of astronomy, and so here we must forbear to pursue this most attractive subject, merely recommending all who are interested in the biography of a most remarkable woman to read the "Memoirs of Caroline Herschel."*

It has, indeed, been well remarked that the most important event in connection with the discovery of Uranus was the discovery of Herschel's unrivalled powers of observation. Uranus must, sooner or later, have been found. Had Herschel not lived, we would still, in all probability, have known Uranus long ere this. The really important point for science was that Herschel's genius should be given full scope, by setting him free from the engrossing details of an ordinary professional calling. The discovery of Uranus secured all this, and accordingly obtained for astronomy all Herschel's future labours.

Of Uranus itself as a planet we have but little to say. It is plunged into space to such a remote distance that even the best of our modern telescopes cannot make of it a striking picture. We can see, as Herschel did, that it has a measurable disc, and from measurements of that disc we conclude that the diameter of the planet is about 31,700 miles. This is about four times as great as the diameter of the earth, and we accordingly see that the volume of Uranus must be about sixty-four times as great as that of the earth. We also find that, like the other giant planets, Uranus seems to be composed of materials much lighter, on the whole, than those we find here; so that though sixty-four times as large as the earth, Uranus is only fifteen times as heavy. If we may trust to the analogies of what we see everywhere else in our system, we can feel but little doubt that Uranus must rotate about an axis. The ordinary means of demonstrating this rotation can be hardly available in a body whose surface appears so small and so faint, that we cannot conclude with any certainty as to the existence of markings on that surface definite enough to be watched during rotation. The period of rotation is accordingly unknown.

There is, however, one feature about Uranus which presents many points of interest to those astronomers who are possessed of telescopes of unusual size and perfection. Uranus is accompanied

* By Mrs. John Herschel. London, 1876.

by a system of satellites, some of which are so faint as to require the closest scrutiny for their detection. The discovery of these satellites was one of the subsequent achievements of Herschel. It is, however, remarkable that even his penetration and care did not preserve him from errors with regard to these very delicate objects. Some of the points which he thought to be satellites must, it would now seem, have been merely stars enormously more distant, which happened to lie in the field of view. It has been finally ascertained that the known satellites of Uranus are four in number, and their movements have been made the subject of prolonged and interesting telescopic research. The four satellites bear the names of Ariel, Umbriel, Titania, and Oberon. Arranged in order of their distance from the central body, Ariel, the nearest, accomplishes its journey in 2·52 days. Oberon, the most distant satellite, completes its journey in 13·46 days.

The law of Kepler declares that the path of a satellite around its primary, no less than of the primary around the sun, must be an ellipse. It leaves us, however, boundless latitude in the choice of the shape of the ellipse. The ellipse may be nearly a circle, it may be absolutely a circle, or it may be something quite different from a circle. The paths pursued by the planets are, generally speaking, nearly circles; but we meet with no exact circle among planetary orbits. So far as we at present know, the closest approach made to a perfectly circular movement is that by which the satellites of Uranus revolve around their primary. We are not prepared to say that these paths are absolutely circular. All that can be said is that our telescopes fail to show any measurable departure therefrom. It is also to be noted as an unparalleled circumstance that the orbits of the satellites of Uranus all lie in the same plane. This is not true of the orbits of the planets around the sun, nor is it true of the orbits of any other system of satellites around their primary. The most singular circumstance attending the whole Uranian system is, however, found in the position which this plane occupies. This is indeed almost as great an anomaly in our system as are the rings of Saturn themselves. We have already had occasion to notice that the plane in which the earth revolves around the sun is very nearly coincident with the planes in which all the other great planets revolve. The same is true, to a large extent, of the orbits of the minor planets; though here, no doubt, we meet with a few cases in which the plane of the orbit is inclined at no inconsiderable angle to the plane in which the earth moves. The plane in which the moon revolves also lies very close to this system of planetary planes. So, too, do the orbits of the satellites of Saturn and of Jupiter, while even the more recently discovered satellites of Mars form no exception to the rule. The whole solar system—at least so far as the great planets are concerned—would require but comparatively little alteration if the orbits were to be entirely flattened down into one plane. There is, however, one notable exception to this all but universal rule. The satellites of Uranus revolve in a plane which is far from coinciding with the plane to which all other orbits approximate. In fact, the paths of the satellites of Uranus lie in a plane nearly at right angles to the orbit of Uranus. We are not in a position to give any satisfactory explanation of this circumstance. It is, however, evident that in the genesis of the Uranian system there must have been some influence of a quite exceptional and local character.

Soon after the discovery of the planet Uranus, in 1781, sufficient observations were accumulated to enable the orbit it follows to be determined. When the orbit was known, it was then a mere matter of mathematical calculation to ascertain where the planet was situated at any past time, and where it would be situated at any future time. A very interesting inquiry was thus suggested as to how far it might be possible to find any observations of the planet made previously to its discovery by Herschel. Uranus looks like a star of the sixth magnitude. Not many astronomers were provided with telescopes of the perfection attained by Herschel, and the personal delicacy of perception characteristic of Herschel was a still more rare possession. It was therefore, to be expected that, if such previous observations existed, they would merely record Uranus as a star visible, and indeed bright, in a moderate telescope, but still not claiming any exceptional attention over thousands of apparently similar stars. Many of the early astronomers had devoted themselves to the useful and laborious work of forming catalogues of stars. In the preparation of a star catalogue, the telescope was directed to the heavens, the stars were observed, their places were carefully measured, the brightness of the star was also estimated, and thus the catalogue was gradually compiled in which each star had its place faithfully recorded, so that at any future time it could be identified. The stars were thus inventoried by hundreds and by thousands, at various dates from the birth of accurate astronomy till the present time. The suggestion was then made that, as Uranus looked so like a star, and as it was quite bright enough to have engaged the attention of astronomers possessed of even very moderate instrumental powers, there was a possibility that it had already been observed, and thus actually lay recorded as a star in some of the older catalogues. This was indeed a suggestion worthy of every attention, and pregnant with the most important consequences in connection with the immortal discovery to be discussed in our next chapter. But how was such an examination of the catalogues to be conducted? Uranus is constantly moving about; does it not seem that there is every element of uncertainty in such an investigation? Let us consider a memorable example.

The great national observatory at Greenwich was founded in 1675, and the first Astronomer-Royal was the illustrious Flamsteed, who in 1676 commenced that series of observations of the heavenly bodies which has been continued to the present day with such incalculable benefits to science. At first the instruments were of the most primitive description, but in the course of some years Flamsteed succeeded in procuring instruments adequate to the production of a catalogue of stars, and he devoted himself with extraordinary zeal to the undertaking. It is in this memorable work, the "Historia Cœlestis" of Flamsteed, that we have found the first observation of Uranus. In the first place, it was known that the orbit of Uranus, like the orbit of every other great planet, was inclined at a very small angle to the ecliptic. It hence follows that Uranus is at all times only to be met with along the ecliptic, and it is possible to calculate where the planet was in each year. It was thus seen that in the year 1690 the planet was situated in that part of the ecliptic where Flamsteed was at the same date making his observations. It was natural to search the observations of Flamsteed, and see whether any of the so-called stars could have been Uranus. An object was found in the "Historia Cœlestis" which occupied a position identical with that which Uranus must have filled on the same date. Could this be Uranus? A decisive test was at once available. The telescope was directed to the spot in the heavens where Flamsteed saw a sixth-magnitude star. If that were really a star, then would it still be visible. The trial was made: no such star could be found, and hence the presumption that this was really Uranus could hardly be for a moment doubted. Speedily other confirmations flowed in. It was shown that Uranus had been unconsciously observed by Bradley and by Tobias Mayer, and it also became apparent that Flamsteed had observed Uranus not only once, but that he had actually measured its place four or five times between the years 1690 and 1715. Yet Flamsteed was never conscious of the discovery that lay so nearly in his grasp. He was of course under the impression that all these observations related to different stars. A still more remarkable case is found among the observations of Lemonnier. After Uranus had been discovered, Lemonnier turned to examine his former observations, and among them he found that he had really observed Uranus no less than three times, on each occasion, of course, recording it as a distinct star. But Uranus had in reality a still narrower escape from being detected by Lemonnier. Another astronomer on going again over Lemonnier's work found no fewer than nine additional observations of Uranus, and of these four had been made on consecutive nights. How close, indeed, was Lemonnier to the discovery which would have immortalised him! During the intervals between those four nights, the planet of course moved, and was of course taken for a different star each night. If Lemonnier had only carefully looked over his own work; if he had perceived, as he might have done, how the star he observed yesterday was gone to-day, while the star visible to-day had moved away by to-morrow, there is no doubt that Uranus would have been discovered, and William Herschel would have been anticipated. Would Lemonnier have made as good use of his fame as Herschel did? ' This is a question hard, perhaps, to answer; but those who estimate Herschel as the present writer thinks he ought to be estimated, will probably agree in thinking that it was most fortunate for science that Lemonnier did *not* compare his observations.*

These early accidental observations of Uranus are not merely to be regarded as matters of historical interest or curiosity. That they are of the deepest importance with regard to the science itself a few words will enable us to show. It is to be remembered that Uranus requires no less than eighty-four years to accomplish his mighty revolution around the sun. The planet has completed one entire revolution, since its discovery, and up to the present time (1885) has accomplished nearly one-third of another. For the careful study of the nature of the orbit, it was desirable to have as many observations as possible, and extending over the widest possible interval. This was in a great measure secured by the identification of the early observations of Uranus. An approximate knowledge of the orbit was quite capable of giving the places of the planet with sufficient accuracy to identify it when met with in the catalogues. But when by their aid the actual observations have been discovered, they tell us precisely the place of Uranus; and hence, instead of our knowledge of the planet being limited to but little more than one revolution, we have at the present time information with regard to it extending over considerably more than two revolutions.

From the observations of the planet the ellipse in which it moves can be ascertained. We can compute this ellipse from the observations made during the time since the discovery. We can also compute the ellipse from the early observations made

* Arago says that "Lemonnier's records were the image of chaos." Bouvard showed to Arago one of the observations of Uranus which was written on a paper bag that in its time had contained hair-powder.

before the discovery. If Kepler's laws were rigorously verified, then, of course, the ellipse performed in the present revolution must differ in no respect from the ellipse performed in the preceding, or indeed in any other revolution. We can test this point in an interesting manner by comparing the ellipse derived from the ancient observations with that deduced from the modern ones. These ellipses closely resemble each other; they are nearly the same; but it is most important to observe that they are not *exactly* the same, even when allowance has been made for every known source of disturbance in accordance with the principles explained in the next chapter. The law of Kepler seems thus not absolutely true in the case of Uranus. Here is, indeed, a matter demanding our most earnest and careful attention. Have we not repeatedly laid down the universality of the laws of Kepler in controlling the planetary motions? How then can we reconcile this law with the irregularities proved beyond a doubt to exist in the motions of Uranus?

Let us look a little more closely into the matter. We know that the laws of Kepler are a consequence of the laws of gravitation. We know that it is in virtue of the sun's attraction that the planet moves in an elliptic path around the sun, and we know that the ellipse will be preserved without the minutest alteration if the sun and the planet be left to their mutual attractions, and no other force intervene to disturb them. The conclusion is irresistible. Uranus does not move solely in consequence of the sun's attraction and that of the planets of our system interior to Uranus; there must therefore be some further influence acting upon Uranus besides those already known. To the development of this subject the next chapter will be devoted.

CHAPTER XV.

NEPTUNE.

Discovery of Neptune—A Mathematical Achievement—The Sun's Attraction—All Bodies Attract—Jupiter and Saturn—The Planetary Perturbations—Three Bodies—Nature has Simplified the Problem—Approximate Solution—The Sources of Success—The Problem Stated for the Earth—The Discoveries of Lagrange—The Eccentricity—Necessity that all the Planets Revolve in the same Direction—Lagrange's Discoveries have not the Dramatic Interest of the more Recent Achievements—The Irregularities of Uranus—The Unknown Planet must Revolve outside the Path of Uranus—The Data for the Problem—Le Verrier and Adams both Investigate the Question—Adams Indicates the Place of the Planet—How the Search was to be Conducted—Le Verrier also Solves the Problem—The Telescopic Discovery of the Planet—The Rival Claims—Early Observation of Neptune—Difficulty of the Telescopic Study of Neptune—Numerical Details of the Orbit—Is there any Outer Planet?—Contrast between Mercury and Neptune.

We enter in this chapter into a discovery so extraordinary that the whole annals of science may be searched in vain for a parallel. We are not here concerned with technicalities of practical astronomy. Neptune was first revealed to us by profound mathematical research rather than by telescopic investigation. We must develop the account of this striking epoch in the history of science with the fulness of detail which is commensurate with its importance; and it will accordingly be necessary, at the outset of our narrative, to make an excursion into a difficult but most attractive region of astronomy, to which we have as yet made little reference.

The supreme controlling power in the solar system is the attraction of the sun. Each planet of the system experiences that attraction, and, in virtue of it, the planet is constrained to move around the sun in an elliptic path. The efficiency of the sun as an attractive agent is directly proportional to its mass, and as its mass is more than a thousand times as great as the mass of Jupiter, which, itself, exceeds that of all the other planets collectively, the attraction of the sun is necessarily the chief determining force of all the movements in our system. The law of gravitation, however, does not merely say that the sun attracts each planet. Gravitation is a doctrine much more general, for it asserts that every body in the universe attracts every other body. In obedience to this law each planet must be attracted, not only by the sun, but by innumerable bodies, and the movement of the planet must be the joint effect of all the attractions. As to the influence of the stars on our solar system, it may be at once set aside as inappreciable. The stars are no doubt enormous bodies, in many cases possibly transcending our sun in magnitude, but the law of gravitation tells us that the intensity of the attraction decreases with the square of the distance. Most of the stars are a million times as remote as the sun, and consequently their attraction is so slender as to be absolutely inappreciable in the discussion of this question. The only attractions necessary to consider are those which arise from the action of one body of the system upon another. Let us take, for instance, the two largest planets of our system, Jupiter and Saturn. Each of these planets moves mainly in consequence of the sun's attraction, but each planet also attracts the other, and the consequence is that each planet is slightly drawn away from the position it would otherwise have occupied. In the language of astronomy we would say that the path of Jupiter is perturbed by the attraction of Saturn; and, conversely, that the path of Saturn is perturbed by the attraction of Jupiter.

For many years these irregularities of the planetary motions presented problems which astronomers were not able to solve. Gradually, however, one difficulty after another has been vanquished, and though there are no doubt some small irregularities still outstanding which have not been completely explained, yet all the larger and more important phenomena of the kind are well understood. The subject is one of the most difficult which the astronomer has to encounter in the whole range of his science. He has here to calculate what effect one planet is capable of producing on another planet. Such calculations bristle with the most formidable difficulties, which can only be overcome by consummate skill in the loftiest branches of mathematics. Let us state what the problem really is.

When two bodies move in virtue of their mutual attraction, each of these bodies revolves in a path which admits of being exactly calculated. Each path is, in fact, an ellipse, and the focus of each ellipse is at the centre of gravity of the two bodies. In the case of a sun and a planet, in which the mass of the sun preponderates so enormously over the mass of the planet, the centre of gravity of the two lies very near the centre of the sun; the actual path of the sun is in such case very small in comparison with the path of the planet; and hence we are justified in most cases in regarding the sun as at rest, and the planet revolving around it. All these matters admit of perfectly rigid calculation of a somewhat elementary character. But now let us add a third body to the system. This third body attracts each of the others, and is attracted by them. By this attraction, the third body is displaced, and accordingly its influence on the others is modified; they in turn act upon it, and these actions and reactions introduce endless complexity into the system. Such is the famous "problem of three bodies," which has engaged the attention of almost every great mathematician since the time of Newton. Stated in its mathematical aspect, and without having its rigour abated by any modifying circumstances, the problem is one that defies solution. No one has yet been able to construct the analysis which shall be powerful enough to cope with the mutual attractions of three bodies moving freely in space. If the number of bodies be greater than three, as is actually the case in the solar system, the problem becomes still more hopeless.

Nature, however, has in this matter dealt kindly with us. She has, it is true, proposed a problem which cannot be rigorously solved; but she has introduced into the problem, as proposed in the solar system, certain special features which materially reduce the difficulty. We are still unable to make what a mathematician would describe as a rigorous solution of the question; we cannot solve it with the completeness of a sum in arithmetic; but we can do what is nearly if not quite as useful. We can solve the problem approximately; we can find out what the effect of one planet on the other is *very nearly*, and we can by additional labour reduce the limits of uncertainty to as low a point as may be desired. We thus have really a practical solution of the problem adequate for all the purposes of science. It avails us little to know the place of a planet with absolute mathematical accuracy. If we can calculate the place with so close a degree of approximation to the true position that no telescope could possibly disclose the difference, then every practical end will have been attained. The reason why in this case we are enabled to get round the difficulties which we cannot surmount, lies in the exceptional character of the problem of three bodies as exhibited in the solar system. In the first place the sun is of such pre-eminent mass, that many matters may be overlooked which would be of moment were he rivalled in mass by any of the planets. Another great source of success arises from the small inclinations of the planetary orbits to each other; while the fact that the orbits are nearly circular also greatly facilitates the work. The mathematicians who reside in some of the other parts of the universe are not equally favoured. Among the sidereal systems we find not a few cases where the problem of three bodies, or even of more than three, would have to be faced without any of the alleviating circumstances which our system presents. In such groups as the marvellous quadruple star of Orion, we have four or more bodies comparable in size, which must produce movements of the utmost complexity. Even if terrestrial mathematicians had the hardihood to face such problems, there is no likelihood of their being able to do so for ages to come; such researches must repose on accurate observations as their foundation; and the observations of these distant systems are at present utterly inadequate for the purpose.

The elliptic revolution of a planet around the sun, in conformity with Kepler's law, would assure for that planet permanent conditions of climate. Our earth, for instance, if guided solely by Kepler's laws, would at each day of the year return exactly to the same position which it had on the same day of last year. From age to age the quantity of heat received by our earth would remain constant if the sun himself remained constant, and our climate would thus be preserved indefinitely. But since the existence of planetary perturbation has become recognised, questions arise of the deepest importance with reference to the possible effects which such perturbations may have. We now see that the path of the earth is not absolutely fixed. That path is deranged by Venus and by Mars; it is deranged, it must be deranged, by every planet in our system. It is true that in a year, or even in a century, the amount of derangement produced is not very great; the ellipse which represents very nearly the path of our earth this year, does not differ much from the ellipse which represented the movement of the earth one hundred years ago. But the important question arises as to whether the slight difference which does exist may not be constantly increasing, and may not ultimately assume such proportions as to profoundly modify our climates, or even to render life utterly impossible. Indeed, if we look at the subject without attentive calculation, nothing would seem more probable than that such should be the fate of our system. Our earth revolves in a path inside that of the mighty Jupiter. The earth is constantly attracted by Jupiter, and when it overtakes Jupiter, and comes between Jupiter and the sun, then the two planets are comparatively close together, and the earth is pulled outwards by Jupiter. It might, therefore, be supposed that the tendency of Jupiter would be gradually to draw the earth away from the sun, and thus to cause the earth to describe a path ever growing wider and wider. It is, however, not possible to decide a dynamical question by merely superficial reasoning of this character. The question has to be brought before the tribunal of mathematical analysis, where every element in the case is duly taken into account. Such an inquiry is by no means a simple one. It worthily occupied the splendid talents of Lagrange, and his discoveries in the theory of planetary perturbation are some of the most remarkable contributions ever made to theoretical astronomy.

We cannot here attempt to give even the slightest sketch of the reasoning which Lagrange employed. It can only be expressed by the formulæ of the mathematician, and would then be hardly intelligible without previous years of mathematical study. It fortunately happens, however, that the results to which Lagrange was conducted, and which have been abundantly confirmed by the labours of other mathematicians, admit of being described in simple language.

Let us suppose the case of the sun, and of two planets circulating around him. These two planets are mutually disturbing each other, but the amount of the disturbance is small in comparison with the effect of the sun on each of them. Lagrange demonstrated that, though the ellipse in which each planet moved was gradually altered in some respects by the attraction of the other planet, yet, there is one feature of the ellipse which the perturbation is powerless to alter permanently: the longest axis of the ellipse—that is to say, a line equal to double the mean distance of the planet from the sun—must remain unchanged. This is really a discovery as important as it was unexpected. It at once removes all fear as to the effect which the perturbations can produce on the stability of the system. It shows in particular, so far as the earth is concerned, that, notwithstanding the attractions of Mars and of Venus, of Jupiter and of Saturn, our earth will for ever continue to revolve at the same mean distance from the sun, and thus the succession of the seasons and the length of the year will, so far as this cause at least is concerned, remain for ever unchanged.

But Lagrange went further into the inquiry. He saw that the mean distance did not alter, but it remained to be seen whether the eccentricity of the ellipse described by the earth could not be affected by the perturbations. This is a matter of hardly less consequence than that just referred to. It might be that, while the earth still preserved the same average distance from the sun, yet that the greatest and least distance might change enormously: the earth might pass very close to the sun at one part of its orbit, and then recede to a very great distance at the opposite part. So far as the welfare of this earth and its inhabitants is concerned, this is quite as important as the question of the mean distance; too much heat in one half of the year would be indifferent compensation for too little during the other half. Lagrange submitted this question also to his powerful analysis. Again he vanquished the mathematical difficulties, and again he was able to give assurance of the permanence of our system. It is true that he was not this time able to say that the eccentricity of each path will remain constant; this is not the case. What he does assert, and what he has abundantly proved, is that the eccentricity of each orbit will always remain small. Here, then, we see that the shape of the earth's orbit gradually swells and gradually contracts; the length of the ellipse is invariable, but sometimes it approaches more to a circle, and sometimes becomes more elliptical. These changes are comprised within narrow limits; so that, though they may probably correspond with measurable climatic changes, yet still the safety of the system is not imperilled, as it would be if the eccentricity could increase indefinitely. Once again Lagrange applied the resources of his calculus to study the effect which perturbations can have on the inclination of the path in which the planet moves. The result in this case was similar to that obtained with respect to the eccentricities. If we commence with the assumption that the mutual inclinations of the planets are small, then Lagrange's formulæ tell us that they must always remain small. We are thus finally led to the conclusion that the planetary perturbations are unable to threaten the stability of the solar system.

We shall perhaps appreciate the more fully the importance of these memorable researches, if we consider how easily matters might have been otherwise. Let us suppose a system resembling ours in every respect save one. Let that system have a sun, as ours has; a system of planets and of satellites like ours. Let the masses of all the bodies in this hypothetical system be identical with the masses in our system, and let even the distances and the periodic times be the same in the two cases. Let all the planes of the orbits be similarly placed; and yet this hypothetical system might have the seeds of decay from which ours is exempt. There is one point in the hypothetical system which we have not specified. In our system all the planets revolve in the *same direction* around the sun. Let us, in the hypothetical system, merely suppose this law violated by reversing one planet on its path. That slight change alone will turn the system into one doomed to destruction by the planetary perturbations. Here, then, we find the explanation of that remarkable feature in our system, the uniformity of the directions in which the planets revolve around the sun. Had that not been the case, our system must, in all probability, have perished ages ago, and we should not be here to discuss perturbations or any other subject.

Great as was the success of the eminent French mathematician who made these beautiful discoveries, it was left for this century to witness the crowning triumph of mathematical analysis applied to the law of gravitation. The work of Lagrange lacks the dramatic interest of the discovery made by Le Verrier and Adams, which gave still wider extent to the solar system by the discovery of the planet Neptune revolving far outside Uranus.

We have already alluded to the difficulties which were experienced when it was sought to reconcile the early observations of Uranus with those made since its discovery. We have shown that the path in which Uranus revolved experienced change, and that consequently Uranus was submitted to the action of some other force besides the sun's attraction.

The question, then, arises as to the nature of these disturbing forces. From what we have already learned as to the mutual disturbing influence of one planet upon another, it seems natural to inquire whether the irregularities of Uranus could not be accounted for by the attraction of the other planets. Uranus revolves just outside Saturn. The mass of Saturn is much larger than the mass of Uranus. Could it not be that Saturn draws Uranus aside, and thus causes the changes? This is a question to be decided by the mathematician. He can compute what Saturn is able to do, and he finds, no doubt, that Saturn is capable of producing some displacement of Uranus. In a similar manner Jupiter, with his mighty mass, acts on Uranus, and produces some disturbance which the mathematician calculates. When these calculations had been made for all the known planets they were applied to

Uranus, and we might expect to find that they would fully account for the observed irregularities of his path. This was, however, not the case. After every known source of disturbance had been carefully allowed for, Uranus was still shown to be influenced by some further disturbance; and hence the conclusion was established that Uranus must be affected by some unknown body. What could this unknown body be, and where must it be situated? Analogy was here the guide of those who speculated on this matter. We know no cause of disturbance of a planet's motion except it be the attraction of another planet. Could it then be that Uranus was really attracted by some other planet at that time utterly unknown? This suggestion was made by many astronomers, and it was easy to see some of the conditions which the unknown body should fulfil. Its orbit must, in the first place, lie outside the orbit of Uranus. This seemed obvious, because the unknown planet must be a large and massive one to produce the observed disturbances. If, therefore, it were nearer than Uranus, it would be a conspicuous object, and must have been discovered long ago. Other reasonings were also available to show, without doubt, that if the disturbances of Uranus were caused by the attraction of a planet, that planet must revolve outside Uranus. The general analogies of the planetary system might also be invoked in support of the hypothesis that the path of the unknown planet, though really elliptic, did not differ very widely from a circle, and that the plane in which it moved must also be nearly coincident with the plane of the earth's orbit.

The sole data available for the discovery of the planet were the measured deviations of Uranus at the different points of its orbit. We have, as it were, to fit the orbit of the unknown planet, as well as the mass of that planet itself, in such a way as to account for the various perturbations. Let us, for instance, assume a certain distance for the hypothetical body, and try whether we can assign an orbit and a mass for the planet, with that distance, which shall account for the perturbations. The distance is perhaps too great. We try again with a lesser distance. We can now represent the observations with greater accuracy. A third attempt will give the result closer still, until at length the distance of the unknown planet is determined. In a similar way the weight of the body can be also determined. We assume a weight, and calculate all the perturbations. If the results seem greater than those obtained by observations, then the assumed weight is too great. We amend the assumption, and recompute with a lesser weight, and so on until at length we obtain a mass for the planet which harmonises with the results of actual measurement. The other elements of the unknown orbit—its eccentricity and the position of its axis—are all to be determined in a similar manner. At length it appeared that the perturbations of Uranus could be completely explained if the unknown planet had a certain definite mass, and moved in an orbit which had a certain definite position, while it was also manifest that no other orbit or greatly different mass would explain the observed facts.

These very remarkable computations were undertaken quite independently by two astronomers—one in England and one in France. Each of them attacked, and each of them succeeded in solving, the great problem. The scientific men of England and the scientific men of France joined issue on the question as to the claims of their respective champions to the great discovery; but in the forty years which have elapsed since these memorable researches the question has gradually become settled. It is the impartial verdict of the scientific world outside England and France, that the merits of this splendid triumph of science must be divided equally between this present distinguished Professor J. C. Adams, of Cambridge, and the late U. J. J. Le Verrier, the director of the Paris Observatory.

Shortly after Mr. Adams had taken his degree at Cambridge, in 1843, when he obtained the distinction of Senior Wrangler, he turned his attention to the perturbations of Uranus, and, guided by these perturbations alone, commenced his search for the unknown planet. Long and arduous was the inquiry—demanding an enormous amount of arithmetical calculation, no less than consummate mathematical resource; but gradually Mr. Adams overcame the difficulties. Gradually, as the subject developed, he saw how the perturbations of Uranus could be fully explained by the existence of an exterior planet, and at length he had ascertained, not alone the orbit of this exterior planet, but he was even able to indicate the part of the heavens in which this planet was placed. With his researches in this advanced condition, Mr. Adams, in October of 1845, called on the Astronomer-Royal, Sir George Airy, at Greenwich, and placed in his hands the computations which indicated with marvellous accuracy the place of the unknown planet. It thus appears that seven months before any one else had solved this problem Mr. Adams had effected the solution, and had actually located the planet in a position but little more than a degree distant from the spot which it is now known to have occupied. All that was wanted to complete the achievement, and to gain for Professor Adams and for English science the undivided glory of this discovery, was a strict telescopic search of the heavens in the neighbourhood indicated.

Why, it may be said, was not such an inquiry instituted at once? No doubt it would have been done at once, if the observatories forty years ago had been furnished generally with those elaborate star-charts which are now available. In the absence of such charts (and it does not seem that any English astronomers then possessed them) the search for the planet was a most laborious matter. It had been suggested that it could be detected by its possessing a visible disc; but it must be remembered that even Uranus, so much closer to us, had a disc so small that it was observed nearly a score of times without particular notice, though it did not escape the eagle glance of Herschel. There then remained only one available method of finding Neptune. It was to construct a chart of the heavens in the neighbourhood indicated, and then to compare this chart night after night with the stars in the heavens. Before recommending the commencement of a labour so onerous, the Astronomer-Royal thought it right to submit Mr. Adams's researches to a crucial preliminary test. Mr. Adams had shown how his theory rendered exact account of the perturbations of Uranus in longitude. The Astronomer-Royal asked Mr. Adams whether his theory would give an equally clear explanation of the notable variations in the distance of Uranus. There can be no doubt that his theory would render a satisfactory account of these variations also; but, unfortunately, Mr. Adams seems not to have thought the matter of sufficient importance to give the Astronomer-Royal any speedy reply, and hence it happened that no less than nine months elapsed between the time when Mr. Adams first communicated his results to the Astronomer-Royal, and the time when the telescopic search for the planet was systematically commenced. Up to this time also no account of Mr. Adams's researches had been published. His labours were known to but few besides the Astronomer-Royal and Professor Challis of Cambridge, to whom the duty of making the search was afterwards entrusted.

In the meantime the great French mathematician and astronomer Le Verrier had had his attention specially invited by Arago to the problem of the perturbations of Uranus. With the most exhaustive analysis he investigated every possible known source of disturbance. The influences of the older planets were estimated with the most elaborate detail, but only to confirm the conclusion already arrived at as to their inadequacy to account for the perturbations. Le Verrier then, in complete ignorance of the labours of Adams, commenced the search for the unknown planet by the aid of mathematical investigation. On the 1st of June, 1846, Le Verrier's results appeared; and then the Astronomer-Royal perceived that the results obtained by him coincided practically with those of Adams, insomuch that the places assigned to the unknown planet by the two astronomers were not more than a degree apart! This was, indeed, a most remarkable result. Here was a planet unknown to human sight, yet felt, as it were, by mathematical analysis with a certainty so great that two astronomers, each in absolute ignorance of the other's labours, actually concurred in locating the planet in almost the same spot of the heavens. The existence of the planet was thus raised nearly to a certainty, and it was incumbent on practical astronomers at once to commence the search. In June, 1846, the Astronomer-Royal announced to the visitors at Greenwich Observatory the close coincidence between the calculations of Le Verrier and of Adams, and he urged the importance of the proposal that a search should be at once commenced. Professor Challis, having the command of the great Northumberland equatorial telescope at Cambridge, was induced to undertake the work, and on the 29th of July, 1846, he began his labours.

The plan of search adopted by Professor Challis was an onerous one. He first took the theoretical place of the planet, as given by Mr. Adams, and after allowing a margin for the necessary uncertainties of a calculation so recondite, he marked out a certain region of the heavens, near the ecliptic, in which it might be anticipated that the unknown planet must be found. He then determined to observe all the stars in this region and measure their relative distances. When this work was once done it was to be repeated a second time. His scheme even contemplated a third complete set of measures of all the stars contained within this limited region. There can be no doubt that this process would determine the planet if it should be bright enough to come within the limits of stellar magnitude which Professor Challis adopted. The planet would be detected by its motion relatively to the stars, when the three series of measures came to be compared together. The scheme was organised so thoroughly that it must have led to the discovery of the planet—in fact it afterwards appeared that Professor Challis actually did observe the planet more than once, and the subsequent comparison of its places must infallibly have led to its detection.

In the meantime Le Verrier was maturing his no less elaborate investigations in the same direction. He felt confident of the existence of the planet, and he even went so far as to predict not only the place of the planet but its actual appearance. He thought the planet would be large enough (though still of course only a telescopic object) to be distinguished from the stars by the possession of a disc. These definite predictions strengthened the belief that we were on the verge of another great discovery in the solar system, so much so that when Sir John Herschel addressed the British Association on the 10th of September, 1846, he introduced the following words:—"The past year has given to us the new planet Astræa—it has done more, it has given us the probable prospect of another. We see it as Columbus saw America from the shores of Spain. Its movements have been felt trembling along the far-reaching line of our analysis, with a certainty hardly inferior to ocular demonstration."

The time of the discovery was now rapidly approaching. On the 18th of September, 1846, Le Verrier wrote to the astronomers of the Berlin Observatory describing precisely the place of the planet as indicated by his calculations, and asking their aid in making its telescopic discovery. The request thus preferred was similar to that made on behalf of Adams to Professor Challis. Both at Berlin and at Cambridge the telescopic research was to be made in the same region of the heavens. The Berlin astronomers were, however, most fortunately possessed of an invaluable aid to the research which was not, at the time, in the hands of Professor Challis. We have mentioned how the search for a telescopic planet can be enormously facilitated by the use of a carefully-executed chart of the stars. In fact a mere comparison of the chart with the sky is all that is necessary. It happened that the preparation of a series of star charts had been undertaken by the Berlin Academy of Sciences some years previously. On these charts the place of every star, down even to the tenth magnitude, was faithfully engraved. This work was one of very great utility, but its originators could hardly have anticipated the brilliant discovery which would arise from their herculean labours. It was only possible to accomplish such a vast piece of surveying work by instalments, and accordingly sheet by sheet, as the chart was completed, it issued from the press. It happened that before the news of Le Verrier's labours reached Berlin the chart of that part of the heavens had been engraved and printed, although not actually published.

It was on the 23rd of September that Le Verrier's letter reached the Berlin astronomers. The sky that night was clear, and we can easily understand with what anxiety Dr. Galle directed his telescope to the heavens. The instrument was pointed in accordance with Le Verrier's instructions. The field of view showed, as does every part of the heavens, a multitude of stars. One of these was really the planet. The new chart was unrolled, and, star by star, the heavens were compared with the chart. As the process of identification went on, one object after another was found in the heavens as engraved on the chart, and was of course rejected. At length a star of the eighth magnitude —a brilliant object—was brought into review. The map was examined, but there was no star there. This object could not have been in its present place when the map was formed. The object was therefore a wanderer—a planet. Yet it is necessary to be excessively cautious in such a matter. Many possibilities had to be guarded against. It was, for instance, possible that the object was really a star which, by some mischance, eluded the careful eye of the astronomer who constructed the map. It was even possible that the star might be one of the large class of variables which alternate in brightness, and it might have been conceivable that it was too faint to be seen when the chart was made. Even if neither of these explanations would answer, it was still necessary to show that the object was now moving, and moving with that particular velocity and in that particular direction which the theory of Le Verrier indicated. The lapse of a single day was sufficient to dissipate all doubts. The next night the object was again observed. It had moved, and when its motion was measured it was found to accord precisely with what Le Verrier had foretold. Indeed, as if no circumstance in the confirmation should be wanting, it was ascertained that the diameter of the planet, as measured by the micrometers at Berlin, was practically coincident with that anticipated by Le Verrier.

The world speedily rang with the news of this splendid achievement. Instantly the name of Le Verrier rose to a pinnacle hardly surpassed by that of any astronomer of any age or country. The whole circumstances of the discovery were most dramatic in their character. We picture to ourselves the great astronomer buried in profound meditation for many months; his eyes are bent, not on the stars, but on his calculations and his formulæ. No telescope is in his hand; it is the human intellect which is the instrument he alone uses. With patient labour, guided by the most consummate mathematical artifice, he manipulates his columns of figures. He attempts one solution after another. In each he learns something to avoid; by each he obtains some light to guide him in his future labours. At length he begins to see harmony in those results where before there was but discord. Gradually the clouds disperse, and he sees with a certainty little short of actual vision the planet glittering in the far depths of space. He rises from his desk and invokes the aid of a practical astronomer, and lo! there is the planet in the indicated spot. The annals of science present no such achievement as this. It was the most triumphant proof of the law of universal gravitation. The Newtonian theory had indeed long ere this attained an impregnable position; but, as if to place its truth in the most dazzling and conspicuous light, this discovery of Neptune was accomplished.

For a moment it seemed as if the French nation were to enjoy the undivided honour of this splendid triumph; nor would it, indeed, have been unfitting that the nation which gave birth to Lagrange and to Laplace, and which developed the great Newtonian theory by their immortal labours, should have obtained this distinction. Up to the time of the telescopic discovery of the planet by Dr. Galle at Berlin, no public announcement had been made of the labours of Challis in searching for the planet, or even of the theoretical researches of Adams on which those observations were based. But in the midst of the pæans of triumph with which the enthusiastic French nation hailed the discovery of Le Verrier, there appeared a letter from Sir John Herschel in the *Athenæum* for 3rd October, 1846, in which he announced the researches made by Adams, and claimed for him a participation in the glory of the discovery. Subsequent inquiry has shown that this claim was a just one, and it is now universally admitted by all independent authorities. Yet it will easily be imagined that the French savants, jealous of the fame of their countryman, could not at first be brought to recognise a claim so put forward. They were asked to divide the unparalleled honour between their own illustrious countryman and a young foreigner of whom but few had ever heard, and who had not even published a line of his work or had any claim of his put forward until after the whole work had been completely finished by Le Verrier. The demand made on behalf of Adams was accordingly resented by the French nation; and a somewhat embittered controversy arose on the matter, but point by point the English astronomers succeeded in establishing the claim of their countryman. It was true that Adams had not published his researches to the world, but he had communicated them to the Astronomer-Royal, the official head of astronomy in this country. They were also well known to Professor Challis, the Professor of Astronomy at Cambridge. Then, too, when the work of Adams was examined, it was found to be quite as thorough and quite as successful as that of Le Verrier. It was also found that the method of search adopted by Professor Challis not only must have been eventually successful, but that it actually was in a sense already successful. When the telescopic discovery of the planet had been achieved, Challis turned naturally to see whether he had observed the planet or not. It was on the 1st October that he heard of the success of Dr. Galle, and by that time Challis had accumulated observations in connection with this research of no fewer than 3,150 stars. Among them he speedily found that an object observed on the 12th of August was not in the same place on the 30th of July. This was really the planet; and its discovery would thus have been assured when Challis had had time to compare his measurements. In fact, if he had only discussed his observations at once, there cannot be much doubt that the entire glory of the discovery would have been awarded to Adams. He would then have been first, no less in the theoretical calculations than in the optical discovery of the planet. It may also be remarked that in another way Challis very narrowly missed making the telescopic discovery of Neptune. In his paper, Le Verrier had pointed out the possibility of detecting the planet by its disc. Challis made the attempt, and before the intelligence of the actual discovery at Berlin had reached him, he had made an examination of the region indicated by Le Verrier. About 300 stars passed through the field of view, and among them he selected one on account of its disc; it afterwards appeared that this was indeed the planet.

Even if the researches of Le Verrier and of Adams had never been undertaken it is certain that the distant Neptune must be some time discovered; yet that discovery might have been made in a manner which every true lover of science would now deplore. We hear constantly of the discovery of minor planets, yet no one attaches to such achievements a fraction of the consequence belonging to the discovery of Neptune. The danger to be feared by delay was, that Neptune should have been discovered by simple survey work, just as Uranus was discovered, or just as the hosts of minor planets are now found. In this case the science of Theoretical Astronomy, the great science founded by Newton, and raised to a marvellously interesting and beautiful system by the labours of Lagrange and Laplace, would have been deprived of its most brilliant illustration.

Neptune had, in fact, a very narrow escape on at least one previous occasion of being discovered in a very simple way. This was shown when sufficient observations had been obtained to enable the path of the planet to be calculated. It was then possible to trace back the movements of the planet among the stars and thus to institute a search in the catalogues of earlier astronomers to see whether they contained any record of Neptune. It was soon found that the place of the planet on May 10th, 1795, must have coincided with the place of a star recorded on that day in the "Histoire Céleste" of Lalande. It further appeared by actual examination of the heavens that there was no star in the place indicated by Lalande, so the fact that this was really an observation of Neptune was placed quite beyond doubt. When reference was made to the original observations of Lalande a matter of the very greatest interest was brought to light. It was there found that he had observed the same star (for so he regarded it) both on May 8th and on May 10th; on each day he had determined its position, and both are duly recorded. But when he came to prepare his catalogue and found that the places on the two occasions were different, he discarded the earlier result, and merely printed the later. Had Lalande but had the courage to believe implicitly in his own observations, an immortal discovery lay in his grasp; had he manfully said, "I was right on the 10th of May and I was right on the 8th of May; I made no mistake on either occasion, and the object I saw on the 8th must have moved between that and the 10th," then he must without fail have found Neptune. But had he done so, how great would have been the loss to science! The discovery of Neptune would then merely have been an accidental reward to a laborious worker, instead of being one of the most glorious achievements in the loftiest department of human reason.

With the conclusion of the brief sketch here given of the discovery of Neptune, we nearly conclude all that can be narrated of this planet. If we fail to see in Uranus any of those features which make Mars or Venus, Jupiter or Saturn such attractive telescopic objects, what can we expect to find in Neptune, which is half as far again as Uranus? With a good telescope and a suitable magnifying power we can see that Neptune has a disc, but no features on that disc can be identified. We are consequently not in a position to ascertain the period in which Neptune rotates around its axis, though from the general analogy of the system we must feel assured that it really does rotate. More successful have been the attempts to measure the diameter of Neptune, which is found to be nearly 35,000 miles, or more than four times the diameter of the earth. It would also seem that, like Jupiter and like Saturn, the planet must be enveloped with a vast cloud laden atmosphere, for the density of the globe is only about one-fifth that of the earth. This great globe revolves around the sun at a mean distance of no less than 2,780 millions of miles, being about thirty times as great as the mean distance from the earth to the sun. The journey, though accomplished at a rate of more than three miles a second, is yet so long that Neptune requires almost 155 years to complete one revolution. Since its discovery, some forty years ago, Neptune has moved only through about one quarter of its path, and even since the time it was first casually seen by Lalande, in 1795, it has only had time to traverse half of its mighty circuit.

Neptune, like our earth, is attended by a single satellite; this delicate object was discovered by Mr. Lassell with his colossal reflecting telescope shortly after the planet itself was known. The satellite performs its journey around Neptune in a period of but little less than six days. By observing the motions of the satellite we are enabled to determine the mass of the planet, and thus it appears that Neptune is about one twenty-thousandth part of the weight of the sun.

Our review of the planetary system closes when we have arrived at Neptune; whether any planets revolve around the sun in orbits beyond Neptune it is impossible to say. All we can assert is that no such planets have been seen, nor is there at present any good ground for believing in their existence. The negative evidence on the question is indeed very strong. We have in our chapter on the minor planets entered into a full discussion of the way in which these objects are discovered. It is by minute and diligent comparison of the heavens with elaborate star charts that these planets are found. But these inquiries would be equally efficacious in searching for a trans-Neptunian planet; in fact, we could design no better method to hunt for a trans-Neptunian planet than that which is at this moment in constant practice at many observatories. The labours of those who search for small planets have been abundantly rewarded by discoveries now counted by hundreds. Yet it is a most noteworthy fact that all these planets are limited to one region of the solar system. It is approximately true to say that all their orbits are included between those of Jupiter and of Mars. In one or two cases the orbits just reach beyond the path of Jupiter on the one hand, or inside the path of Mars on the other; but never has a planet been found in this way which goes out even as far as Saturn, and, of course, still less has any trans-Neptunian object been found. It has sometimes been conjectured that time may disclose perturbations in the orbit of Neptune, and that these perturbations may lead to the discovery of a planet still further, even though that planet be so remote and so faint that it actually eludes telescopic research. At present, however, and for generations to come, such an inquiry will hardly be within the range of practical astronomy. Neptune has not yet done more than travel through a quarter of its path since it was found by Le Verrier and Adams. Its movements since then have no doubt been studied minutely, but it must at least describe one whole revolution before it would be feasible to construct from any perturbations of its path the orbit of an unknown and still more remote planet.

We have thus seen the planetary system to be bounded on one side by Mercury and on the other by Neptune. The discovery of Mercury was in itself a brilliant achievement of prehistoric times. The early astronomer who accomplished that feat, when devoid of instrumental assistance and unsupported by accurate theoretical knowledge, merits our hearty admiration for his untutored acuteness and penetration. On the other hand, the discovery of the exterior boundary of the planetary system is worthy of special attention from the fact that it was founded on profound theoretical learning, and verified by consummate instrumental and practical skill.

Though we here close our account of the planets and their satellites, we have still two chapters to add before we shall have completed what is to be said with regard to the solar system. A further and notable class of bodies, neither planets nor satellites, own allegiance to the sun, and revolve around him in conformity to the laws of universal gravitation. These bodies are the comets, and their somewhat more humble associates, the shooting stars. We find, in the study of these objects, many matters of interest, to which we shall proceed in the ensuing chapters.

CHAPTER XVI.

COMETS.

Comets Contrasted with Planets in Nature as well as in their Movements—Coggia's Comet—Periodic Returns—The Law of Gravitation—Parabolic and Elliptic Orbits—Theory in Advance of Observations—Most Cometary Orbits are sensibly Parabolic—The Labours of Halley—The Comet of 1682—Halley's Memorable Prediction—The Retardation Produced by Disturbance—Successive Returns of Halley's Comet—Encke's Comet—Effect of Perturbations—Orbit of Encke's Comet—Attraction of Mercury and of Jupiter—How the Identity of the Comet is secured—How to Weigh Mercury—Distance from the Earth to the Sun found by Encke's Comet—The Disturbing Medium—The Comets of 1843 and 1858—Passage of a Comet between the Earth and the Stars—Comets not composed of Gas of appreciable Density—Can the Comet be Weighed?—Evidence of the Small Mass of the Comet derived from the Theory of Perturbation—The Tail of the Comet—Its Changes—Views as to its Nature—Carbon present in Comets.

In our previous chapters which treated of the sun and the moon, the planets and their satellites, we found in all cases that the celestial bodies with which we were concerned were nearly globular in form, and many are undoubtedly solid bodies. All these objects possess a density which, even if in some cases it be much less than that of the earth, is still hundreds of times greater than the density of merely gaseous substances. We now, however, approach the consideration of a class of objects of a totally different character. We have no longer to deal with globular objects possessing a considerable mass. Comets are of the most irregular shapes, they are in large part, at all events, formed of materials in the utmost state of tenuity, and their masses are so small that no means we possess have enabled them to be measured. Not only are comets different in constitution from planets or from the other more solid bodies of our system, but the movements of comets are quite distinct from the orderly return of the planets at their appointed seasons. The comets appear sometimes with almost startling unexpectedness; they

PLATE XII.

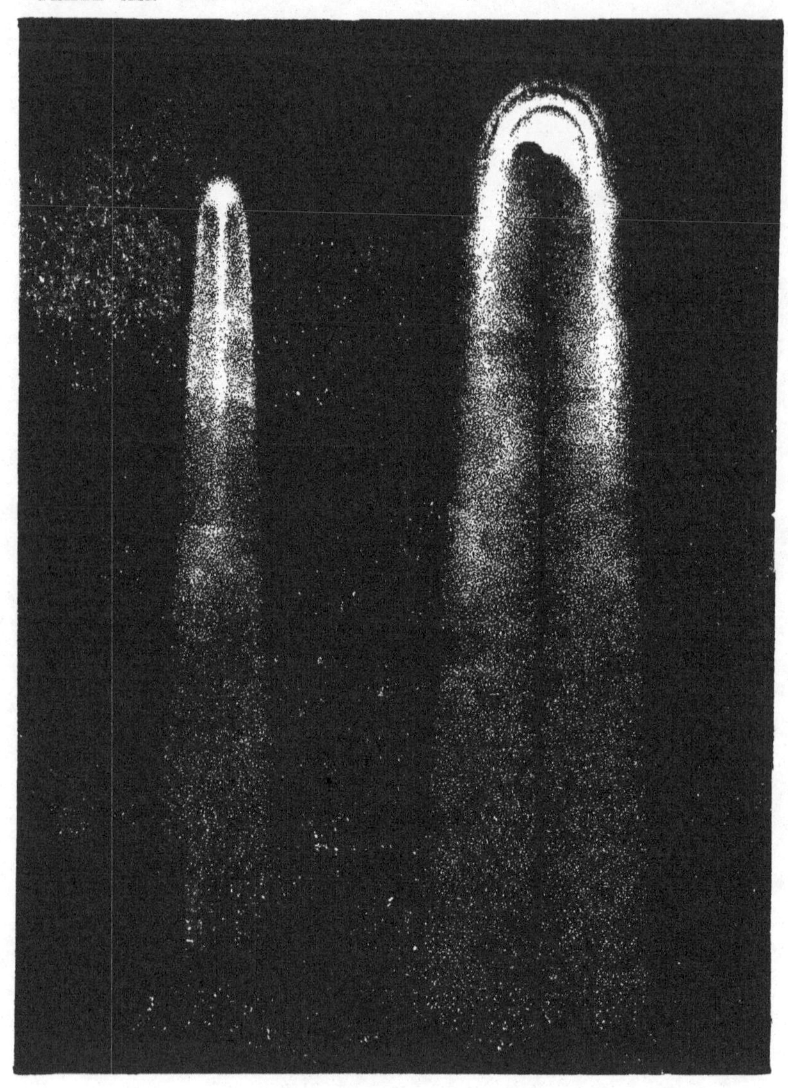

COGGIA'S COMET.

(AS SEEN ON JUNE 10TH AND JULY 9TH, 1874.

rapidly swell in size to an extent that in superstitious ages called forth the utmost terror; again, they disappear, often never again to return. Modern science has, no doubt, removed a great deal of the mystery which once invested the whole subject of comets. Their movements are now to a large extent explained, and some additions have been made to our knowledge of their nature, though we must still confess that what we do know, bears but a very small proportion to what remains unknown.

Let us first describe in general terms the nature of a comet, in so far as its structure is disclosed by the aid of a powerful refracting telescope. We represent in Plate XII. two interesting sketches made at Harvard College of the great comet of 1874, distinguished by the name of its discoverer Coggia.

We see here the head of the comet, containing as its brightest spot what is called the nucleus, and in which the material of the comet seems to be much denser than elsewhere. Surrounding the nucleus we find certain definite layers of luminous material, from which the tail seems to stream away. This view may be regarded as that of a typical comet, but the varieties of structure presented by different comets are almost innumerable. In some cases we find the nucleus absent; in other cases we find the tail absent. The tail is, no doubt, a conspicuous feature in those great comets which receive universal attention; but in the small telescopic comets, of which a few are generally found every year, the absence of the tail is quite an ordinary character. Not only do comets present great variety in appearance, but even the aspect of a single comet undergoes great change. The comet will sometimes increase enormously in bulk; sometimes it will diminish, sometimes it will have a large tail, or sometimes no tail at all. Measurements of a comet's size are almost futile; they may cease to be true even during the few hours in which a comet is observed in the course of a night. It is, in fact, impossible to identify a comet by any description of its personal appearance. Yet the question as to the personal identity of a comet is often of very great consequence. We must provide means by which that identity can be established, entirely apart from what the comet may be like.

It is now well known that several comets make periodic returns. After being invisible for a certain number of years such a comet comes again into view, and again retreats into space to perform another revolution. The question then arises as to how are we to recognise the comet when it comes back? The personal features of its size or brightness, the presence or absence of a tail, large or small, are fleeting characters of no value for such a purpose. Fortunately, however, the law of elliptic motion established by Kepler has suggested the means of defining the identity of a comet with absolute precision.

After Newton had made his immortal discovery of the law of gravitation, and after he had succeeded in demonstrating that the elliptic paths of the planets around the sun were a necessary consequence of that law, he was naturally tempted to apply the same reasoning to explain the movements of comets. Here again he met with marvellous success, and illustrated his theory by completely explaining the movements of the remarkable comet which appeared in the year 1680.

There is a certain very beautiful curve known to geometricians by the name of the parabola. Its form is shown in the adjoining figure; it is a curved line which bends in towards and around a certain point known as the focus. This would not be the occasion for any allusion to the geometrical properties of this curve; these are fully discussed in works on mathematics. It will here only be necessary to point to the connection which exists between the parabola and the ellipse. We have in a former chapter explained the construction of the ellipse, and we have shown how it possesses two foci. Let us suppose that a series of ellipses be drawn, each of which has a greater distance between its foci than the preceding one. Imagine the process carried on until at length the distance between the foci became enormously great in comparison with the distance from each focus to the curve, then each end of this long ellipse will have practically the same form as a parabola. We may thus look on the parabola represented in Fig. 55 as being one end of an ellipse whereof the other end is at an indefinitely great distance. Newton showed that the law of gravitation would permit a body to move in an ellipse of this very extreme type no less than in one of the more ordinary proportions. If an object revolve in a parabolic orbit about the sun at the focus, then the object moves in gradually towards the sun, sweeps around the sun, and then begins to retreat; but there is one great and necessary distinction between parabolic and elliptic motion. In the latter case

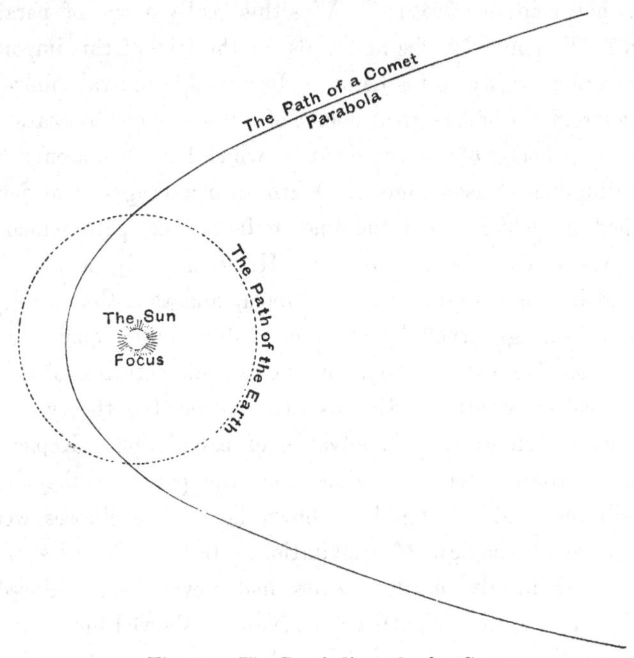

Fig. 55.—The Parabolic path of a Comet.

the body, after its retreat to a certain distance, will turn round and again draw in towards the sun; in fact, it will make periodic returns, as was found to be the case with the planets. But in the case of the true parabola the body can never return; to do so it would have to double the distant focus, and as that is infinitely remote, it could not be reached except in the lapse of infinite time.

The characteristic feature of the movement in a parabola is thus enunciated. The body draws in gradually towards the focus from an indefinitely remote distance on one side, and after passing round the focus gradually recedes to an indefinitely remote distance on the other side, never again to return. When Newton had perceived that parabolic motion of this type could arise from the law of gravitation, it at once occurred to him that by its means the movements of a comet might be explained. He knew that comets must be attracted by the sun; he saw that the usual course of a comet was to appear suddenly, to sweep around the sun and then retreat, never again to return. Was this really a case of parabolic motion? Fortunately, the materials for the trial of this important theory were all ready to his hand. He was able to avail himself of the measurements of the great comet of 1680, and of observations of several other bodies of the same nature which had been accumulated by the diligence of astronomers. With his usual sagacity he devised a method by which, from the known facts, the path which the comet pursues could be determined. He accordingly calculated the path and he found that it was a parabola, and that the velocity of the comet was governed by the law of describing equal areas in equal times. Here was another convincing demonstration of the law of universal gravitation. In this case, indeed, the theory may be said to have been actually in advance of calculation. Kepler had determined from observation alone that the paths of the planets were ellipses, and Newton had shown how these ellipses were a consequence of the law of gravitation. But in the case of the comets their highly erratic orbits had never been reduced to geometrical form until the theory of Newton showed him that they were parabolic, and then he invoked the observations to verify the anticipations of his theory.

The great majority of comets move in orbits which cannot be sensibly discriminated from parabolæ, and any comet whose orbit is of this character can only be seen at a single apparition. The theory of gravitation, though it admits the parabola as a possible orbit for a comet, does not assert that the orbit must be necessarily a parabola. We have pointed out that this curve is really only a very extreme type of ellipse, and it would still be in perfect accordance with the law of gravitation for a comet to pursue a path of any elliptical form, provided only that the sun was placed at the focus, and that the comet obeyed the rule of describing equal areas in equal times. If a comet move in an elliptic path, then it will return to the sun again, and consequently we shall have regular periodical visits fom the same object.

An interesting field of inquiry was here presented to the astronomer. Nor was it long before the discovery of a periodic comet was made which illustrated, in a striking manner, the truth of the laws of universal gravitation. The name of the celebrated astronomer Halley is, perhaps, best known from its association with the great comet, whose periodicity was discovered by his calculations. When Halley learned from the Newtonian theory the possibility that a comet might move in an elliptic orbit, he undertook a most laborious investigation; he collected from various records of observed comets all the reliable particulars that could be obtained, and thus he was enabled to learn, with tolerable accuracy, the nature of the paths pursued by about twenty-four large comets. One of these was the great comet of 1682, which Halley himself observed, and whose path he computed in accordance with the principles of Newton. He then proceeded to the inquiry whether this comet of 1682 could possibly have visited our system at any previous period. To answer this question he turned to the list of recorded comets which he had so carefully elaborated, and he found that his comet very closely resembled, both in appearance and in orbit, a comet observed in 1607, and also another observed in 1531. Could these three bodies be identical? It was only necessary to suppose that a comet, instead of revolving in a parabolic orbit, really revolved in an extremely elongated ellipse, and that it completed each revolution in a period of about seventy-five or seventy-six years. He submitted this hypothesis to every test that he could devise; he found that the orbits, determined on each of the three occasions, were so nearly identical that it would be beyond the wildest probability that the coincidence should be accidental. Accordingly, he decided to submit his theory to the most supreme test known to astronomy. He ventured to make a prediction which posterity would have the opportunity of verifying. If the period of the comet were seventy-five or seventy-six years, as the former observations seemed to show, then Halley estimated that it, if unmolested, ought to return in 1757 or 1758. There were, however, certain sources of disturbance which Halley pointed out, and which would be quite powerful enough to affect materially the time of the return. The comet in its journey passes near the path of Jupiter, and experiences great perturbations from that mighty planet. Halley concluded that the return of the comet might be delayed till the end of 1758 or the beginning of 1759.

This prediction was a memorable event in the history of astronomy, inasmuch as it was the first attempt to foretell the apparition of one of those mysterious bodies whose visits seemed guided by no fixed law, and which were usually regarded as omens of awful import. Halley felt the importance of his announcement. He knew that his earthly course would have run long before the comet had completed its revolution; and, in language almost touching, the great astronomer writes: "Wherefore if it should return according to our prediction about the year 1758, impartial posterity will not refuse to acknowledge that this was first discovered by an Englishman."

As the time drew nigh when this great event was expected, it awakened the liveliest interest among astronomers. The distinguished mathematician Clairaut undertook to compute anew, and by the aid of improved methods, the effect which would be wrought on the comet by the attraction of the planets. His analysis of the perturbations was sufficient to show that the object would be kept back for 100 days by Saturn, and for 518 days by Jupiter. He therefore gave some additional exactness to the prediction of Halley, and finally concluded that this comet would reach the point of its path nearest to the sun about the middle of April, 1759. The sagacious astronomer (who, we must remember, lived long before the discovery of Uranus and of Neptune) further adds that as this comet retreats so far, it may possibly be subject to influences of which we do not know, or to the disturbance even of some planet too remote to be ever perceived. He, accordingly, qualifies his prediction with the statement that, owing to these unknown possibilities, his calculations may be a month wrong one way or the other. Clairaut made this memorable communication to the Academy of Sciences on the 14th of November, 1758. The attention of astronomers was immediately quickened to see whether the visitor, who last appeared seventy-six years previously, was about to return. Night after night the heavens were scanned. On Christmas Day in 1758 the comet was first detected, and it passed closest to the sun about midnight on the 12th of March, just a month earlier than the time announced by Clairaut, but still within the limits of error which he had assigned as being possible.

The verification of this prediction was a further confirmation of the theory of gravitation. Since then, Halley's comet has returned once again, in 1835, under circumstances somewhat similar to those just narrated. Further historical research has also succeeded in identifying Halley's comet with numerous memorable apparitions of comets in former times. It has even been shown that a splendid comet, which appeared eleven years before the commencement of the Christian era, was merely that of Halley in one of its former returns. Among the most celebrated visits of this object was that of 1066, when the apparition attracted universal attention. A picture of the comet on this occasion forms a quaint feature in the Bayeux Tapestry. The next return of Halley's comet is expected about the year 1910.

There are now several comets known which revolve in elliptic paths, and are, accordingly, entitled to be termed periodic. These objects are chiefly telescopic, and are thus in strong contrast to the splendid comet of Halley. Most of the other periodic comets have also periods much shorter than that of Halley. Of these objects, by far the most celebrated is that known as Encke's comet, which merits our careful attention.

The object to which we refer has had a striking career, during which it has suffered much at the hands of the law of gravitation. We are not here concerned with the prosaic routine of a mere planetary orbit. A planet is mainly subordinated to the compelling sway of the sun's gravitation. It is also to some slight extent affected by the attractions which it experiences from the other planets. Mathematicians have long been accustomed to anticipate the movements of the planets by actual calculation. They know how the place of the planet is approximately decided by the sun's attraction, and they can discriminate the different adjustments which that place is to receive, in consequence of the disturbances produced by the other planets. The capabilities of the planets for producing disturbance are greatly increased when the disturbed body follows the eccentric path of a comet. It is frequently found that the path of a comet comes very near to the track of a planet, so that the comet may actually sweep by the very planet itself, even if the two bodies do not actually run into collision. On such an occasion the disturbing effect of the planet is enormously augmented, and we therefore turn to the comets when we desire to illustrate the great theory of planetary perturbations by some striking example.

Having decided to choose a comet, the next question is, *What* comet? There cannot here be much room for hesitation. Those splendid comets which appear so capriciously may be at once excluded. They are visitors apparently coming for the first time, and retreating without any distinct promise that mankind shall ever see them again. A comet of this kind moves in a parabolic path, sweeps once around the sun, and then retreats into the space whence it came. We cannot completely study the effect of perturbations on a comet until it has been watched during successive returns to the sun. Our choice is thus limited to the comparatively small class of objects known as periodic comets; and, from a survey of the entire group, we can select the most suitable to our purpose. One comet, and only one, fulfils all the required conditions. It is the celebrated object best known as Encke's comet, for, though Encke was not the discoverer, yet it is to his calculations that the comet owes its fame. This body is rendered more suitable for our purpose by the memorable researches to which it has recently given rise.

In the year 1818 a comet was discovered by the painstaking astronomer Pons at Marseilles. We are not to suppose that this comet was a splendid spectacle. It was a small telescopic object, not unlike one of those dim nebulæ which are scattered in thousands over the heavens. The comet is, however, readily distinguished from a nebula by its rapid movement relatively to the stars, while the nebula remains at rest. The position of this comet was ascertained by its discoverer, as well as by other astronomers. Encke found from the observations that the comet returned to the sun once in every three years and a few months. This was a startling discovery. At that time no other comet of short period had been detected, so that this new addition to the solar system awakened the liveliest interest. The question was immediately raised as to whether this comet, as it revolved so frequently, may not have been observed during previous returns. The historical records of the apparitions of comets are counted by hundreds, and how among this host are we to select those objects which were identical with the comet discovered by Pons?

We may at once relinquish any hope of identification by drawings of the comet itself, but fortunately, there is one feature of a comet on which we can seize, and which no fluctuations of the actual structure can modify or disguise. The path in which a comet travels through space is independent of the bodily changes of the comet. The shape of that path and its position depend entirely upon those other bodies of the solar system which are specially involved in the theory of Encke's comet. In Fig. 56 we show the orbits of three of the planets. These orbits have been chosen with such proportions as shall make the innermost represent the orbit of Mercury; the next is the orbit of the earth, while the outermost is the orbit of Jupiter. Besides these three, we also perceive a much more elliptical path, representing the orbit of Encke's comet, projected down on the plane of the earth's motion. The sun is situated at the focus of the ellipse The comet is constrained to revolve in this curve by the attraction of the sun, and it requires a little more than three years to accomplish a complete revolution. It passes close to the sun at perihelion, at a point inside the path of Mercury, while at its greatest distance the path approaches to the path of Jupiter. This elliptic path is mainly determined by the attraction of the sun. Whether the comet weighed an ounce, a ton, a thousand tons, or a million tons, whether it was a few miles, or many thousands of miles in diameter, the orbit would still be the same. It is by the shape of this ellipse, by its actual size, and by the position in which it lies that we identify the comet.

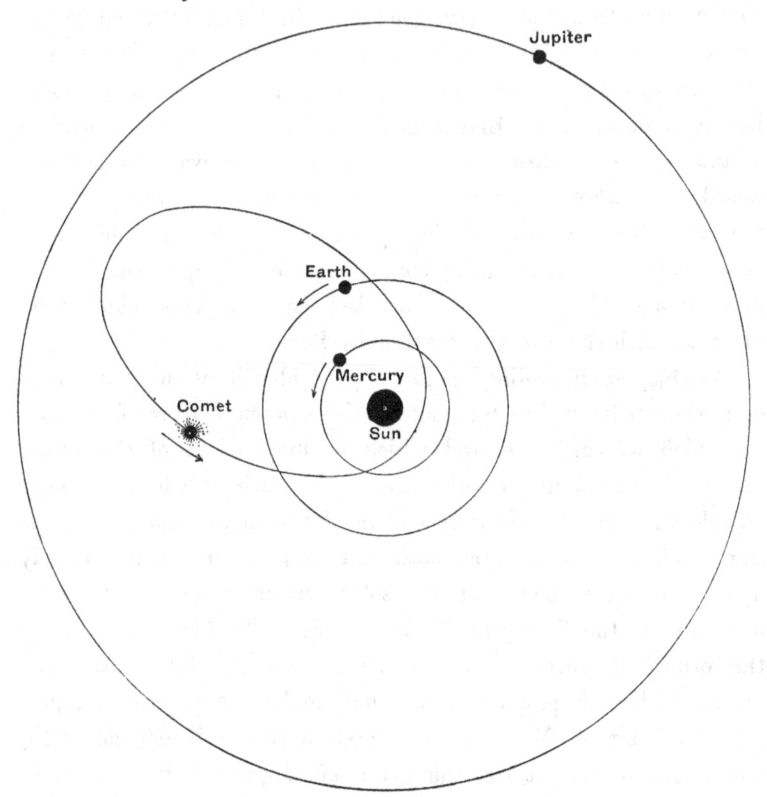

Fig. 56.—The Orbit of Encke's Comet.

Encke's comet is usually invisible, and even the most powerful telescope in the world would not show a trace of it. After one of its periodical visits, the comet withdraws until it passes to the outermost part of its path, then it will turn, and again approach the sun. It would seem that it becomes invigorated by the sun's rays, and commences to dilate under their genial influence. While moving in this part of its path the comet lessens its distance from the earth. It daily increases in splendour, until at length, partly by the intrinsic increase in brightness, and partly by the decrease in distance from the earth, it comes within the range of our telescopes. We can generally anticipate when this will occur, and we can tell to what point of the heavens the telescope is to be pointed which will discern the comet at its next return to perihelion. The comet has eluded the grasp of the artist, but it cannot elude the grasp of the mathematician. The mathematician will tell when and where the comet is to be found, but no one can say what it will be like.

Once recognised as a regular member of our system, Encke's comet becomes a teacher of marvellous power. At one time its periodical voyage brings it close to the sun and close to the path of the planet Mercury. At another time the comet approaches the orbit of Jupiter. Again, the wanderer comes within reach of our telescopes, and we are able to glean many interesting facts from its travels. These movements contain impressive evidence as to the way in which the celestial phenomena are founded on the law of gravitation.

Were all the other bodies of the system removed, then the path of Encke's comet must be for ever performed in the same ellipse and with absolute regularity. The chief interest for our present purpose lies, however, not in the regularity of its path, but in the *irregularities* introduced into that path by the presence of the other bodies of the solar system. Let us, for instance, follow the progress of the comet through its perihelion passage, in which the track lies near that of the planet Mercury. It will usually happen that Mercury is situated in a distant part of its path at the moment the comet is passing, and the influence of the planet will then be comparatively small. It may, however, sometimes happen that the planet and the comet come close together. One of the most memorable instances of a close approach to Mercury took place on the 22nd November, 1848. On that day the comet and the planet were only separated by an interval of about one-thirtieth of the earth's distance from the sun, *i.e.*, about 3,000,000 miles. On two other occasions, viz., 1835, August 23rd, and 1858, October 25th, the distance between Encke's comet and Mercury has been less than 10,000,000 miles—an amount of trifling import in comparison with the dimensions of our system. Approaches so close as this are fraught with serious consequences to the movements of the comet. Mercury, though a small body, is still, as its name would suggest, a massive body. It always attracts the comet, but the efficacy of that attraction is enormously enhanced when the comet in its wanderings comes near the planet. The effect of this attraction is to force the comet to swerve from its path, and to impress certain changes upon its velocity. As the comet recedes, the disturbing influence of Mercury rapidly abates; and ere long becomes insensible. But time cannot efface from the orbit of the comet the effect which the disturbance of Mercury has actually accomplished. The disturbed orbit is a different figure to the undisturbed ellipse which the comet would have occupied had the influence of the sun alone determined its shape. We are able to calculate the movements of the comet as determined by the sun. We could also calculate the effects arising from the disturbance produced by Mercury, provided we know its mass.

Mercury, though one of the smallest of the planets, is perhaps the most troublesome to the astronomer. It lies so close to the sun, that it is seen but seldom in comparison with the other great planets. Its orbit is very eccentric, and experiences disturbances by the attraction of other bodies in a way not yet fully understood. A special difficulty has also been found in the attempt to place Mercury in the weighing scales. We can weigh the whole earth, we can weigh the sun, the moon, and even Jupiter and other planets, but Mercury presents difficulties of a peculiar character. Le Verrier, however, succeeded in devising a method of weighing Mercury. He demonstrated that our earth is attracted by this planet, and he showed how the amount of attraction may be disclosed by observations of the sun, so that by examination of the observations, he made an approximate determination of the mass of Mercury. Le Verrier's result indicated that the weight of the planet was about the fifteenth part of the weight of the earth. In other words, if our earth was placed in a mighty scale, and fifteen globes, each equal to Mercury, were laid in the other, the scales would be evenly balanced. It was necessary that this result should be received with great caution. It depended upon a very delicate interpretation of somewhat precarious measurements. It could only be regarded as a provisional value, to be discarded when a better one should be obtained.

The approach of Encke's comet to Mercury, and the elaborate investigations of Von Asten, by which the observations of the comet have been discussed, have thrown much light on the subject. Von Asten assumed, in the first place, that Mercury was really endowed with the mass which Le Verrier had attributed to it; and he was able to calculate where the comet should be, assuming this determination to be correct. But the comet was not found in the place indicated by the calculations; it was fully an arc of one degree distant. The calculations must, therefore, have some erroneous element. There was only one element which could be erroneous: it was the mass of Mercury. Von Asten then tried a different value for the mass of the planet. He repeated the calculations, and the calculated place became more nearly coincident with the observed place. A third attempt, and this time a final one, showed that by assuming a suitable mass for Mercury the calculated places of the comet can be reconciled with the observed places. Von Asten has thus demonstrated that the weight of Mercury is much less than Le Verrier had supposed. If our earth were divided into twenty-five parts of equal mass, each one of these parts would weigh as much as the planet. No different value of the mass of Mercury can be made consistent with the information obtained from the voyage of Encke's comet.

Do we doubt the accuracy of this delicate weighing-machine? Look, then, at the orbit of Jupiter, to which the comet approaches so nearly when it retreats from the sun. It will sometimes happen that Jupiter and the comet are in close proximity, and then the mighty planet seriously disturbs the pliable orbit of the comet. The path of the latter bears indelible traces of the Jupiter perturbations, as well as of the Mercury perturbations. It might seem a hopeless task to discriminate between the influences of the two planets, overshadowed as they both are by the supreme control of the sun, but the contrivances of mathematical analysis are adequate to deal with the problem. They point out how much is due to Mercury, how much is due to Jupiter; and so the wanderings of Encke's comet can be made to disclose the mass of Jupiter as well as that of Mercury. Here, then, we have a means of testing the precision of our scales. The mass of Jupiter can be accurately measured by his satellites, as we have already mentioned in a previous chapter. As the satellites revolve round and round the planet, they afford the means of measuring his weight by the rapidity of their motion. They tell us that if the sun were placed in one scale of the celestial balance, it would take 1048 bodies equal to Jupiter in the other to weigh him down. Hardly a trace of uncertainty clings to this determination, and it is therefore of great interest to test the theory of Encke's comet by seeing whether it gives an accordant result. The comparison has been made by Von Asten. Encke's comet tells us that the sun is 1050 times as heavy as Jupiter; so the results are practically identical, and the accuracy of the indications of the comet are confirmed.

Encke's comet, by its long wanderings and by its frequent visits, is, indeed, pregnant with information on various other points relating to the solar system. A score of times that comet has returned, and we can only render these returns consistent with each other by making appropriate estimations of the various quantities involved. It is most instructive to see how, in Von Asten's hands, the most recondite information is wrung from the comet. The distance of the earth from the sun is a very important matter to Encke's comet. Our earth contributes its quota of disturbance to the many afflictions of the suffering body. The capacity of our earth to derange the comet depends in a measure upon the distance from the earth to the sun. The investigation is of no little complexity, but still it is easy to see that the earth's distance from the sun must be among the numerous elements of the question. Unless the right distance be chosen in the calculations, the observed place will not correspond with the calculated place. This consideration is sufficient for the mathematician. He can actually disentangle the earth's distance from the other quantities involved. It is somewhat remarkable that Von Asten has found by his calculations a value of the sun's distance *very appreciably less* than that obtained by other methods. Each of these results has good claims to accuracy, and it will be a problem of interest for astronomers to reconcile the discrepancy.

We have hitherto discussed the adventures of Encke's comet in cases where they throw light on questions otherwise more or less known to us. We now approach a celebrated problem, on which Encke's comet is, if not our sole, at all events our principal authority. Every 1,210 days that comet revolves completely around its orbit, and returns again to the neighbourhood of the sun. The return of the comet is, however, open to certain irregularities. We have already explained how irregularities arise from Mercury and from Jupiter. Additional irregularities also arise from the attraction of the earth and of the other remaining planets; but all these can be allowed for, and then we are entitled to expect, if the law of gravitation be universally true, that the comet shall obey the calculations of mathematics. Encke's comet has not justified this anticipation; at each revolution the period is getting steadily shorter! Each time the comet comes back to perihelion in two and a half hours less than on the former occasion. Two and a half hours is, no doubt, a small period of time in comparison with that of an entire revolution; but in the region of its path visible to us the comet is moving so quickly that its motion in two and a half hours is very considerable. This irregularity cannot be overlooked, inasmuch as it has been confirmed by the returns during about twenty revolutions. It has sometimes been thought that the discrepancies might be attributed to some planetary perturbations omitted or not fully accounted for. The masterly analysis of Von Asten has, however, disposed of this explanation. He has minutely studied all the observations down to 1875, but only to confirm the reality of this regular retardation in the movements of Encke's comet.

An explanation of these irregularities was suggested by Encke long ago, and subsequent researches have tended to substantiate his view. Let us briefly attempt to describe this memorable hypothesis. When we say that a body will move in an elliptic path around the sun in virtue of gravitation, it is always assumed that the body has a free course through space. It is assumed that there is no friction, no air, or other source of disturbance. But suppose that this assumption should be incorrect; suppose that there really is some medium pervading space which offers resistance to the comet, in the same way as our air offers resistance to the passage of a rifle bullet, what effect ought such a medium to produce?

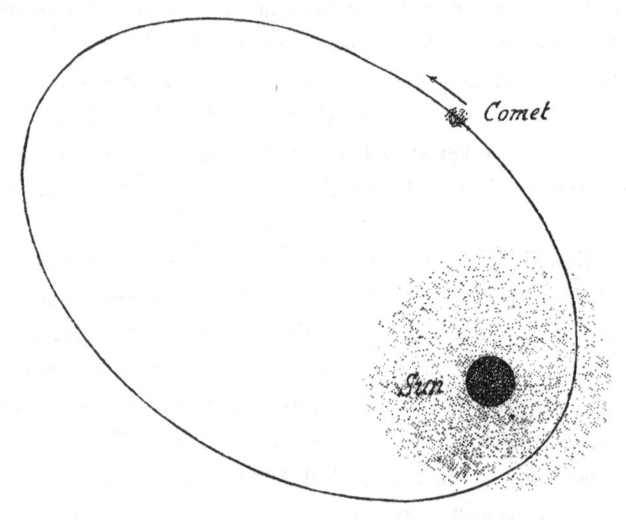

Fig. 57.—The Resisting Medium around the Sun.

The drawing in Fig. 57 represents the theory which Encke put forward, and in support of which we have to add the more recent researches of Von Asten. Even if the greater part of space be utterly void, so that the path of the filmy and almost spiritual comet is incapable of feeling resistance, yet in the neighbourhood of the sun it would seem that there must be some medium of excessive tenuity capable of affecting the comet. It can be demonstrated that the effect of a resisting medium such as we have supposed is to diminish the size of the comet's path, and thus to diminish the periodic time. The existence of the solar corona and the strange phenomenon of the zodiacal light have already shown us, on quite other grounds, the existence of a vast extent of matter of some kind diffused around the sun.

We have selected the comets of Halley and of Encke as illustrations of the class of periodic comets, of which, indeed, they are the most remarkable members. Of the much more numerous class of non-periodic comets, examples in abundance may be cited. We shall mention a few which have appeared during the present century. There is first the splendid comet of 1843, which appeared suddenly in February of that year, and was so brilliant that it could be seen during full daylight. This comet followed a path which could not be certainly distinguished from a parabola, though there is no doubt that it might have been a very elongated ellipse. It is impossible to decide a question of this kind during the brief opportunities available for finding the place of the comet. We can only see the object during a very small arc of its orbit, and even then the comet is not a very well-defined point which admits of being measured with the precision attainable in observations of a star or a planet. This comet of 1843 is, however, especially remarkable for the rapidity with which it moved, and for the close approach which it made to the sun. The heat to which that comet was exposed during its passage around the sun must have been enormously greater than the heat which can be raised in our mightiest furnaces. If the materials had been agate or cornelian, or the most infusible substances known on the earth, they would have been fused and driven into vapour by the appalling intensity of the sun's rays.

The great comet of 1858 was one of the most magnificent comets of modern times. It was first observed on June 2nd of that year by Donati, whose name the comet has subsequently borne; it was then merely a faint nebulous spot, and for about three months it pursued its way across the heavens, without giving any indications of the splendour which it was so soon to attain. Even up to the end of August the comet had hardly become visible to the unaided eye, and was only furnished with a very small tail, but as it gradually drew nearer and nearer to the sun in September, it soon became invested with splendour. A tail of majestic proportions was quickly developed, and by the middle of October, when the maximum brightness was attained, the tail extended over an arc of forty degrees. The beauty and interest of this comet were greatly enhanced by its position in the heavens at a season when the nights were sufficiently dark. Some of the more recent great comets have not been equally fortunate: they have appeared in midsummer, when the nights are not dark, or they have not been well placed in the heavens.

On the 22nd May, 1881, Mr. Tebbutt, of Windsor, in New South Wales, discovered a comet which speedily developed into one of the most interesting celestial objects seen by this generation. About the 22nd of June the comet became visible from these latitudes in the northern sky at midnight. Gradually it ascended higher and higher till it passed around the pole. The nucleus of the comet was as bright as a star of the first magnitude, and its tail was about 20° long. On the 2nd of September it ceased to be visible to the unaided eye, but it remained visible in telescopes until the following February. This was the first comet which was successfully photographed. To this comet also the spectroscope was applied with especial success. We shall in a subsequent chapter describe the use of this remarkable instrument; here it will be sufficient to observe that by its aid the existence of the element carbon in this comet was demonstrated.

Another of the recent comets which received great and deserved attention was that discovered early in September, 1882, in the southern hemisphere. It increased so much in brilliancy that it was seen in daylight by Mr. Common on the 17th September, while on the 18th the astronomers at the Cape of Good Hope were fortunate enough to have observed the comet actually approach up to the sun's limb, where it ceased to be visible. On the same day the spectrum of the comet was observed, and it was shown that the elements sodium and iron were present therein.

As comets are much nearer to the earth than the stars, it will occasionally happen that the comet must arrive at a position directly between the earth and the star. There is a quite similar phenomenon in the movement of the moon. A star is frequently occulted in this way, and the observations of such phenomena are very familiar to astronomers; but when a comet passes in front of a star the circumstances are widely different. The star is indeed seen nearly as well through the comet as it would be if the comet were entirely out of the way. This has often been noticed. One of the most celebrated observations of this kind was made by the late Sir John Herschel on Biela's comet, which is one of the periodic class, and will be alluded to in the next chapter. The illustrious astronomer on one occasion saw this object pass over a star cluster. The cluster consisted of excessively minute stars—they could only be seen by a very powerful telescope, such as the one Sir John was using. The very faintest haze or the merest trace of a cloud would have sufficed completely to obliterate all the stars. It was therefore with no little interest that the astronomer watched the progress of Biela's comet. Gradually the wanderer encroached on the group of stars, so that if it had any appreciable solidity the numerous twinkling points would have been completely screened. But what were the facts? Down to the most minute star in that cluster, down to the smallest point of light which the great telescope could show, every object in the group was distinctly seen to twinkle right through the mass of Biela's comet!

This was a most important observation. We must recollect that the screen interposed between the cluster and the telescope was not merely a thin curtain; it was a mass of cometary substance, very many thousands of miles in thickness. Contrast, then, the almost inconceivable tenuity of a comet with the clouds to which we are accustomed. A mass of cloud a few hundred feet thick will hide not only the stars, but even the great sun himself. The lightest haze that ever floated in a summer sky, would do more to screen the stars from our view than would one hundred thousand miles of such cometary material as was here interposed.

The great comet of Donati passed over many stars which were visible distinctly through its mass. Among these stars was a very bright one—the well-known Arcturus. The comet, fortunately, happened to pass over Arcturus, and though nearly the densest part of the comet was interposed between the earth and the star, yet Arcturus twinkled on with undiminished lustre through the thickness of this stupendous curtain. Recent observations have, however, shown that stars in some cases experience change in lustre when the denser part of the comet passes over them. It is, indeed, difficult to imagine that a star would remain visible if the nucleus of a really large comet passed over it. It does not seem that an opportunity of testing this supposition has yet arisen.

If the comet were made of transparent gaseous material—such, for instance, as our atmosphere—the place of a star would be deranged when the comet approached the star. The refractive power of air is very considerable. When we look at the sunset, we see the sun appearing to pass below the horizon; yet the sun has actually gone entirely below the horizon before it appears to have commenced its descent. The refractive power of the air is so great that it actually bends the luminous rays round and shows the sun, though it is directly screened by the intervening obstacles. If the comet were made of gaseous material, it must act on the places of the stars by refraction, even if it does not alter their brightness. The question has been carefully tested. A comet has been observed to approach two stars; one of those stars was seen through the comet, while the other was seen directly. If the body had any appreciable quantity of gas in its composition the relative places of the two stars would be altered. This question has been more than once submitted to the test of actual measurement. It has sometimes been found that no appreciable change of position could be detected, and that accordingly in such cases the comet has no perceptible density. A series of careful measurements made on the great comet of 1881 showed, however, that in the neighbourhood of the nucleus the comet seemed to possess some refractive power, though still vastly less than the refraction of our atmosphere.

From these considerations it will probably be at once admitted that the *mass* of a comet must be indeed a very small quantity in comparison with its bulk. When, however, we attempt actually to weigh the comet, all our efforts have proved abortive. We have been able to weigh the mighty planets Jupiter and Saturn; we have even been able to weigh the vast sun himself;

the law of gravitation has provided us with a stupendous weighing apparatus, which has been applied in all these cases with success, but when we attempt to apply the same methods to the comets our efforts are speedily seen to be illusory. No weighing machinery known to the astronomer is delicate enough to determine the weight of a comet. All that we can under any circumstances accomplish is to weigh one heavenly body in comparison with another. The comets seem to be almost imponderable when compared with such robust masses as those of the earth, or any of the other great planets, and so we learn nothing except that their weight is inappreciable. Of course it will be understood that when we say the weight of a comet is inappreciable, we mean with regard to the other bodies of our system. Perhaps no one now doubts that a great comet must really weigh tons; though whether those tons are to be reckoned in tens, in hundreds, in thousands, or in millions, the total seems quite insignificant when compared with a body like the earth.

The small weight of comets is also brought before us in a very remarkable way when we recall what has been said in the last chapter on the important subject of the planetary perturbations. We have there treated of the permanence of our system, and we have shown that the permanence of the system depends upon certain laws which the planetary motions must fulfil. The planets move nearly in circles, their orbits are all nearly in the same plane, and they all move in the same direction. The permanence of the system would be imperilled if any one of these conditions was not fulfilled. In that discussion we made no allusion to the comets. Yet they are members of our system, and they far outnumber the planets. And comets laugh to scorn all those rules which the planets so rigorously obey. Their orbits are never like circles; they are, indeed, more usually parabolic, and thus differ as widely as possible from the circular path. Nor do the planes of the orbits of the comets observe any particular locality; they are inclined at all sorts of angles, and the directions in which the comets move seem to be mere matters of caprice. Every article of the planetary convention is violated by the comets, but yet our system lasts; it has lasted for countless ages, and seems destined to last for ages to come. The comets are attracted by the planets, and conversely, the comets must attract the planets, and must perturb the orbits of the planets to some extent; but to what extent? If the comets all moved in orbits guided by the three general laws which characterise planetary motion, then our argument would break down. The planets might experience considerable derangements from cometary attraction, and yet in the lapse of time those disturbances would neutralise each other, and the permanence of the system would be unaffected. But the case is very different when we deal with the actual cometary orbits. If the comets could appreciably disturb the planets, those disturbances would not neutralise each other, and in the lapse of time the system would be wrecked by the continuous accumulation of irregularities. The facts show, however, that the system has lived, and is living, notwithstanding the comets, and hence we are forced to the conclusion that the cometary masses are utterly insignificant in comparison with the great planetary masses of our system.

These considerations exhibit the laws of universal gravitation and their relations to the permanence of our system in a very striking light. If we include the comets, we may say that the solar system includes many thousands of bodies, in orbits of all sizes, shapes, and positions, only agreeing in the fact that the sun is a focus common to all. The great majority of these bodies are imponderable, and their orbits are placed anyhow, so that, although they may suffer much from the perturbations of the other bodies, they can in no case inflict any appreciable disturbance. There are, however, a few great planets capable of producing vast disturbances; and if their orbits were placed anyhow, chaos would sooner or later be the result. By the mutual adaptations of their orbits to a nearly circular form, to a nearly coincident plane, and to a uniformity of direction, a permanent truce has been effected among the great planets. They cannot now permanently disorganise each other, while the slight mass of the comets renders them incompetent to do so. The stability of the great planets is thus assured; but it is to be observed that there is no guarantee of stability for comets. Their eccentric and irregular paths may undergo the most enormous derangements; indeed, the history of astronomy contains many instances of the vicissitudes to which a cometary career is exposed.

Great comets appear in the heavens under circumstances which have but few features in common. There is no part of the heavens, either in the northern hemisphere or the southern hemisphere, no constellation or region, which is not liable to occasional visits from these mysterious bodies. There is no season of the year, no hour of the day or of the night, when comets may not be above the horizon. In like manner, the size and the aspect of the comets are of every character, from the dim spot just visible to an eye fortified by a mighty telescope, up to a gigantic and brilliant object, with a tail stretching across the heavens for a distance which is as far as from the horizon to the zenith. So also the direction of the tail of the comet seems at first to admit of every possible position: it may stand straight up in the heavens, as if the comet were about to plunge below the horizon; it may stream down from the head of the comet, as if the body had been shot up from below the horizon; it may slope down to the right or rise to the right, it may slope down to the left or rise to the left. Amid all this variety and seeming caprice, can we discover any law common to the different phenomena? If we are to succeed in giving any explanation of the tails of comets, we must first discover what features there are—if any there be—which all these tails possess in common; and we shall find that there is a very remarkable law which the tails of comets obey—a law so true and satisfactory, that if we are given the place of a comet in the heavens, it is possible at once to point out in what direction the tail will lie.

A beautiful comet appears in summer in the northern sky. It is near midnight; we are gazing on the faintly luminous tail, which stands up straight and points towards the zenith; perhaps it may be curved a little or possibly curved a good deal, but still, on the whole, it is directed from the horizon to the zenith. We are not here referring to any particular comet. Every comet, large or small, that appears in the north must at midnight have its tail pointed up in a nearly vertical direction. This fact, which has been verified on numerous occasions, is a striking illustration of the law of direction of comets' tails. Think for one moment of the facts of the case. It is summer; the twilight at the north shows the position of the sun, and the tail of the comet points directly away from the twilight and a way from the sun. Take another case. It is evening; the sun has set, the stars have begun to shine, and a long-tailed comet is seen. Let that comet be high or low, north or south, east or west, its tail invariably points *away* from that point in the west where the departing sunlight still lingers. Again, a comet is watched in the early morning, and if the eye be moved from the place where the first streak of dawn is appearing to the head of the comet, then along that direction, streaming away from the sun, is found the tail of the comet. This law is of still more general application. At any season, at any hour of the night, the tail of a comet must be directed away from the sun. The sun is, of course, on some point of the hemisphere below our feet; draw a great circle from the sun, passing above the horizon to the head of the comet, then the direction of the comet's tail will follow the continuation of that circle.

From the earliest times this fact in the movement of comets must have arrested the attention of those who pondered on the movements of the heavenly bodies. It is a fact patent to ordinary observation, it depends on no telescopic research, nor is it the result of any elaborate mathematical discussion. It is the one fact which gives some degree of consistency to the multitudinous phenomena of comets, and it must be made the basis of our inquiries into the structure of the tails.

In the adjoining figure, Fig. 58, we show a portion of the parabolic orbit of a comet, and we also represent the position of the tail of the comet at various points of its path. It would be, perhaps, going too far to assert that throughout the whole vast journey of the comet, its tail must always be directed from the sun. In the first place, it must be recollected that we can only see the comet during that small part of its journey when it is approaching to or receding from the sun. It is also to be remembered that, while actually passing round the sun, the brilliancy of the comet is so overpowered by the sun that the comet often becomes invisible, just as the stars are invisible in daylight. But with these obvious qualifications we can assert that the tail is always directed away from the sun.

In a hasty consideration of the subject it might be thought that as the comet was dashing along with an enormous velocity the tail was merely streaming out behind, just as the shower of sparks from a rocket are strewn along the path which it follows. This

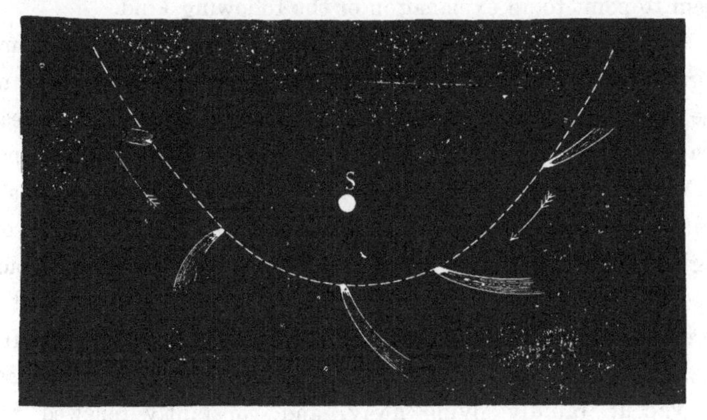

Fig. 58.—The Tail of a Comet directed from the Sun.

would be an entirely erroneous analogy; the comet is moving not through an atmosphere, but through open space, where there is no medium sufficient to sweep the tail into the direction of motion. Another very remarkable feature is the gradual growth of the tail as the comet approaches the sun. While the body is still at a great distance it has usually no perceptible tail, but as it draws in the tail gradually developes, and in some cases reaches the most stupendous dimensions. It is not to be supposed that this increase is a mere optical consequence of the body coming nearer the earth. It can be shown that the growth of the tail is enormously greater than it would be possible to explain merely by the approach of the comet. We are thus led to connect the formation of the tail with the approach to the sun, and we are accordingly in the presence of an enigma which has but little analogy among the other bodies of our system.

That the comet as a whole is attracted by the sun, there can be no doubt whatever. The fact that it moves in an ellipse or in a parabola proves that the two bodies act and re-act on each other in obedience to the law of universal gravitation. But while this is true of the comet as a whole, it is no less certain that the tail of the comet is *repelled* by the sun. It is impossible to speak with certainty as to how this comes about, but the facts of the case seem to point to an explanation of the following kind.

In the materials composing a comet we have some one or more ingredients which give rise to the tail. As the comet draws near the sun and experiences the invigorating effect of increased heat, these ingredients become melted and driven off into vapour. It would seem that though these substances in their solid state are duly attracted to the sun, yet when driven into vapour of a highly rarefied type the heat of the sun exerts on that vapour a repellent power which entirely overcomes the attraction, and accordingly drives the vapour off in a direction pointing away from the sun. We are thus to regard the tail as a stream of smoke or vapour rapidly flying away, and constantly renewed from the evaporation of fresh materials so long as the comet remains sufficiently near to the sun. This explanation will give a plausible reason for the direction of the tail, and, indeed, it seems impossible to believe that the tail could be whirled around the sun with sufficient rapidity if it were really a continuous object.

As to the nature of the repulsive force which we must postulate for the formation of the tail, we are not yet in a position to speak with absolute certainty. It is, however, impossible to omit some account of the remarkable researches of Professor Bredichin, which, when taken in combination with the suggestions of Professor Osborne Reynolds and of others, afford at least a probable explanation of the phenomena. Professor Bredichin has conducted his labours in the strictly philosophical manner which has already led to all the great discoveries in science. He has carefully collaborated the measurements and drawings of the tails of various comets. He has obtained one result from this preliminary part of his inquiry, which will possess a value that cannot be affected even if the ulterior portion of his labours should be found to require qualification. In the examination of the tails of various comets, it is observed that the curvilinear shapes of the outlines fall into one or other of three special types. In the first type we have the straightest tails, which point almost

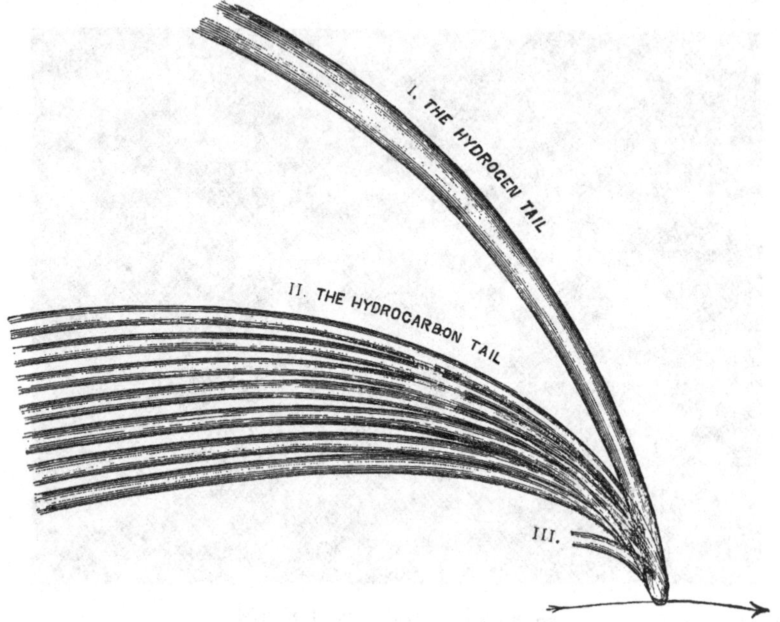

Fig. 59.—Bredichin's Theory of Comets' Tails.

directly away from the sun in a straight line. In the next type we have tails which, while still starting away from the sun, are curved backwards from the direction towards which the comet is moving. In the third type we find the tail curved in still more towards the comet's path. It can be shown that the tails of comets can almost invariably be identified with one or other of these three groups; and in cases where the comet exhibits two tails, as has sometimes happened, then both of these tails will be found to belong to one or other of the types.

The adjoining diagram (Fig. 59) is a sketch of an imaginary comet furnished with tails of the three different types. The direction in which the comet is moving is shown by the arrow-head on the line passing through the nucleus. The straightest of the three tails, marked as Type I., is most probably due to the element hydrogen; the tails of the second type are due to the presence of some of the hydro-carbons in the body of the comet; while the small tails of the third type may be due to iron or to chlorine, or to some other

Fig. 60.—Tails of the Comet of 1858.

element with a high atomic weight. It will of course be understood that this is not the view of any actual comet.

A very beautiful illustration of this theory is afforded in the case of the celebrated comet of 1858 already referred to, of which a drawing is shown in Fig. 60. We see here, besides the great tail, which is the characteristic feature of the comet, two other faint streaks of light. These are really the edges of the hollow cone which forms a tail of Type I., and when we look through the central regions it will be easily understood that the light is not sufficiently intense to be visible; at the edges, however, a good thickness of the cometary matter is looked through, and thus we have the appearance shown in this figure. It would seem that Donati's comet possessed one tail due to hydrogen, and another due to

Fig. 61.—Cheseaux's Comet of 1744.

some of the compounds of carbon. The carbon compounds involved seem to be of considerable variety, and there is, in consequence, a disposition in the tails of the second type to a more indefinite outline than in the hydrogen tails. Cases have been recorded in which several tails have been seen simultaneously on the same comet. The most celebrated of these is that which appeared in the year 1744, and is now known as Cheseaux's comet. Professor Bredichin has devoted special attention to the theory of this marvellous object, and he has shown with a high degree of probability how the multiform tail could be accounted for on his theory.

It is possible to submit some of the questions involved to the test of calculation, and it can be shown that the repulsive force adequate to produce the very straightest tail of Type I. need only be about twelve times as large as the attraction of gravitation. Tails of the second type could be produced by a repulsive force which was equal to gravitation, while tails of the third type would only require a repulsive force of one-quarter the power of gravitation. The chief repulsive force known in nature is derived from electricity, and it has naturally been surmised that the phenomena of comets' tails are really due to the electric condition of the sun and of the comet. It can be shown that when the cometary substances are driven into vapour by the heat of the sun, the electric repulsion may equal—may even greatly exceed—gravitation, and thus produce the phenomena which are observed. It would be premature to assert that the electric character of the comet's tail has been absolutely demonstrated; all that can be said is that, as it seems to account for the observed facts, it would be undesirable to introduce some mere hypothetical repulsive force. It must be remembered that on quite other grounds it is known that the sun is the seat of certain electric phenomena.

As the comet gradually recedes from the sun the heat abates, the evaporation ceases, the repulsion has no longer any object on which to exercise its power, and accordingly we find the tail of the comet declines. If the comet be a periodic one, the same series of changes may take place at its next return to perihelion. A new tail is formed, which also gradually disappears as the comet regains the depths of space. If we may employ the analogy of terrestrial vapours to guide us in our reasoning, then it would seem that as the comet retreats its tail would condense into myriads of small particles. Over these small particles the law of gravitation would resume its sway, no longer obscured by the superior efficiency of the repulsion produced by heat acting on the vapour. The mass of the comet is, however, so extremely small that it would not be able to recall these particles by the mere force of attraction. It follows that, as the comet at each perihelion passage makes a tail, it must on each occasion sacrifice a corresponding quantity of tail-making material. Let us suppose that the comet was endowed in the beginning with a certain capital of those particular materials which are adapted for the production of tails. Each perihelion passage witnesses the production of a tail, and the expenditure of a corresponding proportion of the capital. It is obvious that this operation cannot go on indefinitely. In the case of the great majority of comets the visits to perihelion are so extremely rare that the consequences of the extravagance are not very apparent; but in the case of those periodic comets which have short periods and frequent returns, the consequences are precisely what might have been anticipated: the tail-making capital has been gradually squandered, and thus at length we have the frequent spectacle of a comet without any tail at all.

It is not a little remarkable that some of the materials present in a comet are identical with substances which are familiar on the earth. The most notable instance is the element carbon. This is especially interesting when we reflect on the significance of carbon on the earth. We know that carbon is one of the most abundant and the most important of the elements. We see it as the chief constituent of all vegetable life; we find it to be invariably present in animal life—and, indeed, in all organic substances. It is an interesting fact that this element, of such transcendent importance on the earth, should now have been proved to be present in these wandering bodies. Although no one suggests that a comet could be the abode of life, yet it is certainly a significant discovery that the most important factor in the material substratum of life has now been shown not to be limited to the surface of the earth, but to be as widespread as comets themselves.

To say that carbon is a constituent of comets is, indeed, to attribute to that element a presence throughout the length and breadth, the depth and the height, of the material universe. There is no quarter of the heavens from which comets may not appear. They exist in such profusion that, as Kepler said, there may be more comets in the sky than there are fishes in the ocean; and as to the distance to which comets recede in their wanderings, or from which they may be drawn into our system, its calculation baffles all astronomical measurement. We are in the habit of regarding the orbits of most comets as sensibly of a parabolic shape; and assuming this to be the case, let us ask the law of gravitation to describe the history of one of these wanderers.

Far away in the depths of space, at a distance so remote that from it our earth and the other planets are all invisible, and from which even our sun has been dwarfed to the magnitude of a mere star, lies the future comet. Throughout the mighty abyss which intervenes between the comet and the sun, the law of gravitation reigns. Under its influence the comet begins to approach the sun. The force is so enfeebled by the effect of distance that at first the movements of the comet are very slow. Years roll on; the comet is still found to be approaching, and the motion, though still extremely slow, has made some slight improvement. Centuries roll on, and thousands of years roll on, and now the comet is seen to be rapidly approaching. The distance decreases and the speed increases. The attraction of the sun has now increased greatly, owing to the diminished distance; and the comet at length, with an appalling velocity, amounting to hundreds of miles a second, whirls around the sun, and in a few hours commences its retreat. The mighty velocity of the body now tends to carry it away from the sun, while the whole efficiency of the solar attraction is expended in the effort to recall it; but the comet—if its orbit be such as we are here considering—will not be recalled. The effect of the sun's attraction will speedily abate the velocity of the motion. The comet retreats precisely as it approached. The lapse of a year after it has left the sun finds the comet as far as it was a year before it reached the sun. The same is true for 100 years—for 1,000 years. More and more slowly will the comet retreat as it gradually sinks into the depths of space. It never can entirely escape from the sun's attraction. With ever-abating velocity it will recede further and further, but never again to return to our system.

Such is a brief outline of the principal facts known with regard to these most interesting, but most perplexing bodies. As to the origin of comets, it is impossible to offer more than a conjecture. It has been supposed that they may have been merely drawn into our system from the depths of infinite space. It has also been suggested that possibly comets may have originated in our system —that they may, in fact, have been merely fragments driven off from the sun himself. We must be content with the mere recital of what we know, rather than with attempting guesses about matters beyond our reach. We have confined our remarks to the humbler, but more certain process of indicating the part which comets play in the solar system. We see that comets are obedient to the great laws of gravitation, and afford the most striking illustration of their truth. We have seen how modern science has dissipated the fear and the superstition with which, in earlier ages, the advent of a comet was regarded. We no longer regard a comet as a sign of impending calamity; we may rather look upon it as an interesting and a beautiful visitor, which comes to please us and to instruct us, but never to threaten or to destroy.

CHAPTER XVII.

SHOOTING STARS.

Small Bodies of our System—Their Numbers—How they are Observed—The Shooting Star—The Theory of Heat—A great Shooting Star—The November Meteors—Their Ancient History—The Route followed by the Shoal—Diagram of the Shoal of Meteors—How the Shoal becomes Spread out along its Path—Absorption of Meteors by the Earth—The Discovery of the Relation between Meteors and Comets—The remarkable Investigations concerning the November Meteors—Two Showers in Successive Years—No Particles have ever been Identified from the great Shooting Star Showers—Meteoric Stones—Chladni's Researches—Early Cases of Stonefalls—The Meteorite at Ensisheim—Collections of Meteorites—The Rowton Siderite—Relative frequency of Iron and Stony Meteorites—Fragmentary Character of Meteorites—No Reason to connect Meteorites with Comets—Tschermak's Theory—Effects of Gravitation on a Missile ejected from a Volcano—Can they have come from the Moon?—The Claims of the Minor Planets to the Parentage of Meteorites—Possible Terrestrial Origin—The Ovifak Iron.

In the preceding chapters we have dealt with the gigantic bodies which form the chief objects in what we know as the solar system. We have studied mighty planets measuring thousands of miles in diameter, and we have followed the movements of comets, whose dimensions are to be told by millions of miles. Once, indeed, in a previous chapter, we have made a descent to objects much lower in the scale of magnitude, and we have examined that numerous class of small bodies which we call the minor planets. It is now, however, our duty to make a still further, and this time a very long step, downwards in the scale of magnitude. Even the minor planets must be regarded as colossal objects, when compared with those little bodies whose presence is revealed to us in a most interesting, and sometimes in a most striking manner.

These small bodies compensate in some degree for their minute size, by the enormous profusion in which they exist. No attempt, indeed, could be made to tell in figures the myriads in which they swarm throughout space. They are probably of very varied dimensions, some of them being many pounds or perhaps tons in weight, while others seem to be not larger than pebbles, or even than grains of sand. Yet, insignificant as these bodies may seem, the great sun himself does not disdain to accept their control. Each particle, whether it be as small as the mote in a sunbeam or as mighty as the planet Jupiter, will perform its path around the sun in conformity with the laws of Kepler.

Who does not know that very beautiful occurrence which we call a shooting star, or which, in its more splendid forms, is sometimes called a meteor or fire-ball? It is to objects of this class that we are now to to direct our attention.

A small body, perhaps as large as a paving-stone or larger, more often perhaps not as large as a marble, is moving round the sun. Just as a mighty planet revolves in an ellipse, so this small object will move round and round in an ellipse, with the sun in the focus. There are, at the present moment, inconceivable myriads of such meteors moving in this manner. They are too small and too distant for our telescopes, and we can never see them except under extraordinary circumstances.

At the time we see the meteor it is usually moving with enormous velocity, so that it often traverses a distance of more than twenty miles in a second of time. Such a velocity is almost impossible near the earth's surface: the resistance of the air would prevent it. Aloft, in the emptiness of space, there is no air to resist the meteor. It may have been moving round and round the sun for thousands, perhaps for millions of years, without let or hindrance; but the supreme moment arrives, and the meteor perishes in a streak of splendour.

In the course of its wanderings the body comes near the earth, and within a few hundred miles of its surface of course begins to encounter the upper surface of the atmosphere with which the earth is enclosed. To a body moving with the appalling velocity of a meteor, a plunge into the atmosphere is usually fatal. Even though the upper layers of air are excessively attenuated, yet they suddenly check the velocity, almost as a rifle bullet would be checked when fired into water. As the meteor rushes through the atmosphere the friction warms the surface of the meteor; gradually it becomes red-hot, becomes white-hot, and is usually driven off into vapour with a brilliant light, while we on the earth, one or two hundred miles below, exclaim: "Oh, look! there is a shooting star."

We have here on a very striking scale an experiment illustrating the mechanical theory of heat. It may seem incredible that mere friction should be sufficient to generate heat enough to produce so brilliant a display, but we must recollect two facts: first, that the velocity of the meteor is, perhaps, one hundred times that of a rifle bullet; and second, that the efficiency of friction in developing heat is proportioned to the square of the velocity. The meteor in passing through the air must therefore be heated by the friction of the air about ten thousand times as much as the rifle bullet is heated. We do not make an exaggerated estimate in supposing that by its rush through the air the rifle bullet becomes heated ten degrees; yet if this be admitted, we must grant that there would be such an enormous development of heat attending the flight of the meteor that even a fraction of it would be sufficient to drive the object into vapour.

Let us first consider the circumstances under which these external bodies are manifested to us, and for the sake of illustration, we may take a remarkable fire-ball which occurred on November 6th, 1869. This fire-ball was extensively seen from different parts of England; and by combining and comparing these observations, we obtain accurate information as to the height of the object and the velocity with which it travelled.

It appears that this meteor commenced to be visible at a point ninety miles above Frome, in Somersetshire, and that it disappeared at a point twenty-seven miles over the sea, near St. Ives, in Cornwall, both the path, its height, and the principal localities from which it was observed being shown in the map (Fig. 62). The whole length of its course was about 170 miles, which was performed in a period of five seconds, thus giving an average velocity of thirty-four miles per second. A remarkable feature in the appearance which this fire-ball presented was the long persistent streak of luminous cloud, about fifty miles long and four miles wide, which remained in sight for fully fifty minutes. We have in this example an illustration of the chief features of the

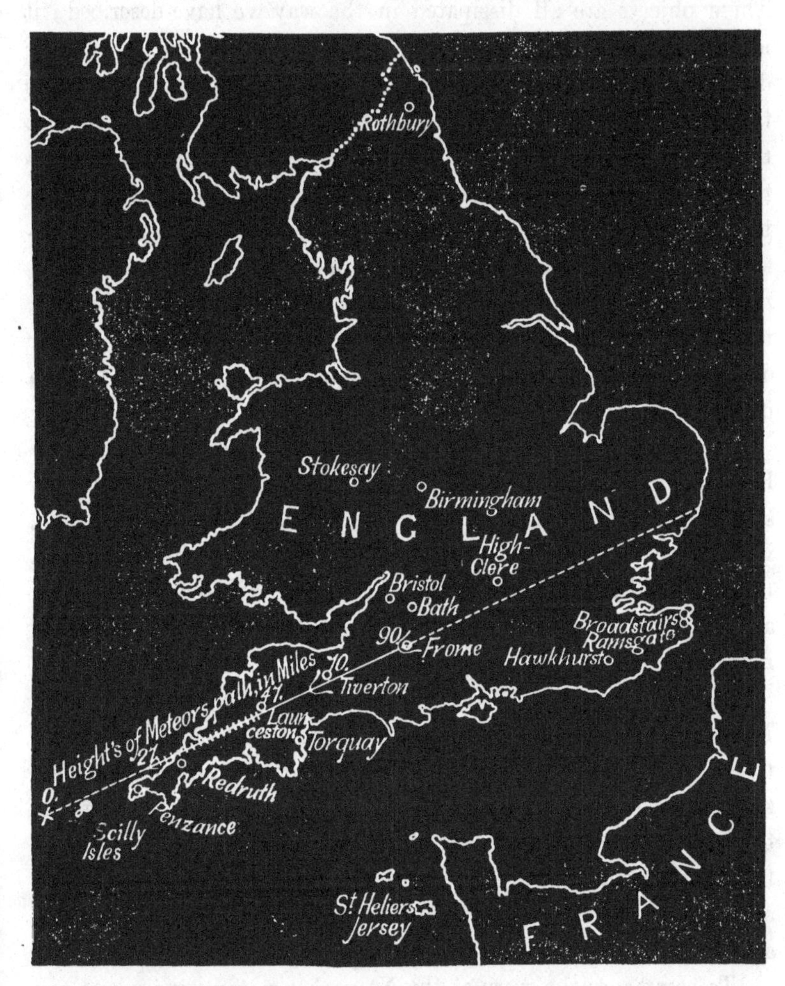

Fig. 62.—The Path of the Fireball of November 6th, 1869.

phenomena of a shooting star presented on a very grand scale. It is, however, to be observed that the persistent luminous streak is not a universal, nor, indeed, a very common characteristic of a shooting star.

The small objects which occasionally flash across the field of the telescope show us that there are innumerable telescopic shooting stars, too small and too faint to be visible to the unaided eye. These objects are all dissipated in the way we have described; it is, in fact, only at the moment, and during the process of their dissolution, that we become aware of their existence. Small as these missiles probably are, their velocity is so prodigious, that they would render the earth uninhabitable were they permitted to rain down unimpeded on its surface. We must, therefore, among the other good qualities of our atmosphere, not forget that it constitutes a kindly screen, which shields us from a tempest of missiles, the velocity of which no artillery could equal. It is, in fact, the very fury of these missiles which is the cause of their utter destruction. Their anxiety to strike us is so great, that friction dissolves them into harmless vapour.

Next to the splendid event of a mighty meteor such as that we have just described, the most striking occurrence in connection with shooting stars is what is known as a shower. These showers have, within the last century, attracted a great deal of attention, and they have abundantly rewarded the labour devoted to them by affording some of the most interesting astronomical discoveries of modern times.

The showers of shooting stars do not occur very frequently. No doubt the quickened perception of those who especially attend to meteors will detect a shower when others see only a few straggling shooting stars; but speaking generally, we may say that the present generation can hardly have witnessed more than two or three such occurrences. I have myself seen two great showers, one of which, in November, 1866, has impressed itself on my memory as a most glorious astronomical spectacle.

To commence the story of the November meteors it is necessary to look back for nearly a thousand years. In the year 902 occurred the death of a Moorish king, and in connection with this event an old chronicler relates how "that night there were seen, as it were lances, an infinite number of stars, which scattered themselves like rain to right and left, and that year was called the Year of the Stars."

No one now believes that the heavens intended to commemorate the death of the king by that display. The record is, however, of considerable importance, for it indicates the year 902 as one in which a great shower of shooting stars occurred. It was with the greatest interest that astronomers perceived that this was the first recorded instance of the well-known periodical shower, one of whose regular returns we saw in 1866. Further diligent literary research has revealed here and there records of startling appearances in the heavens, which fit in with the successive returns of the November meteors. From the first instance, in 902, to the present day there have been twenty-nine visits of the shower; and it is not unlikely that these twenty-nine showers were all more or less seen in some parts of the earth. Sometimes they may have been witnessed by savages, who had neither the inclination nor the means to place on record an apparition which to them was a source of terror. Sometimes, however, these showers were witnessed by civilised communities. Their nature was not understood, but the records were made; and in some cases, at all events, these records have withstood the corrosion of time, and have now been brought together to illustrate this curious subject. We have altogether historical notices of twelve of these showers, mainly collected by the industry of Professor Newton, whose labours have contributed so much to the advancement of our knowledge of shooting stars.

Let us imagine an enormous shoal of shooting stars moving in space. Think of a shoal of herrings in the ocean, extending over many square miles, and containing countless myriads of individuals; or think of those enormous flocks of wild pigeons in the United States of which Audubon has told us. The shoal of shooting stars is perhaps much more numerous than the herrings or the pigeons. The shooting stars are, however, not very close together; they are, on an average, probably some few miles apart. The actual bulk of the shooting star shoal is therefore prodigious; and its dimensions are to be measured by hundreds of thousands of miles.

The meteors cannot choose their own track, like the shoal of herrings, for they are compelled to follow the route which is prescribed to them by the sun. Each one pursues its own ellipse in complete independence of its neighbour, and accomplishes its mighty journey, thousands of millions of miles in length, every thirty-three years. We cannot see the meteors in the greater part of their orbit. There are countless myriads of the meteors at this very moment coursing round their path. The great shoal is at present (1885) far out from the sun, but we know they are there, though we have not seen them. We know it precisely in the same way as we know that there are herrings in the sea. We have not seen the herrings at present in the sea: they are not often seen until they are caught; so it is with the meteors; we never see them till the earth catches them. Every thirty-three years the earth catches these meteors just as successfully as the fisherman catches herrings, and in very much the same way, for while the fisherman spreads his net in which the fishes meet their doom, so the earth spreads out the atmosphere wherein the meteors perish. We are told that there is no fear of the herrings getting exhausted, for all the fishermen catch are as nothing compared to the boundless profusion in which they exist. We may hold the same language true with regard to the meteors. They exist in such myriads, that though the earth swallows up millions each thirty-three years, yet plenty are still left for future showers. The diagram (Fig. 63) will explain the way in which the earth spreads her net. We there see the orbit in which the earth moves around the sun, as well as the elliptic path of the meteors, though it should be remarked that it is not convenient to draw the figure exactly to scale, so that the path of the meteors is relatively much larger than here represented. Once a year the earth accomplishes its revolution, and between the 12th and the 14th of November, crosses the track in which the meteors move. It will usually happen that the great shoal is not at this point when the earth is passing. There are, however, some stragglers all along the path, so that the earth usually obtains a few of these every year at this date. They dart into our atmosphere as shooting stars, and form what we usually speak of as the November meteors.

It will, however, occasionally happen that when the earth is in the act of crossing the track it encounters the dense body of meteors. Through the mighty shoal the earth then plunges, enveloped, of course, with the surrounding coat of air. Into this net the meteors dash in countless myriads, never again to emerge. In a few hours' time, the earth, moving at the rate of eighteen miles a second, has crossed the track and emerges on the other side, bearing with it the spoils of the encounter. The remaining

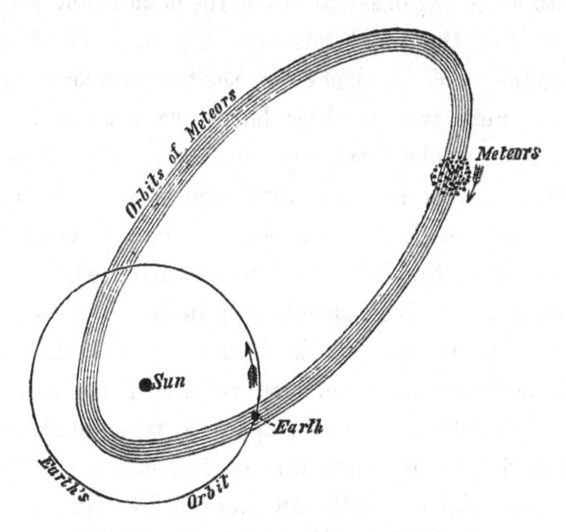

Fig. 63.—The Orbit of a Shoal of Meteors.

meteors of the shoal continue their journey without interruption; perhaps millions have been taken, but probably hundreds of millions have been left.

Such was the occurrence which astonished the world on the night between November 13th and 14th, 1866. We then plunged into the middle of the shoal. The night was fine; the moon was absent. The meteors were distinguished not only by their enormous multitude, but by their intrinsic magnificence. I shall never forget that night. On the memorable evening I was engaged in my usual duty at that time of observing nebulæ with Lord Rosse's great reflecting telescope. I was of course aware that a shower of meteors had been predicted, but nothing that I had heard prepared me for the splendid spectacle so soon to be unfolded. It was about ten o'clock at night when an exclamation from an attendant by my side made me look up from the telescope, just in time to see a fine meteor dash across the sky. It was presently followed by another, and then again by others in twos and in threes, which showed that the prediction of a great shower was likely to be verified. At this time the Earl of Rosse (then Lord Oxmantown) joined me at the telescope, and, after a brief interval, we decided to cease our observations of the nebulæ and ascend to the top of the wall of the great telescope (Fig. 6, p. 17) from which a clear view of the whole hemisphere of the heavens could be obtained. There, for the next two or three hours, we witnessed a spectacle which can never fade from my memory. The shooting stars gradually increased in number until sometimes several were seen at once. Sometimes they swept over our heads, sometimes to the right, sometimes to the left, but they all diverged from the east. As the night wore on the constellation of Leo ascended above the horizon, and then the remarkable character of the shower was disclosed. All the tracks of the meteors radiated from Leo. (See Fig. 74, p. 383). Sometimes a meteor appeared to come almost directly towards us, and then its path was so fore-shortened that it had hardly any appreciable length, and looked like an ordinary fixed star swelling into brilliancy and then as rapidly vanishing. Occasionally luminous trains would linger on for many minutes after the meteor had flashed across, but the great majority of the trains in this shower were evanescent. It would be impossible to say how many thousands of meteors were seen, each one of which was bright enough to have elicited a note of admiration on any ordinary night.

The adjoining figure (Fig. 64) shows the remarkable manner in which the shooting stars of this shower diverged from a point. It is not to be supposed that all these objects were in view at the same moment. The observer of a shower is provided with a map of that part of the heavens in which the shooting stars appear. He then fixes his attention on one particular shooting star, and observes carefully its track with respect to the fixed stars in its vicinity. He then draws a line upon his map in the direction in which the shooting star moved. Repeating the same observation for several other shooting stars belonging to the shower, his map will hardly fail to show that the different tracks of almost all the shooting stars tend from one point or region of the figure. There are, it is true, a few erratic ones, but the majority observe this law. It certainly looks, at first sight, as if all the shooting stars did actually dart from this point; but a little reflection will

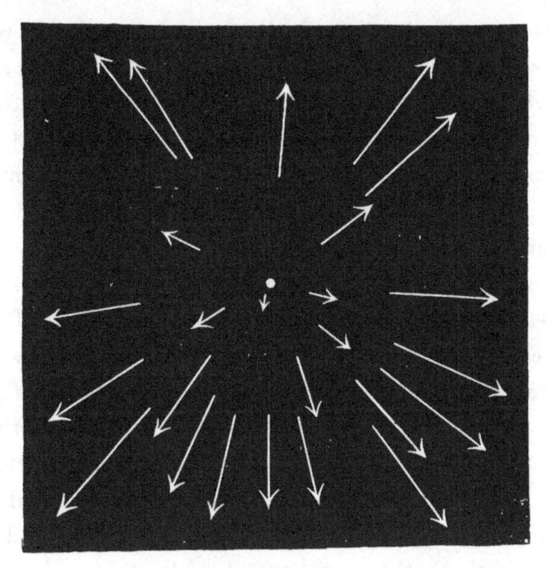

Fig. 64.—Radiant Point of Shooting Stars.

show that this is a case in which the real motion is different from the apparent. If there actually were a point from which these meteors diverged, then from different parts of the earth the point would be seen in different positions with respect to the fixed stars; but this is not the case. The radiant, as this point is called, is seen in the same part of the heavens from whatever station the shower is visible.

We are, therefore, led to accept the simple explanation afforded by the theory of perspective. Those who are acquainted with the principles of this science know, that when a number of parallel lines in an object have to be represented in a drawing they must all be made to pass through the same point in the plane of the picture, unless in the exceptional case where they are also lines parallel to the picture. When we are looking at the shooting stars, what we really see are only the projections of their paths upon the surface of the heavens. From the fact that all those paths pass through the same point, we are to infer that the shooting stars belonging to the same shower are all moving in parallel lines.

We are now able to ascertain the actual direction in which the shooting stars are moving, because a line drawn from the eye of the observer to the radiant point must be parallel to that direction. Of course, it is not intended to convey the idea that throughout all space the shooting stars of one shower are moving in parallel lines; all we mean is, that during the short time in which we see them the motion of each of the shooting stars is sensibly a straight line, and that all these straight lines are parallel.

At present the great meteor shoal of the Leonids (for so this shower is called) has accomplished about half its mighty journey. In the year 1885 it attains its greatest distance from the sun, and then commences to return. Each year the earth crosses the orbit of the meteors, but the shoal is not met with, and no shower of stars is perceived. Every succeeding year will find the meteors approaching the critical point, and the year 1899 will bring the shoal to the earth's track. In that year, therefore, another brilliant meteoric shower may be looked for. The shoal of meteors is of such enormous length that it takes more than a year for the mighty procession to pass through the critical portion of its orbit which lies across the track of the earth. We thus see that the meteors cannot escape the earth. It may be that when the shoal begins to reach this neighbourhood the earth will have just left this part of its path, and a year will have elapsed before the earth gets round again. Those meteors that have the good fortune to be in the front of the shoal will thus escape the net, but some of those behind will not be so fortunate, and the earth will again devour an incredible host. It has sometimes happened thar casts into the shoal have been obtained in two consecutive years.

If the earth happened to pass through the front part in one year, then the shoal is so long that the earth will have moved right round its orbit of 600,000,000 miles, and will again dash through the critical spot before the entire number have passed. History contains records of cases when, in two consecutive Novembers, brilliant showers of Leonids have been seen.

As the earth consumes such myriads of Leonids each thirty-three years, it follows that the total number must be decreasing. The splendour of the showers in future ages will, no doubt, be affected by this circumstance. They cannot be always as bright as they have been. It is also of interest to notice that the shape of the shoal is gradually changing. Each meteor of the shoal moves in its own ellipse around the sun, and is quite independent of the rest of these bodies. Each one has thus a special period of revolution which depends upon the length of the ellipse in which it happens to revolve. Two meteors will move around the sun in the same time if the lengths of their ellipses are exactly equal, but not otherwise. The lengths of these ellipses are many hundreds of millions of miles, and it is impossible that they can be all absolutely equal. In this may be detected the origin of a gradual change in the character of the shower. Suppose two meteors A and B be such that A travels completely round in thirty-three years, while B takes thirty-four years. If the two start together, then when A has finished the first round B will be a year behind; the next time B will be two years behind, and so on. The case is exactly parallel to that of a number of boys who start for a long race, in which they have to run several times round the course before the distance has been accomplished. At first they all start in a cluster, and perhaps for the first round or two they may remain in comparative proximity; gradually, however, the faster runners get ahead and the slower ones lag behind, so the cluster becomes elongated. As the race continues, the cluster becomes dispersed around the entire course, and perhaps the first boy will even overtake the last. Such seems the destiny of the November meteors in future ages. The cluster will in time to come be spread out around the whole of this mighty track, and no longer will a superb display have to be recorded every thirty-three years.

It was in connection with the shower of November meteors in 1866 that a very interesting and beautiful discovery in mathematical astronomy was made by Professor Adams. We have seen that the Leonids must move in an elliptic path, and that they return every thirty-three years, but the telescope cannot follow them during their wanderings. All that we know by observation is the date of their occurrence, the point of the heavens from which they radiate, and the great return every thirty-three years. Putting these various facts together, it is possible, not exactly to determine the ellipse in which the meteors move—the facts do not go so far—they only tell us that the ellipse must be one of five possible orbits. These five possible orbits are—firstly, the immense ellipse in which we now know the meteorites do revolve, and for which they require the whole thirty-three years to complete a revolution; secondly, a nearly circular orbit, very little larger than the earth's path, which the meteors would traverse in a few days more than a year; another similar orbit, in which the time would be a few days short of the year; and two other small elliptical orbits lying inside the earth's orbit. It was clearly demonstrated by Professor Newton that the observed facts would be explained if the meteors moved in any one of these paths, but that they could not be explained by any other hypothesis. It remained to see which of these orbits was the true one. Professor Newton himself made the suggestion of a possible method of solving the problem. The test he proposed was one of very great difficulty, and involved the most intricate calculations. Fortunately, however, Professor Adams undertook the inquiry, and by his successful labours the path of the Leonids has been completely ascertained.

When the ancient records of the appearance of great Leonid showers are examined, it is found that the date of their occurrence undergoes a gradual and continuous change. It follows as a necessary consequence that the point where the path of the meteors crosses the earth's track is not fixed, but that at each successive return the meteors pass at a point about half a degree further on in the direction in which the earth is travelling. It follows that the orbit in which the meteors are revolving is undergoing change; the path they follow in one revolution varies slightly from that pursued in the next. As, however, these changes proceed in the same direction, they may gradually attain considerable dimensions; and the amount of change which is produced in the path of the meteors in the lapse of centuries may be estimated by the two ellipses shown in Fig. 65. The continuous line represents the orbit in A.D. 126; the dotted line represents it at present.

This very definite change in the orbit is one that astronomers attribute to what we have already spoken of as perturbation. It is certain that the elliptic motion of these bodies is due to the sun, and that if they were only acted on by the sun the ellipse would remain absolutely unaltered. We see, then, in this gradual change of the ellipse the influence of the attractions of the planets. It was shown that if the meteors moved in the large orbit, this shifting of the path must be due to the attraction of the planets Jupiter, Saturn, Uranus, and the Earth; while if the meteors followed one of the smaller orbits, the planets that would be near enough, and large enough to act sensibly on them, would be the Earth, Venus, and Jupiter. Here, then, we see how the question may be answered by calculation. It is difficult, but it is possible, to calculate what the attraction of the planets would be capable of producing for each of the five different suppositions as to the orbit. This is what Adams has done. He found that if the meteors moved in the great orbit, then the attraction of Jupiter would account for two-thirds of the observed change, while the remaining third was due to the influence of Saturn, supplemented by a small addition on account of Uranus. In this way the calculation showed that the large orbit was a possible one. Professor Adams also computed the amount of displacement in the path that could be produced if the meteors revolved in any of the four smaller ellipses. This investigation was one of an extremely arduous character, but the results amply compensated for the labour. It was shown that in this case it would be impossible to obtain by the aid of perturbations a displacement even one-half of that which was observed. These four orbits were, therefore, impossible ones. Thus the demonstration was complete as to the character of the path in which the meteors revolve.

The movements in each revolution are guided by Kepler's laws. When at the part of its path most distant from the sun the velocity of a meteor is at its lowest, being then but little more than a mile a second; as it draws in the speed gradually increases, until when the meteor crosses the earth's track its velocity is no less than twenty-six miles a second. The earth is moving very nearly in the opposite direction at the rate of eighteen miles a second, so that if the meteor happen to strike the earth's atmosphere, it does so with an enormous velocity of nearly forty-four miles a second. If the meteor escape the earth, then it resumes its outward journey with gradually declining velocity, and by the time it has again completed its mighty journey a period of thirty-three years years and a quarter will have elapsed.

The innumerable meteors which form the Leonids are arranged in an enormous stream, of a breadth very small in comparison with its length. If we represent the orbit by an ellipse whose length is seven feet, then the meteor stream will be represented by a thread of the finest sewing-silk, about a foot and a half or two feet long, creeping along the orbit.* The size of this stream may be estimated from the consideration that even its width cannot be less than 100,000 miles. Its length may be estimated from the circumstance that, although it is moving about twenty-six miles a second, yet the stream takes about two years to pass the point where its orbit crosses the earth's track. On the memorable occasion of the night between the 13th and 14th of November, 1866, the earth plunged into this stream near its head, and did not emerge on the other side until five hours after. During that time it happened that the hemisphere of the earth which was in front contained the

* This illustration, as well as the figure of the path of the meteors, has been derived from Mr. G. J. Stoney's interesting lecture on "The Story of the November Meteors," at the Royal Institution in 1879.

continents of Europe, Asia, and Africa, and consequently it was in the Old World that the great shower was seen. On that day twelvemonth, when the earth had regained the same spot, the shoal had not entirely passed, and the earth made another plunge. This time the American continent was in the van, and consequently it

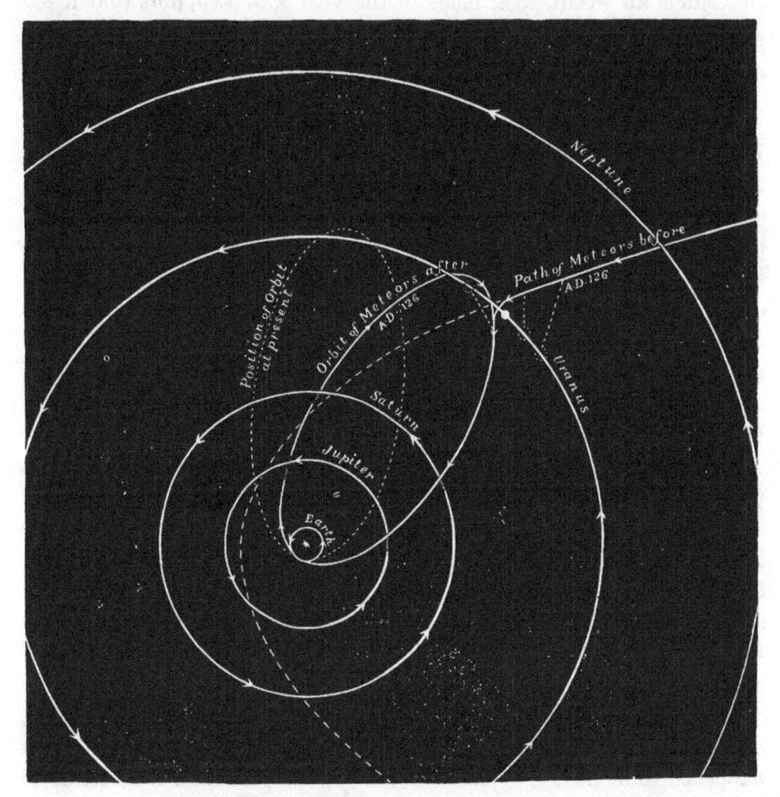

Fig. 65.—The History of the Leonids.

was there that the shower of 1867 was seen. Even in the following year the great shoal had not entirely passed, and a few stragglers along the route are still occasionally encountered at each annual transit of the earth across the great meteoric highway.

The diagram is also designed to indicate a remarkable speculation which was put forward on the high authority of Le Verrier, with the view of explaining how the shoal came to be introduced into the solar system. The orbit in which the meteors revolve does not intersect the paths of Jupiter, Saturn, or Mars, but it does intersect the orbit of Uranus. It must, therefore, sometimes happen that Uranus will be passing through this point of its path just as the shoal has arrived there. Le Verrier has demonstrated that such an event took place in the year A.D. 126, but that it has not happened since. We thus seem to have a clue to a very wonderful past history :—

"All would be explained if we may suppose that before the year 126, the meteors had been moving beyond the solar system; and that in that year they chanced to cross the path of the planet Uranus, travelling along some such path as that represented in the diagram. Had it not been for the planet, they would have kept on the course marked out with a dotted line, and after having passed the sun, would have withdrawn on the other side into the depths of space, to the same measureless distance from which they had originally come. But their stumbling on the planet changed their whole destiny. Even so great a planet would not sensibly affect them until they got within a distance, which would look very short indeed upon our diagram. But they seem to have almost grazed his surface, and while they were very close to such a planet, he would be able to drag them quite out of their former course. This the planet Uranus seems to have done; and when, pursuing his own course, he again got too far off to influence them sensibly, they found themselves moving slowly backwards and slowly inwards; and accordingly began the new orbit round the sun, which corresponds to the situation into which they had been brought, and the direction and moderate speed of their new motion.

"They seem to have passed Uranus while they were still a small, compact cluster. Nevertheless, those members of the group which happened to be next the planet as they swept past, would be attracted with somewhat more force than the rest; the farthest members of the group with the least. The result of this must inevitably have been that when the group were soon after abandoned to themselves, they did not find themselves so closely compacted as before, nor moving with an absolutely identical motion, but with motions which differed, although perhaps very little, from one another. These are conditions which would have started them in those slightly differing orbits round the sun, which, as we have seen, would cause them, as time wore on, to be drawn out into the long stream in which we now, after seventeen centuries, find them.

"What is here certain is, that there was a definite time when the meteors entered upon the path they are now pursuing—that this time was the end of February or beginning of March in the year 126 is still a matter of probability only. It is, however, *highly* probable, because it explains all the phenomena at present known; but astronomers are not yet in a position to assert that it is ascertained, since one link in the complete chain of proof is wanting. We who live now should be in possession of this link if our ancestors had made sufficiently full observations; and our posterity will have it when they compare the observations they can make with those which we are now carefully placing on record for their use. They will then know whether the rate at which the stream is lengthening out is such as to indicate that A.D. 126 was the year in which this process began. If so, Le Verrier's hypothesis will be fully proved." *

The determination of the true elliptic path of the Leonids led Professor Schiaparelli to a discovery of rather an unexpected character, which revealed in a manner almost startling, a mysterious connection between the shooting star showers and periodic comets.

In the year 1866 a comet appeared. This comet was observed, and from these observations the path in which the comet moved could be calculated. What was the astonishment of astronomers, when it became known that the path of this comet was also the path followed by the Leonids! To emphasize this coincidence, it is desirable to think separately of the various points which it implies. Take, for instance, the plane in which the comet moves: that plane may have any position, subject only to the condition that it shall pass through the centre of the sun. So, too, the orbit of the meteors may lie in any one of the myriad planes passing through the sun's centre. That these two planes should absolutely coincide is a matter which would be almost infinitely improbable if it were merely a question of chance. The coincidence does not rest even here; not only do the planes coincide, but the directions of the axes of the orbits in these planes are also identical: this, again, would be infinitely unlikely were the question one merely of chance. But we

* G. J. Stoney: "The Story of the November Meteors."

have not yet exhausted the subject, for it further appears that the shortest distance from the sun to the orbit is also in each case identical. We have here a concentration of evidence which proves to demonstration that there must be some profound physical connection between comets and swarms of meteors. The observed facts would be accounted for if we suppose the comet to be preceded or followed in its path by a meteor stream; but this, of course, implies some very intimate connection between the comet and the meteors, though at the present time it is not possible to state the real nature of that connection.

The remarkable association between comets and meteors has been noticed in the case of several other meteoric streams besides the Leonids. We may, for instance, cite one of the showers which is most constant in its annual appearance, and is called the Perseids, from the circumstance that the radiant point lies in the constellation of Perseus. This shower is known also as that of the August meteors, which fall from the 9th to the 11th of that month in each year. Their orbit has been carefully determined, and it has been found to be also the path of a comet. There are several other showers of which the same statement may be made. We shall instance one in particular which has attracted great and deserved attention.

A copious meteoric shower took place on the night of the 27th November, 1872. On this occasion the shooting stars diverged from a radiant point in the constellation of Andromeda. I estimated at the time that there were at least three hundred meteors per hour. It is true that these shooting stars were not bright, nor were their trains very long. As a spectacle, it was, in my judgment, unquestionably inferior to the magnificent display of 1866. But though, from a picturesque point of view, the shower of 1872 is inferior to that of 1866, yet it is difficult to say which of the two showers is of greater scientific importance.

It surely is a remarkable coincidence that the earth should encounter the Andromedes (for so this shower is called) at the very moment when it is crossing the track of Biela's comet. We have observed the direction from which the Andromedes come when they plunge into the atmosphere; we can ascertain also the direction in which Biela's comet is moving when it passes the earth's track, and we find that the direction in which the comet moves and the direction in which the meteors move are identical. We have therefore learned that the orbit in which the Andromedes move and the orbit in which Biela's comet moves are coincident. This is, in itself, a strong and almost overwhelming presumption that the comet and the shooting stars are connected; but this is not all. This comet was observed in 1772, and again in 1805-6, before its periodic return in seven years was discovered. It was again discovered by Biela in 1826, and was observed in 1832. In 1846 the astronomical world was startled to find that there were now two comets in place of one, and the two fragments were again perceived at the return in 1852. No trace of Biela's comet was seen in 1859, nor in 1865-66, when its return was also due. It therefore happens that at the end of 1872 the time had arrived for the return of Biela's comet, and thus the occurrence of the great shower of the Andromedes took place about the time when Biela's comet was actually due. The inference is irresistible that the shooting stars, if not actually a part of the comet itself, are at all events most intimately connected therewith. This shower will also be memorable by reason of the telegram sent from Professor Klinkerfues to Mr. Pogson at Madras. The telegram ran as follows:—"Biela touched earth last night. Search near Theta Centauri." Pogson did search and did find a comet, though whether this was really the comet of Biela is still a matter which some astronomers doubt.

We have only mentioned three of the periodic showers; they are, indeed, the only three which produce a display worthy of note, from a picturesque point of view. There are, however, many others well known to those who have studied the subject; and the annual reports of the British Association are replete with information sedulously collected from every available source on the entire subject of luminous meteors.

It is a noticeable circumstance that the great meteoric showers seem never yet to have succeeded in projecting a missile which has reached the earth's surface. Out of the myriads of Leonids, of Perseids, or of Andromedes, not one particle has ever been seized and identified. Those bodies which do fall from the sky to the earth, and which we call meteorites, never come from the great showers, so far as we know. They seem, indeed, to be phenomena of quite a different character to the periodic meteors.

It is somewhat curious that the belief in the celestial origin of meteorites is of modern growth. In ancient times there were, no doubt, rumours of wonderful stones which had fallen down from the heavens on the earth, but these reports seem to have obtained but little credit. They were a century ago regarded as perfectly fabulous, though there was abundant testimony on the subject. Eye-witnesses averred that they had seen the stones fall. The stones themselves were unlike other objects in the neighbourhood, and cases were even authenticated where men had been killed by blows from these celestial visitors.

The objects were, however, generally seen by ignorant and illiterate persons. The true parts of the record were so mixed up with imaginary additions, that science refused to adopt the belief that such objects really fell from the sky. Even at the present day it is often extremely difficult to obtain accurate testimony as to the circumstances attending the fall of a meteorite. For instance, the fall of a meteorite was observed by a Hindoo in the jungle. The stone was there, its meteoric character was undoubted, and the witness was duly examined as to the details of the occurrence; but he was so frightened by the noise and by the danger he had narrowly escaped, that he could tell little or nothing. The only fact of which he felt certain was that the meteorite had hunted him for some time through the jungle before it fell to the earth!

In the year 1794 Chladni published an account of the remarkable mass of iron which the traveller Pallas had discovered in Siberia. It was then for the first time that the important step was taken of recognising that this object and others similar to it were really of celestial origin. But even Chladni's reputation and the arguments he brought forward failed to procure universal assent. Shortly afterwards, in 1795, a stone of fifty-six pounds was exhibited in London, which several witnesses declared they had seen fall at Wold Cottage, in Yorkshire; and it was subsequently deposited in the collection at the British Museum. The evidence then began to pour in from other quarters; portions of stone from Italy and from Benares were found to be identical with the Yorkshire stone. The incredulity of those who had doubted the celestial origin of these objects began to give way. A careful memoir on the Benares stone, by Howard, was published in the "Philosophical Transactions" for 1802, while, as if to complete the demonstration, a great shower of falling stones took place in the following year at L'Aigle, in Normandy. The French Academy deputed the physicist Biot to make a detailed examination on the spot of all the circumstances attending this memorable shower. His inquiry completely removed every trace of doubt, and the meteoric stones have accordingly been transferred from the dominions of geology to those of astronomy. It may be noted as a coincidence that the recognition of the celestial origin of meteorites was simultaneous with that of the discovery of the first of the series of minor planets. In each case our knowledge of the solar system has been extended by the addition of numerous minute bodies, which, notwithstanding their insignificant dimensions, are pregnant with information. Once the reality of these stone-falls has been admitted, we can turn with great interest to the ancient records, and restore to them the credit which many of them undoubtedly merit, and which was for so many centuries withheld. Perhaps the earliest of all the records which can be said to have much pretension to historical accuracy, is that of the shower of stones which Livy describes as having fallen, about the year 654 B.C., on the Alban Mount, near Rome. Among the more modern instances, we may mention one which was authenticated in a very emphatic manner. It occurred in the year 1492 at Ensisheim, in Alsace. The Emperor Maximilian ordered a minute narrative of the circumstances to be drawn up and deposited with the stone in the church. The stone was suspended in the church for three centuries, until in the French

Revolution it was carried off to Colmar, and pieces were broken from it, one of which is now in the British Museum. Fortunately, this interesting object has been restored to its ancient position in the church at Ensisheim, where it remains an attraction to sight-seers to this day. The account is as follows:—"In the year of the Lord 1492, on the Wednesday before St. Martin's Day, November 7th, a singular miracle occurred, for between eleven o'clock and noon there was a loud clap of thunder and a prolonged confused noise, which was heard at a great distance, and a stone fell from the air in the jurisdiction of Ensisheim which weighed 260 pounds, and the confused noise was at other places much louder than here. Then a child saw it strike on ploughed ground in the upper field towards the Rhine and the Ill, near the district of Gisgang, which was sown with wheat, and it did no harm, except that it made a hole there; and then they conveyed it from the spot, and many pieces were broken from it, which the Land Vogt forbade. They therefore caused it to be placed in the church, with the intention of suspending it as a miracle, and there came here many people to see this stone, so there were many remarkable conversations about this stone; the learned said they knew not what it was, for it was beyond the ordinary course of nature that such a large stone should smite from the height of the air, but that it was really a miracle from God, for before that time never was anything heard like it, nor seen, nor written. When they found that stone, it had entered into the earth to half the depth of a man's stature, which everybody explained to be the will of God that it should be found, and the noise of it was heard at Lucerne, at Villingen, and at many other places, so loud that the people thought that houses had been overturned; and as the King Maximilian was here, the Monday after St. Catherine's Day of the same year, his Royal Excellency ordered the stone which had fallen to be brought to the castle, and after having conversed a long time about it with the noblemen, he said that the people of Ensisheim should take it and order it to be hung up in the church, and not to allow anybody to take anything from it. His Excellency, however, took two pieces of it, of which he kept one, and sent the other to Duke Sigismund of Austria, and they spoke a great deal about the stone, which they suspended in the choir, where it still is, and a great many people came to see it."

Once we have recognised the celestial origin of the meteorites, they claim our closest attention. They afford the only direct method we possess of obtaining a knowledge of the materials of bodies exterior to our planet. We can take a meteorite in our hands, we can analyse it, and find the materials of which it is composed. We shall not attempt to enter into any very detailed account of the structure of meteorites; it is rather a matter for the consideration of chemists and mineralogists than for astronomers. A few of the more obvious features will be all that we require. They will serve as a preliminary to the discussion of the probable origin of these bodies.

In the British Museum we may examine a superb collection of meteorites. They have been brought together from all parts of the earth, and vary in size from bodies not much larger than a pin's head up to vast masses weighing many hundredweights. There are also many models of celebrated meteorites, of which the originals are dispersed through various other museums.

Many of these objects have nothing very remarkable in their external appearance. If they were met with on the sea beach, they would be passed by without more notice than would be given to any other stone. Yet, what a history such a stone might tell us if we could only manage to obtain it! It fell; it was seen to fall from the sky; but what was its course anterior to that movement? Where was it 100 years ago, 1,000 years ago? Through what regions of space has it wandered? Why did it never fall before? Why has it actually now fallen? Such are some of the questions which crowd upon us as we ponder over these most interesting bodies. Some of these objects are composed of very characteristic materials; take, for example, one of the more recent meteorites, known as the Rowton siderite. This body differs very much from the more ordinary kind of stony meteorite. It is an object which even a casual passer-by would hardly pass without notice. Its great weight would also attract attention, while if it be scratched or rubbed with a file, it would be found that it was not in any sense a stone, but that it was a mass of nearly pure iron. We know the circumstances under which that piece of iron fell to the earth. It was on the 20th of April, 1876, about 3.40 p.m., that a strange rumbling noise, followed by a startling explosion, was heard over an area of several miles in extent among the villages in Shropshire, eight or ten miles north of the Wrekin. About an hour after this occurrence, a farmer noticed the ground in one of his grass-fields to have been disturbed, and he probed the hole which the meteorite had made, and found it, still warm, about eighteen inches below the surface. Some men working at no great distance had actually heard the noise of its descent, but without being able to indicate the exact locality. This remarkable object weighs 7¾ lbs. It is an irregular angular mass of iron, though all its edges seem to have been rounded by fusion in its transit through the air, and it is covered with a thick black pellicle of the magnetic oxide of iron, except at the point where it first struck the ground. It was first exhibited publicly at a bazaar in Wolverhampton, and by the Duke of Cleveland, on whose property it fell, it was presented to the British Museum, where, as the Rowton siderite, it attracts the attention of every one who is interested in these wonderful bodies.

This siderite is specially interesting on account of its distinctly metallic character. Falls of the siderites (as they are called) are not so common as those of the stony meteorites; in fact, there are only a few known instances of meteoric irons having been actually seen to fall, while the falls of stony meteorites are to be counted in scores or in hundreds. The inference is that the iron meteorites are much less frequent than the stony ones. This is, however, not the impression that the visitor to the British Museum would be likely to receive. In that extensive collection the meteoric irons are by far the most striking objects. The explanation is not difficult. Those gigantic masses of iron are unquestionably meteoric: no one doubts that this is the case. Yet the vast majority of them have never been seen to fall; they have simply been found, under circumstances which point unmistakably to their meteoric nature. Suppose, for instance, that a traveller out on one of the plains of Siberia or of Central America finds a mass of metallic iron lying on the surface of the ground, what explanation can be rendered of such an occurrence? No one has brought the iron there, and there is no iron within hundreds of miles. Man never fashioned that object, and the iron is found to be alloyed with nickel in a manner that is always observed in known meteorites, and is generally regarded as a sure indication of a meteoric origin. Observe also, that as iron perishes by corrosion in our atmosphere, that great mass of iron cannot have lain there for indefinite ages; it must have been placed there at some finite time. There is only one source of such an object conceivable; it must have fallen from the sky. On those boundless plains the stony meteorites also have, doubtless, fallen in hundreds and in thousands, but the stony ones crumble away, and in any case would not arrest the attention of the traveller as the iron meteorites do. Hence it follows, that although the stony meteorites seem to fall much more frequently, yet, unless they are actually observed in the moment of descent, they are much more liable to be overlooked than the meteoric irons. Hence it is, that the great meteorites which are the characteristic feature of the British Museum collection are mainly the meteoric irons.

We have said that a loud noise accompanied the descent of the Rowton siderite, and the Emperor Maximilian has recorded the loud explosion which took place when the meteorite fell at Ensisheim. In this we have a characteristic feature of the phenomenon. Nearly all the descents of meteorites that have been observed, seem to have been ushered in by a loud detonating explosion. We do not, however, assert that this is a quite invariable character; and it is also the case that meteors often detonate without throwing down any stones which are observed. The violence associated with the phenomenon is forcibly illustrated by the Butsura meteorite. This object fell in India about twenty years ago. A loud explosion was heard, several fragments of stone were collected at distances miles apart; and when brought together, they were found to fit, so as to enable the primitive form of the meteorite to be re-constructed. A few of the pieces are wanting (they were, no doubt, lost by falling unobserved into localities from which they could not be recovered), but we have obtained pieces quite numerous enough to permit us to form a very good idea of the irregular shape of the object before the explosion occurred which shattered it into fragments. This is one of the ordinary stony meteorites, and is thus contrasted with the Rowton siderite which we have just been considering. There are also other types of meteorites. The Breitenbach iron, as it is called, is a good representation of a class of meteorites which are intermediate

Fig. 66.—Section of the Chaco Meteorite.

between the meteoric irons and the stones. It consists of a coarse cellular structure of iron, the cavities being filled with mineral substances. In the British Museum will be seen sections of intermediate forms in which this structure is exhibited.

Look first at the most obvious characteristic of these meteorites. We do not now allude to their chemical composition, but to their external appearance. What is the most remarkable feature in the shape of these objects?—surely it is that they are fragments. They are evidently pieces that are *broken* from some larger object. This is apparent by merely looking at their form; it is still more manifest when we examine their mechanical structure. It is often found that meteorites are themselves composed of smaller fragments, united together in the manner of certain well-known volcanic rocks; others are, again, composed of very small particles, analogous to what are known as volcanic tufas. The structure of the meteorites may be illustrated by a section of an aërolite found on the Sierra of Chaco, weighing about 30 lbs. (Fig. 66).

The section here represented shows the composite structure of this object, which belongs to the class of stony meteorites. Its shape shows that it was really a fragment with angular edges and corners. No doubt it may have been much more considerable when it first dashed into the atmosphere. The angular edges now seen on the exterior may be due to an explosion which then occurred; but this will not account for the structure of the interior. We there see irregular pieces of the most varied form and material conglomerated into a single mass. If we would seek for any analogous objects on the earth, we must look to some of the volcanic rocks, where we have multitudes of irregular angular fragments adhering together by a cement in which they are imbedded. The evidence presented by this meteorite is conclusive as to one circumstance, with regard to the origin of these objects. They must have come as fragments, from some body of considerable, if not of vast, dimensions. In this meteorite there are numerous small grains of iron mingled with mineral substances. The iron in many meteorites has, indeed, characters resembling those produced by the actual blasting of iron by dynamite. Thus, a large meteoric iron from Brazil has been found to have been actually shivered into fragments at some time anterior to its fall on the earth. These fragments have been cemented together again by irregular veins of mineral substances.

For an aërolite of a very different type we may refer to the carbonaceous meteorite of Orgueil, which fell in France on the 14th May, 1864. On the occasion of its descent a splendid meteor was seen, rivalling the full moon in size. The actual diameter of this globe of fire must have been some hundreds of yards. There were actually found nearly a hundred fragments of the body scattered about over a tract of country fifteen miles long. This object is of particular interest, inasmuch as it belongs to a rare group of aërolites, from which metallic iron is absent. It contains many of the same minerals which are met with in other meteorites, but in these objects they are *associated with carbon,* and with substances of a white or yellowish crystallisable material, soluble in ether, and resembling some of the hydro-carbons. Such a substance, if found on the earth, would probably be deemed a product resulting from animal or vegetable life!

We have pointed out how a body moving with great velocity and impinging upon the air may become red-hot and white-hot, or even driven off into vapour. How, then, does it happen that meteorites escape this fiery ordeal, and fall down to the earth, no doubt with a great velocity, but still, with a velocity very much less than that which would have sufficed to drive them off into vapour? Had that Rowton siderite, for instance, struck our atmosphere with a velocity of twenty miles a second, it seems unquestionable that it would have been entirely dissipated into vapour, though, no doubt, the particles would ultimately coalesce together, and slowly descend to the earth in microscopic beads of iron. How has the meteorite escaped this fate? It must be remembered that our earth is also moving with a velocity of about eighteen miles per second, and that the velocity with which the meteorite plunges into the air is to be obtained by compounding the velocity of the meteorite with the velocity of the earth. If the meteorite come into direct collision with the earth, the velocity of the collision will be extremely great; but it may happen that though the actual velocities of the two bodies are both enormously great, yet the relative velocity may be comparatively small. This is, at all events, one conceivable explanation of the arrival of a meteorite on the surface of the earth.

We have shown in the earlier parts of this chapter that the well-known star showers are all intimately connected with comets. In fact, each star shower revolves in the path pursued by a comet, and the shooting star particles have, in all probability, been themselves derived from the comet. Showers of shooting stars and comets have, therefore, an intimate connection, but there is no ground for supposing that meteorites have any connection with comets—the facts, indeed, all seem to me to point in the opposite direction. It has already been remarked that meteorites have never been known to fall from the great star showers. No particle of a meteorite was ever dropped from the countless host of the Leonids or of the Perseids; the Lyraids never dropped a meteorite, nor did the Quarantids, the Geminids, or the many other showers with which every astronomer is familiar. There is no reason to connect meteorites with these showers, and there is, accordingly, no reason to connect meteorites with comets. Indeed, the appearance of a comet and the history of its movements and its changes seem entirely at variance with the supposition that it is composed of materials resembling those in meteorites.

With reference to the origin of meteorites it is difficult to speak with any great degree of confidence. Every theory of meteorites is in itself improbable, so it seems that the only course open to us is to choose that view of their origin which seems least improbable. It appears to me that this condition is best fulfilled in the theory entertained by the Austrian mineralogist, Tschermak. He has made a study of the meteorites in the rich collection at Vienna, and he has come to the conclusion that the "meteorites have had a volcanic source on some celestial body." Let us attempt to pursue this hypothesis and discuss the problem, which may be thus stated:—Assuming that meteorites have been ejected from volcanoes, on what body or bodies in the universe must these volcanoes be situated? This is really a question for astronomers and mathematicians. Once the mineralogists assure us that these bodies are volcanic, the question becomes one of calculation and of the balance of probabilities.

The first step in the inquiry is to realise distinctly the dynamical conditions of the problem. Conceive a volcano to be located on a planet. The volcano is supposed to be in a state of eruption, and in one of its mighty throes projects a missile aloft: this missile will ascend, it will turn round, and fall down again. Such is the case at present in the eruptions of terrestrial volcanoes. Cotopaxi has been known to hurl prodigious stones to a vast height, but these stones assuredly return to earth. The gravitation of the earth has gradually overcome the velocity produced by the explosion, and down the body falls. But let us suppose that the eruption is still more violent, and that the stones are projected from the planet to a still greater height above its surface. Suppose, for instance, that the stone should be shot up to a height equal to the planet's radius, the attraction of gravitation will then be reduced to one-fourth of what it was at the surface, and hence the planet will find greater difficulty in pulling back the stone the higher it ascends. Not only is the distance through which the stone has to be pulled back increased as the height increases, but the efficiency of gravitation is weakened, so that in a two-fold way the difficulty of recalling the stone is increased. Once again, let us suppose the stone projected with still greater speed, it will rise up and up; the gravitation is still tending to pull the stone back, but as it ascends the gravitation is lessened; when the stone has attained a height of ten times the radius of the planet, the attraction is reduced to one hundredth part of its original intensity, and thus it may happen that if the velocity be sufficiently great, the stone will rise higher and higher, and gravitation will never be able to recall it. There is thus a certain critical velocity appropriate to each planet, and depending on its mass. If the missile be projected upwards with a velocity equal to this, then it will ascend never to return. We all recollect Jules Verne's voyage to the moon, in which he described the Columbiad, an imaginary cannon, capable of shooting out a projectile with a velocity of six miles a second. This is the critical velocity for the earth. If we could imagine the air removed, then a cannon of six-mile power would project a body upwards which would not fall down.

The great difficulty about Tschermak's view of the volcanic origin of the meteorites, lies in the tremendous initial velocity which is required. The Columbiad is a myth, and we know no agent, natural or artificial, at the present time on the earth, adequate to the production of a velocity so appalling. The thunders of Cotopaxi are said to be heard one hundred and fifty miles away, but in its mightiest throes, it discharges no missiles at the present time with a velocity of six miles a second. We are therefore led to inquire whether any of the other celestial bodies are entitled to the parentage of the meteorites. We cannot see volcanoes on any other body except the moon; all the other bodies are too remote for an inspection so minute. Does it seem likely that volcanoes on the moon can ever launch forth missiles which fall upon the earth?

This was a belief sustained by eminent authority. The mass of the moon is very small as compared with that of the earth. It would take over eighty moons rolled into one, to equal the mass of the earth. A body could therefore be projected from the surface of the moon, so as to leave the moon altogether, with much less velocity than is required to project a body entirely away from the surface of the earth. It would not be true to assert that the velocity of projection varies directly as the masses. The correct law is, that it varies directly as the square root of the mass, and inversely as the square root of the radius. It is hence easy to see that the velocity required to project a missile from the moon is only about one-sixth of that which would be required to project a missile from the earth. If the moon had on its surface volcanoes of one mile power, it is quite conceivable that these might be the source of meteorites. We have seen how the surface of the moon shows traces of intense volcanic activity. There are vast craters, which must have been once the seat of mighty eruptions, so that the whole surface of the moon can only be compared with the most intensely volcanic regions on the earth. A missile thus projected from the moon could undoubtedly fall on the earth, and it is not impossible that some of the meteorites may really have come from this source. There is, however, one great difficulty about the volcanoes on the moon which must be borne in mind. Suppose an object were so projected, it would, under the attraction of the earth, in accordance with Kepler's laws, move around the earth as a focus. If we set aside the disturbances produced by all other bodies, as well as the disturbance produced by the moon itself, we see that the meteorite if it once misses the earth can never fall thereon. It would be necessary that the shortest distance of the earth's centre from the orbit of the projectile shall be less than the radius of the earth, so that if a lunar meteorite is to fall on the earth, it must do so the first time it goes round. The journey of a meteorite from the moon to the earth is only a matter of days, and therefore, as meteorites are still falling, it would follow that they must still be constantly ejected from the moon. The volcanoes on the moon are, however, not now active; observers have long studied its surface, and they find no reliable traces of volcanic activity at the present day. It is utterly out of the question, whatever the moon may once have been able to do, that at the present date she should still continue to launch forth meteorites. It is just possible that a meteorite expelled from the moon in remote antiquity, when its volcanoes were active, may, under the influence of the disturbances previously excepted, have its orbit so altered, that at length it comes within the reach of the atmosphere and falls to the earth, but under no other circumstances could the moon send us a meteorite at present. We are therefore compelled to look elsewhere in our search for volcanoes fulfilling the conditions of the problem.

Let us now direct our attention to the planets, and examine the circumstances under which volcanoes located thereon could eject a meteorite which should ultimately tumble on the earth. We cannot see the planets well enough to tell whether they have or ever had any volcanoes; but the almost universal presence of heat in the large celestial masses seems to leave us in little doubt that some form of volcanic action may be found in the planets. We may at once dismiss the giant planets, such as Jupiter or Saturn: their appearance is very unlike a volcanic surface; while their great mass would render it necessary to suppose that the meteorites were expelled with terrific velocity if they are to succeed in escaping from the gravitation of the planet. Applying the rule already given, a volcano on Jupiter would have to be five or six times as powerful as the volcano on the earth. To avoid this difficulty, we naturally turn to the smaller planets of the system; take, for instance, one of that innumerable host of minor planets, and let us inquire how far this body is likely to have ejected a missile which should fall upon the earth. Some of these planets are only a few miles in diameter. There are bodies in the solar system so small that a very moderate velocity would be sufficient to project a missile away from them altogether. We have, indeed, already illustrated this point in discussing the minor planets. In a cricket match, on a planet twenty miles in diameter, a successful hit might actually drive the ball away never to return; the field would wait in vain for its descent, the ball would have started off in a conic section around the sun in its focus, and would never return to the planet from which it had been driven. It has been supposed that a volcano placed on one of the minor planets might be quite powerful enough to drive off the meteorites so as to wander through space until the chapter of accidents brought them into collision with the earth. There is but little difficulty in supposing that there should be such volcanoes, and that they should be sufficiently powerful to drive bodies from the surface of the planet; but we must remember that the bodies are to fall down on the earth, and certain dynamical considerations here come in which merit our close attention. To concentrate our ideas, we shall consider one of the minor planets in particular, and for this purpose let us take Ceres, which is a small object revolving in an orbit exterior to Mars. If a meteorite is to fall upon the earth, it must pass through the narrow ring some 8,000 miles wide, which marks the earth's path; it will not do for the missile to pass through the ecliptic on the inside or on the outside of the ring, it must be actually through this narrow strip, and then if the earth happens to be there at the same moment the meteorite will fall. The first condition to be secured is, therefore, that the path of the meteorite shall traverse this narrow ring. This is to be effected by projection from some point in the orbit of Ceres. The missile on leaving Ceres will ascend in a conic section around the sun in the focus, and will then descend so as to cross the ecliptic again, at a point within the narrow strip referred to. Ceres is itself moving in its orbit with a high velocity of about eleven miles a second, so that the actual velocity of the projectile will be the resultant obtained by compounding the volcanic velocity with the orbital velocity. It can be shown on purely dynamical grounds that a meteorite projected from a point in the orbit of Ceres could not intersect the earth's orbit unless the velocity perpendicular to the radius at the instant of projection were about eight miles a second. But the velocity of the planet is about eleven miles a second in this direction. It would therefore be necessary for the volcanic velocity to abate the velocity perpendicular to the radius by at least three miles a second. This is a significant result; the volcanic energy sufficient to remove the projectile from Ceres may be small enough, but if that projectile is ever to cross the earth's track, the dynamical requirements of the case show that we must have a volcano on Ceres at the very least of three-mile power. We have thus gained but little by the suggestion of a minor planet, for we have not found that a moderate volcanic power would be adequate. But there is another difficulty in the case of Ceres, inasmuch as the ring on the ecliptic is very narrow in comparison with the other dimensions of the problem. Ceres is a long way off, and it would require very great accuracy in volcanic practice on Ceres to project a missile so that it should just traverse this ring and fall neither inside nor outside. There must be a great many misses for every hit. We have attempted to make the calculation by the aid of the theory of probabilities, but it is a question not very easy to state definitely. If the total velocity with which the projectile leaves the orbit of Ceres be less than eight miles a second, then the projectile will fall short of the earth's track. On the other hand, if the total initial velocity exceed sixteen miles a second the orbit in which the projectile passes will be hyperbolic, and though it may cross the earth's track once, it will never do so again. Taking a mean between these extreme velocities, we may investigate the following problem. Suppose that a projectile is discharged in a random direction from Ceres with the total initial velocity of twelve miles a second (resulting, of course, from the composition of the velocity of projection with the actual orbital velocity of Ceres), what is the probability that this projectile shall cross the earth's track? Solved by the theory of probabilities, we find that the chances against this occurrence are about 50,000 to 1, so that out of every 50,000 projectiles hurled from a point in the orbit of Ceres only a single one can be expected to cross the track of the earth. It is thus evident that there are two objections to Ceres (and the same may be said of the other minor planets) as a possible source of the meteorites. Firstly, that notwithstanding the small mass of the planet a very powerful volcano would still be required ; and secondly, that we are obliged to assume that for every meteorite which could ever fall on the earth at least 50,000 must have been ejected. It is thus plain that if the meteorites have really been driven from some planet of the solar system, large or small, the volcano must, from one cause or another, be a very powerful one. As we must have a very powerful volcano in any case, we are led to inquire which planet possesses on other grounds the greatest probability in its favour.

We have already spoken of the volcanoes on the earth, and we admit of course that at the present time they are utterly devoid of the necessary energy; but were the terrestrial volcanoes always so feeble as they are at present? Grounds are not wanting for the belief that in the very early days of geological time the volcanic energy on the earth was much greater than at present. We admit fully the difficulties of the view that the meteorites have really come from the earth; but they must have some origin, and it is reasonable to indicate the source which seems to have most probability in its favour. Grant for a moment that in the primæval days of volcanic activity there were some mighty throes which hurled forth missiles with the adequate velocity : these missiles would ascend, they would pass from the gravitation of the earth, they would be seized by the gravitation of the sun, and they would be compelled to revolve around the sun for ever after. No doubt the resistance of the air would be a very great difficulty, but this resistance would be greatly lessened were the crater at a very high elevation above the sea level. Some of these objects might perhaps revolve in hyperbolic orbits, and retreat never to return, while others would be driven into elliptic paths. Round the sun these objects would revolve for ages, but at each revolution—and here is the important point—they would traverse the point from which they were originally launched. In other words, every object so projected from the earth, would at each revolution cross the track of the earth. We have in this fact an enormous probability in favour of the earth as contrasted with Ceres. Only one Ceres-ejected meteorite out of every 50,000 could possibly cross the earth's track, while every earth-projected meteorite would necessarily do so.

If this view be true, then there must be countless hosts of meteorites traversing space in elliptic orbits around the sun. All these orbits have one feature in common: they all intersect the track of the earth. It will sometimes happen that the earth is found at this point at the moment the meteorite is crossing; when this is the case the long travels of the meteorite are at an end, and it tumbles back on the earth from which it parted so many ages ago.

It is well to emphasise the contrast between the lunar theory of meteorites (which we think improbable), and the terrestrial theory (which appears to be probable). For the lunar theory it would, as we have seen, be necessary that some of the lunar volcanoes should be still active. In the terrestrial theory it is only necessary to suppose that the volcanoes on the earth once possessed sufficient explosive energy. No one supposes that the volcanoes at present on the earth eject now the fragments which are to form future meteorites; but it seems probable that the earth may be now slowly gathering back, in these quiet times, the fragments she ejected in an early stage of her history. Assuming, therefore, with Tschermak that the meteorites have had a volcanic origin on some considerable celestial body, we are led to agree with those who think that most probably that body is the earth.

We cannot forbear to mention one or two circumstances which seem to corroborate the view that the meteorites are really of ancient terrestrial origin. The most characteristic constituent of meteorites is the alloy of iron and nickel, which is almost universally present. Sometimes, as in the Rowton siderite, the whole object consists of little else, sometimes this alloy is in grains distributed through the mass. When Nordenskjöld discovered in Greenland a mass of native iron containing nickel, this was at once regarded as a celestial visitor. It was called the Ovifak meteorite, and large pieces of the iron were conveyed to our museums. There is, for instance, in the British Museum a most interesting exhibit of the Ovifak substance. Close examination shows that this so-called meteorite lies in a bed of basalt which has been vomited from the interior of the earth. Those who believe in the meteoric origin of the Ovifak iron are constrained to admit that shortly after the eruption of the basalt, and while it was still soft, this stupendous iron meteorite of gigantic mass and bulk happened to fall into this particular soft bed. The view is, however, steadily gaining ground that this great iron mass was no celestial visitor at all, but that it simply came forth from the interior of the earth with the basalt itself. The beautiful specimens in the British Museum show how the iron graduates into the basalt in such a way as to make it highly probable that the source of the iron is really to be sought in the earth and not external thereto. Should further research establish this, as now seems probable, a most important step will have been taken in proving the terrestrial origin of meteorites. If the Ovifak iron be really associated with the basalt, we have a proof that the iron-nickel alloy is indeed a terrestrial substance, found deep in the interior of the earth, and associated with volcanic phenomena. This being so, it will be no longer difficult to account for the iron in undoubted meteorites. When the vast volcanoes were in activity they ejected masses of this iron-alloy, which, having circulated round the sun for ages, have at last come back again. As if to confirm this view, Professor Andrews discovered particles of native iron in the basalt of the Giant's Causeway, while the probability that large masses of iron are there associated with the basaltic formation was proved by the researches on magnetism of the late Provost Lloyd.

Besides the more solid meteorites which seem to be terrestrial, there can be no doubt that the debris of the ordinary shooting stars must rain down upon the earth in gentle showers of celestial dust. The evidence on this point is overwhelming. The snow in the Arctic regions has often been found stained with traces of dust which contains particles of iron. Similar particles have been found in the towers of cathedrals and under many other circumstances. There can be hardly a doubt that some of the motes in the sunbeam, and many of the particles which good housekeepers abhor as dust, are really of a cosmical origin. In the famous cruise of the *Challenger* the dredges brought up from the depths of the Atlantic no "wedges of gold, great anchors, heaps of pearl," but among the mud which they raised are to be found numerous magnetic particles which there is every reason to believe fell from the sky, and thence subsided to the depths of the ocean. Sand from the deserts of Africa, when examined under the microscope, yields traces of minute iron particles which bear the marks of having experienced a high temperature.

The earth draws in this cosmic dust continuously, but the earth now never parts with a particle of its mass. The consequence is inevitable; the mass of the earth must be growing, and though the change may be a small one, yet to those who have studied Darwin's treatise on "Earth-worms," or to those who are acquainted with the modern theory of evolution, it will be manifest that stupendous results can be achieved by slight causes which tend in one direction. It is quite probable that an appreciable part of the solid substance of the earth may thus have been derived from meteoric matter, which in perennial showers descends upon its surface.

CHAPTER XVIII.

THE STARRY HEAVENS.

Whence the Importance of the Solar System?—Home—View in Space—Other Stellar Systems—The Sun a Star—Stars are Self-Luminous—We see the Points of Light, but nothing else—The Constellations—The Great Bear and the Pointers—The Pole Star—Cassiopeia—Andromeda, Pegasus, and Perseus—The Pleiades: Auriga, Capella, Aldebaran—Taurus, Orion, Sirius; Castor and Pollux—The Lion—Boötes, Corona, and Hercules—Virgo and Spica—Vega and Lyra—The Swan.

In the previous chapters of this work we have considered the sun and the system of planets and other bodies which revolve in obedience to his potent sway. We have found in the survey of this system much to impress us, and much which is calculated to awaken our conceptions to the stupendous scale on which the heavens are constructed. It is, however, desirable that we should not disguise from our thoughts the circumstances which lead us to attribute to the solar system a position of such importance in the scheme of the universe. It is the fact of our residence on a planet belonging to the solar system, which gives the solar system its great importance in our eyes. The solar system is our abode in the universe. The other planets, Mercury, Venus, Mars, Jupiter, and Saturn are our neighbours; the moon is our inseparable attendant; while the sun is the lamp which gives us light, and the fire which gives us warmth. It thus happens that in our eyes the solar system has the interest and the familiarity of home. We are surrounded by it; its welfare is our welfare. We live our lives without perhaps bestowing many thoughts as to whether there may not be other systems besides ours, or as to whether our solar system may not be far surpassed in size and in splendour by many, perhaps we might say by countless, other systems in the universe.

When we become aware of the existence of such systems, a multitude of questions arise. We long to know their details; we long to know the sizes of those great suns, and the dimensions of the planets which circulate around them; we long to see the configuration of their planets, to learn what their surface is like, and, it may be, to speculate on the possibility of their being inhabited. We long to compare or to contrast such systems with our own—to ask whether those planets have satellites, as ours have; whether those suns show the remarkable features which characterise our sun; whether they are attended like our own sun by innumerable comets. Have they showers of shooting stars, as we have? Such are a few of the questions which occur to us when we pass from a review of the solar system to the study of the other systems in space.

These questions can be asked, but can they be answered? In great part they cannot. These other systems are plunged into space to a distance so appalling, that all those details which we are so anxious to learn become utterly lost by distance. Of the planets surrounding these systems we can see nothing, though here and there their presence can be, as it were, felt, and thus their existence is made probable. Of these systems, all that we see are what we call the stars—those gems of light that stud the midnight sky. These stars are the suns of the distant systems, but their distance is such that they have entirely lost the glory which we attach to the word sun. Yet that they really are suns can be easily shown. We shall merely allude to the subject now, and we shall discuss it more fully hereafter. Suppose that a traveller, endowed with a miraculous power of voyaging through space, were to start from the solar system and soar away into the stellar regions. As he receded he would find that the earth and the other planets dwindled down and gradually became invisible. Still further on he would find that the sun began to lose its splendour, and even its pre-eminence, among the celestial host. On and on the traveller wings his way, until, by the time he has travelled over a distance comparable with that which intervenes between the sun and the stars, he would find that our sun had dwindled down to a star not so bright as many of those that twinkle in our skies every night. We are thus led to realise that our solar system is a little island group, situated at the most appalling distances from the stars. The solar system is isolated from its neighbours just as a rock a few yards square in the middle of the Atlantic would be isolated from the coasts of Europe and of America.

It is very essential to observe that the stars we see are bodies which shine by their own light. Aldebaran shines with a light not very unlike that which comes from Mars. The planets are, indeed, not unfrequently mistaken for stars, and stars are often mistaken for planets; yet no language can be too strong to emphasise the vast difference between bodies belonging to these two classes. If Aldebaran looks like Mars, the resemblance is quite a fortuitous one. Aldebaran is probably thousands of times as large as Mars, and it is certainly hundreds of thousands of times as far off. But this is by no means all the difference; it is not even the chief difference. Aldebaran is a sun. It has its own light, and is a body intensely heated and glowing, just like our sun; but Mars has no light of its own. Mars is as dark as an ordinary stone, and is only visible from the fact that the rays of light from our sun shine upon it.

We are thus led to perceive, that although the points of light which we see in stellar space are one and all suns on their own account, yet we do not see—we need not expect to see—any of the dark bodies in their neighbourhood which are illuminated by their light. We can hardly avoid acquiring by this thought a greatly enlarged conception of the importance and the extent of the stellar regions. As a traveller at night looks down on the distant city from the height which he has attained, he sees in that city, not the houses or the monuments, not even the churches or the greatest buildings, all he sees are the bright points of light scattered here and there in the gloom. Some, perhaps, may be the lights which serve to guide the wayfarer through the streets; some may be the lights proceeding from the houses where rejoicing and feasting are going on; some will be the lights where patient watchers minister to the wants of the sick and the dying; but of the purport of these lights, of what these lights illuminate, the distant traveller sees nothing and knows nothing—he sees the lights, but he can see nothing else. So it is when we look on the starry host—we see the bright points of light, but we see nothing else; of all the dark objects illuminated by those lights we see absolutely nothing. We cannot resist the conjecture that this unseen universe is of great interest and complexity, though we are unable to see anything more than the system of lights by which it is illuminated.

But though our acquaintance with the sidereal universe is so narrowly limited that only objects of colossal dimensions can be seen in our greatest telescopes, yet the field within our reach is replete with interest. Let us in this chapter make a general survey of the sidereal heavens, and we shall subsequently develop the subject in its different aspects.

The student of astronomy should make himself acquainted with the principal constellations in the heavens. This is a most pleasing acquirement, and should form a part of the education of every child in the kingdom. We shall commence our discussion of the sidereal system with a brief account of the principal constellations visible in the northern hemisphere, and we shall accompany our description with such outline maps of the stars as will enable the beginner to identify the objects we shall name.

We have in an earlier chapter directed the attention of the student to the remarkable constellation of stars which are known to astronomers as Ursa Major, or the Great Bear. It is the most conspicuous group in the northern skies, and in northern latitudes it never sets. At eleven p.m. in the month of April the Great Bear is directly overhead; at the same hour in the month of September it is low down in the north; in July it will be seen in the west; at Christmas it is in the east. From the remotest antiquity this group of stars has attracted attention. The stars in the Great Bear were comprised in the great catalogue of stars, made two thousand years ago, which has been handed down to us. From the positions of the stars given in this catalogue it is possible to reconstruct the Great Bear as it appeared two thousand years ago. This has been done, and it appears that the seven principal stars have not changed in this lapse of time to any large extent, so that the configura-

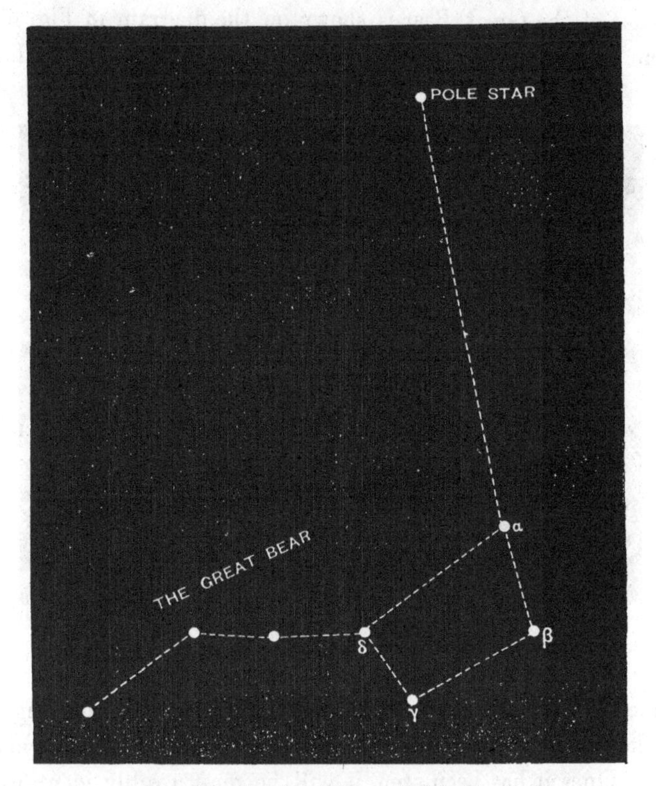

Fig. 67.—The Great Bear and Pole Star.

tion of the Great Bear remains practically the same now as it was then. The beginner must first obtain an acquaintance with this group of seven stars, and then his further progress in this branch of astronomy will be greatly facilitated. The Great Bear is, indeed, a splendid constellation, and its only rival is to be found in Orion, which contains more brilliant stars, though it does not occupy so large a region in the heavens.

In the first place, we observe how the Great Bear enables the Pole Star, which is the most important object in the northern heavens, to be readily found. The Pole Star is very conveniently indicated by the direction of the two stars, β and a, of the Great Bear, which are, accordingly, generally known as the "pointers." This use of the Great Bear is shown on the diagram in Fig. 67, in which the line β a, produced onwards and slightly curved, will

Fig. 68.—The Great Bear and Cassiopeia.

conduct to the Pole Star. There is no likelihood of making any mistake in this star, as it is the only bright one in the neighbourhood. Once it has been seen, it will be most readily identified on future occasions, and the observer will not fail to notice how constant is the position which it preserves in the heavens. The other stars either rise or set, or, like the Great Bear, they dip down low in the north without actually setting, but the Pole Star exhibits no changes of this magnitude. In summer or winter, by night or by day, the Pole Star is ever found in the same place—at least, so far as ordinary observation is concerned. No doubt, when we use the accurate instruments of the observatory the notion of the fixity of the Pole Star is dissipated; we then see that it has a slow motion, and that it describes a circle every twenty-four hours around the true pole of the heavens, which is not coincident with the Pole Star, though closely adjacent thereto. This distance is at present somewhat less than a degree and a half, and it is gradually lessening, until, in the year A.D. 2095, the distance will be under half a degree.

The Pole Star itself belongs to another not inconspicuous group of stars, known as the Little Bear. The two principal stars of this group, next in brightness to the Pole Star, are sometimes called the "Guards." The Great Bear and the Little Bear, with their cynosure the Pole Star, form a group in the northern sky not paralleled, either in beauty or in utility, by anything in the southern heavens. At the South Pole there is no conspicuous star to indicate its position approximately—a circumstance disadvantageous to astronomers and to navigators in the southern hemisphere.

It will now be easy to add a third constellation to the two already acquired. On the opposite side of the Pole Star to the Great Bear, and at about the same distance, lies a very pleasing group of five bright stars, forming a W. These are the more conspicuous members of the constellation Cassiopeia, which contains altogether about fifty-five lucid stars. When the Great Bear is low down in the north, then Cassiopeia, on the contrary, is high overhead; when the Great Bear is high overhead, then Cassiopeia is to be looked for low down in the north. The configuration of the leading stars is so striking, that once the eye has recognised them, future identification will be very easy—the more so when it is borne in mind that the Pole Star lies midway between Cassiopeia and the Great Bear (Fig. 68). These important constellations will serve as guides to all the rest. We shall accordingly proceed step by step in showing how the learner is to distinguish the various other groups visible from the British Islands or similar northern latitudes.

The next constellation to be recognised is the imposing group which contains the Great Square of Pegasus. This is not, like Ursa Major, or like Cassiopeia, said to be "circumpolar." The Great Square of

Pegasus sets and rises daily. It cannot be seen conveniently during the spring and the summer, but in autumn and in winter the four stars which mark the corners of the square can be very easily perceived. There are certain small stars within the square; perhaps about thirty can be counted by an unaided eye of ordinary power in these latitudes. In the south of Europe, where purer and brighter skies are to be found, the number of small stars appears to be

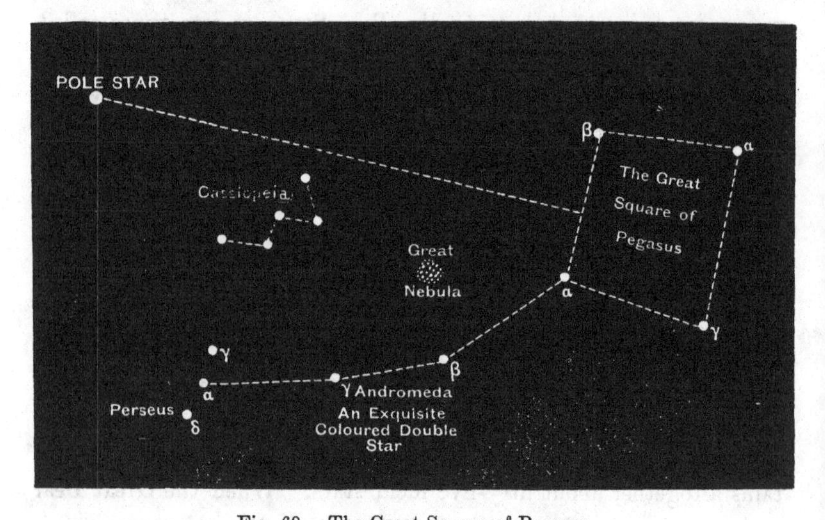

Fig. 69.—The Great Square of Pegasus.

greatly increased. An acute observer at Athens has counted 102 in the same region.

The Great Square of Pegasus can be readily identified by imagining a line from the Pole Star over the end of Cassiopeia. This line produced about as far again will conduct the eye to the centre of the Great Square of Pegasus (Fig. 69).

The line through β and a in Pegasus produced 45° to the south points out the important star Fomalhaut in the mouth of the southern fish. To the right of this line, nearly half-way down, is the rather vague constellation of Aquarius, where a small equilateral triangle with a star in the centre may be noticed.

The Square of Pegasus is one illustration of the absurd way in which the boundaries of the constellations are defined. There can be no more completely associated group than the four stars of this square, and all four ought surely to belong to the same constellation. Three of the corners—marked *a*, *β*, *γ*—do belong to Pegasus, but the fourth corner—also marked *a*—is sometimes placed in a different constellation, known as Andromeda, whereof it is, indeed, the brightest member. The remaining part of the constellation Andromeda—or, at all events, its brightest stars—can now be readily found. They are marked *β* and *γ*, and are readily identified by producing one side of the Square of Pegasus in a curved direction. We have now a remarkable group of seven stars, which it is easy to identify and easy to remember, although formed out of parts belonging to three different constellations. They are respectively *a*, *β*, and *γ* from Pegasus, *a*, *β*, and *γ* from Andromeda, and *a* from Perseus. The three form a sort of handle, as it were, extending to a great length on one side of the square, and are a group both striking in appearance, and very useful in the further identification of the celestial objects. *β* Andromedæ with two smaller stars form the girdle of the unfortunate heroine.

a Persei lies between two other stars (*γ* and *δ*) of the same constellation. The three form a curve, and if we prolong that curve, we are conducted to one of the gems of the northern heavens—the beautiful reddish star Capella, in the constellation of Auriga (Fig. 7). Close to Capella are three small stars forming an isosceles triangle, these are the Hœdi or Kids. Capella and Vega are the two most brilliant stars in the northern heavens; and though Vega, with its whiter ray, is generally considered the more lustrous of the two, yet the opposite opinion has been entertained. Different eyes will frequently form various estimates of the relative brilliancy of stars which approach each other in brightness. The difficulty of making a satisfactory comparison between Vega and Capella is greatly increased by the wide distance in the heavens by which they are separated, as well as by a slight difference in colour, for Vega is distinctly whiter than Capella. This contrast between the colour of stars is often another source of uncertainty in the attempt to compare their relative brilliancy; so that when actual measurements are to be effected by instrumental means, it is necessary to compare the two stars alternately with some object of intermediate colour.

On the opposite side of the pole to Capella, but not quite so far away, will be found four stars in a quadrilateral. They form

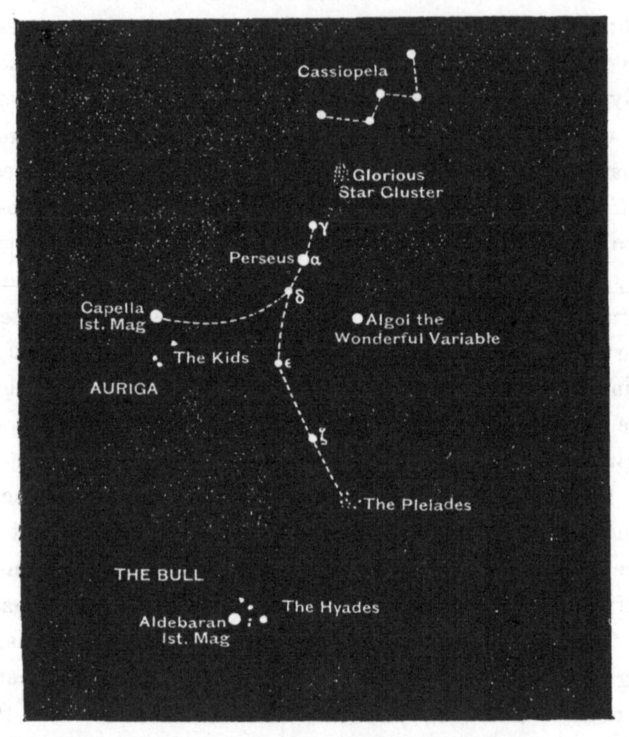

Fig. 70.—Perseus and its neighbouring Stars.

the head of the dragon, the rest of whose form coils right round the pole.

If we continue the curve formed by the three stars γ, a, and δ in Perseus, and if we bend round this curve gracefully into one of an opposite flexion, in the manner shown in Fig. 70, we are first conducted to two other principal stars in Perseus, marked ϵ and ζ. Perseus is one of the richest regions of the heavens. We have here a most splendid portion of the Milky Way, and the fields of the telescope are crowded with stars beyond number. Even a small telescope or an opera-glass directed to this teeming constellation cannot fail to delight the observer, and convey to him a profound impression of the extent and majesty of the sidereal heavens. We shall return later on to a brief enumeration of some of the remarkable telescopic objects in Perseus. Pursuing in the same figure the line ϵ and ζ, we are conducted to the remarkable little group known as the Pleiades. We have here a suggested resemblance to the Great Bear on a much smaller scale; indeed, in many ways the designation of Little Bear would be applied with greater

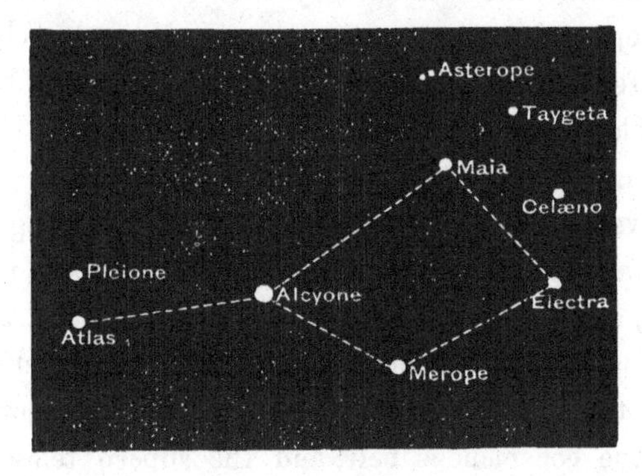

Fig. 71.—The Pleiades.

propriety to the Pleiades than to the constellation which is actually so called.

The Pleiades form a group so universally known and so easily identified, that it hardly seems necessary to give any further specific instructions for their discovery. It may, however, be observed that in these latitudes they cannot be seen during the summer. Let us suppose that the search is made at about 11 p.m. at night: on the 1st of January the Pleiades will be found high up in the sky in the south-west; on the 1st of March, at the same hour, they will be seen to be setting in the west. On the 1st of May they are not visible; on the 1st of July they are not visible; on the 1st of September they will be seen low down in the east. On the 1st of November they will be high in the heavens in the south-east. On the ensuing 1st of January the Pleiades will be in the same position as they were on the same date in the previous year, and so on throughout the cycle. It need, perhaps, hardly be explained here that these changes are not really due to movements of the constellations; they are due, of course, to the apparent annual motion of the sun among the stars.

The Pleiades are shown in the figure on the previous page, a group of ten, being about the number visible with the unaided eye to those who are gifted with very acute vision. The lowest telescopic power will increase the number of stars up to thirty or forty (Galileo saw more than forty with his first telescope), while with telescopes of greater power the number is largely increased; indeed, no fewer than 625 have been counted with the aid of a powerful telescope. The group is, however, rather too large and coarse for an effective telescopic object, except with a large field and low power. In an opera-glass it is a very pleasing spectacle.

If we draw a ray from the Pole Star to Capella, a point already determined, and produce this ray on sufficiently far, as shown in the adjoining figure, we come to the glorious constellation of our winter sky, the splendid group of Orion. The brilliancy of the stars in Orion, the conspicuous belt, and the superb telescopic objects which it contains, alike render this group remarkable, and place it perhaps at the head of the constellations. The leading star in Orion is known either as α Orionis, or as Betelgueze, by which name it is designated in the figure. It lies above the three stars, δ, ϵ, and ζ, which form the belt. Betelgueze is a star of the first magnitude, and so also is Rigel, on the opposite side of the belt. Orion thus enjoys the distinction of containing two stars of the first magnitude in its group, while the five other stars here shown are of the second magnitude.

The neighbourhood of Orion contains some of the most important stars. If we carry on the line of the belt upwards to the right, we are conducted to another star of the first magnitude, Aldebaran, which strongly resembles Betelgueze in its ruddy colour. Aldebaran is the brightest star in the constellation of Taurus. It is this constellation which contains the Pleiades already referred

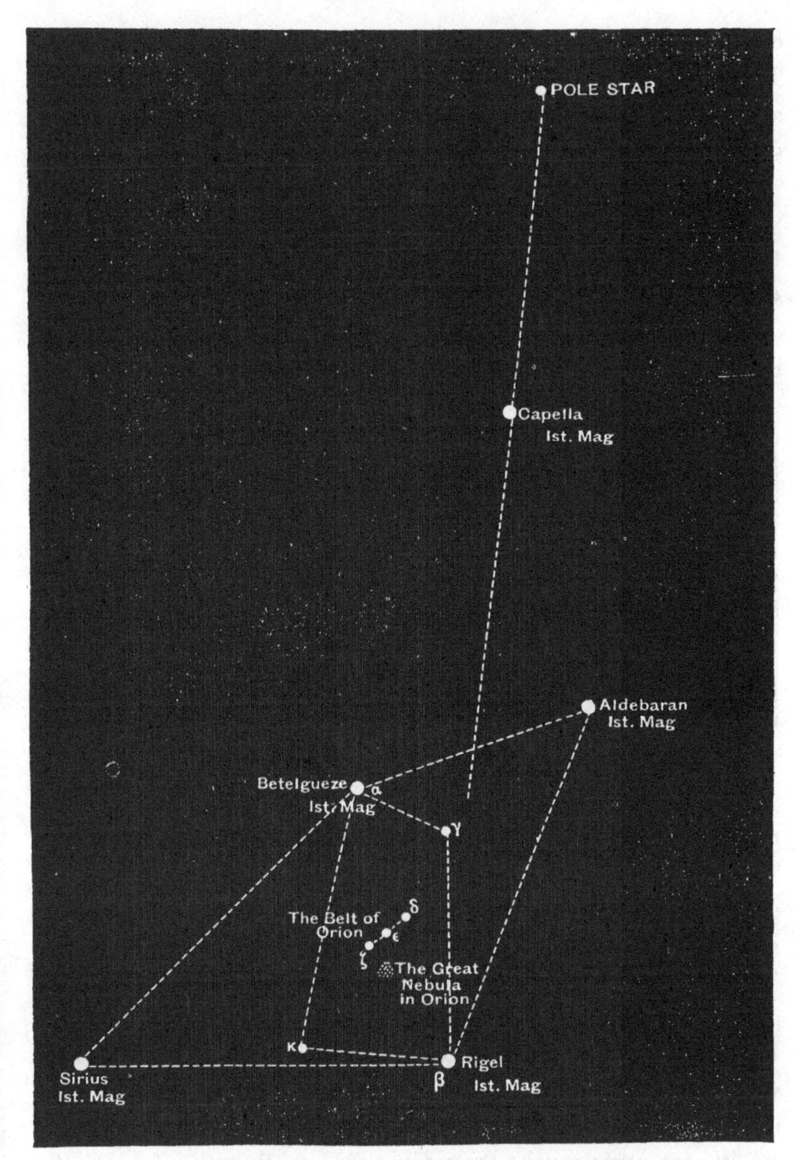

Fig. 72.—Orion, Sirius, and the neighbouring Stars.

to, and another more scattered group known as the Hyades, which will be easily discovered near Aldebaran.

The line of the belt of Orion continued downwards to the left conducts the eye to the gem of the sky, the splendid star Sirius. This is an object which is beyond all controversy the most brilliant star in the heavens. It has, indeed, been necessary to create a special order of magnitude wherein to place Sirius by himself; all the other first magnitude stars, such as Vega and Capella, Betelgueze and Aldebaran, coming a long way behind. Sirius, with a few other stars of much less lustre, form the constellation of Canis Major.

It is useful for the learner to note the large configuration, of an irregular lozenge shape, of which the four corners are the first

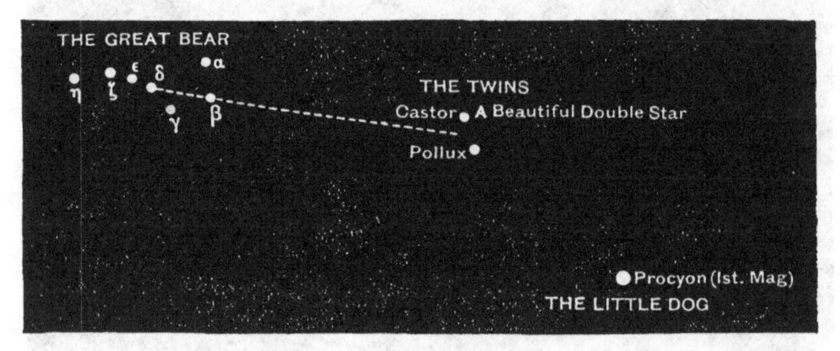

Fig. 73.—Castor and Pollux.

magnitude stars, Aldebaran, Betelgueze, Sirius, and Rigel (Fig. 72). The belt of Orion is placed symmetrically in the centre of the group, and the whole figure is so striking that once perceived it is not likely to be forgotten.

One-third of the way from the Square of Pegasus to Aldebaran a bright star of the second magnitude is the chief star in the Ram; with two others it forms a curve, at the other end of which will be found γ of the same constellation, which was the first double star ever noticed.

We can again invoke the aid of the Great Bear to point out the stars in the constellation of Gemini (Fig. 73). If the diagonal joining the stars δ and β of the body of the Bear be produced in the opposite direction to the tail, it will lead to Castor and Pollux, two remarkable stars of the second magnitude. This same line carried a little further on passes near the star Procyon, of the first magnitude, which is the only conspicuous object in the constellation of the Little Dog.

The pointers in the Great Bear marked *a*, *β* will also serve to indicate the constellation of the Lion. If we produce the line joining them in the opposite direction from that used in finding the

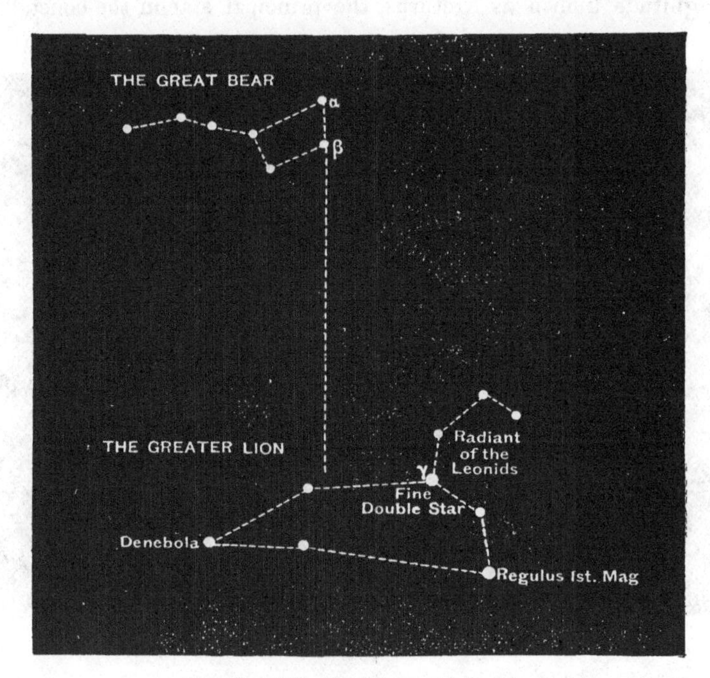

Fig. 74.—The Great Bear and the Lion.

Pole, we are brought into the body of the Lion. This group will be recognised by the star of the first magnitude called Regulus. It is one of a series of stars forming an object somewhat resembling a sickle; three of the group are of the second magnitude. The Sickle is specially famous in astronomy as containing the radiant point from which the periodic shooting star shower known as the Leonids diverge, while Regulus lies alongside the sun's highway through the stars, at a point which he passes on the 21st of August every year.

Between Gemini and Leo lies the inconspicuous constellation of the Crab; the most striking object in it is the misty patch called Præsepe or the Bee-Hive, which the smallest opera-glass will resolve into its component stars.

The tail of the Great Bear, when prolonged with a continuation of the curve which it possesses, leads to a brilliant star of the first magnitude known as Arcturus, the principal star in the constella-

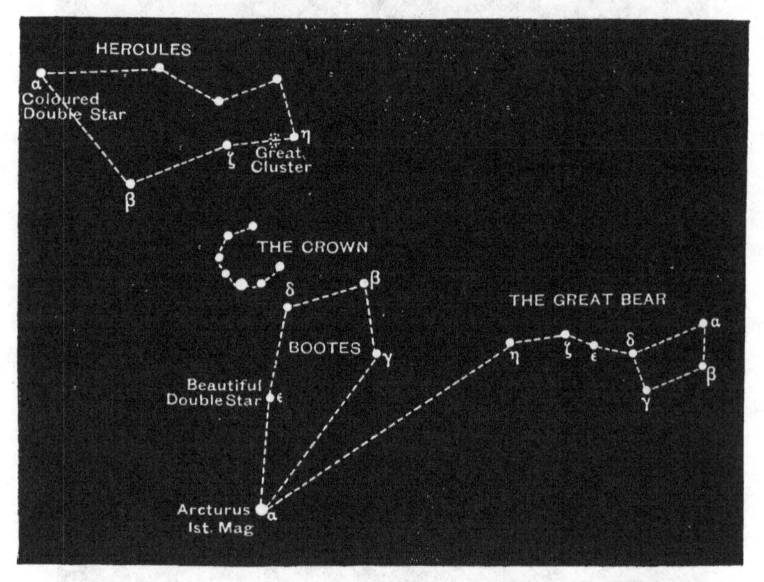

Fig. 75.—Boötes and the Crown.

tion of Boötes (Fig. 75). A few other stars, marked β, γ, δ, and ϵ in the same constellation, are also shown in the figure. It seems an open question whether among the stars visible in these latitudes Arcturus is not to be placed next to Sirius in point of brightness, the other two candidates for the place being Vega and Capella. Two stars in the southern hemisphere invisible in these latitudes termed *a* Centauri and Canopus, are both much brighter than Arcturus, Vega, or Capella.

In the immediate neighbourhood of Boötes is a striking semi-circular group known as The Crown or Corona Borealis. It will be readily found from its position as indicated in the figure, or it may be identified by following the curved line indicated by β, δ, ϵ and ζ in the Great Bear.

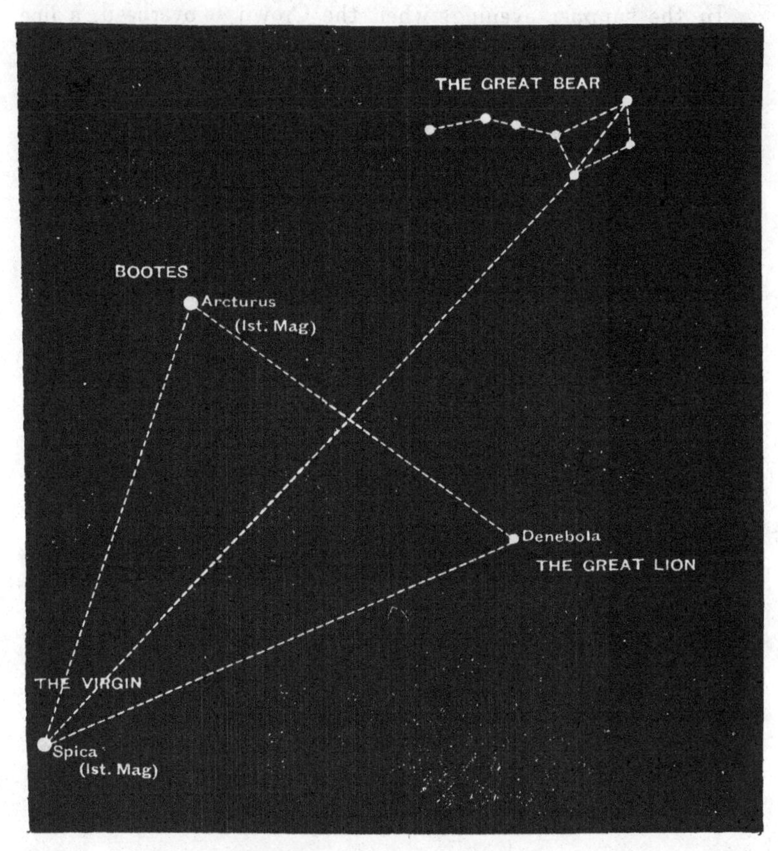

Fig. 76.—Virgo and the neighbouring Constellations.

The constellation of Virgo is principally characterised by the first magnitude star called Spica, or a Virginis. This may be found from the Great Bear; for if the line joining the two stars a and γ in that constellation be prolonged with a slight curve, it will conduct the eye to Spica. We may here notice another of those large configurations which are of great assistance in the study of the stars. There is a fine equilateral triangle, whereof Arcturus and Spica form two of the corners, while the third is indicated by Denebola, the bright star near the tail of the Lion (Fig. 76).

In the summer evenings when the Crown is overhead, a line from the Pole Star through its fainter edge continued nearly to the

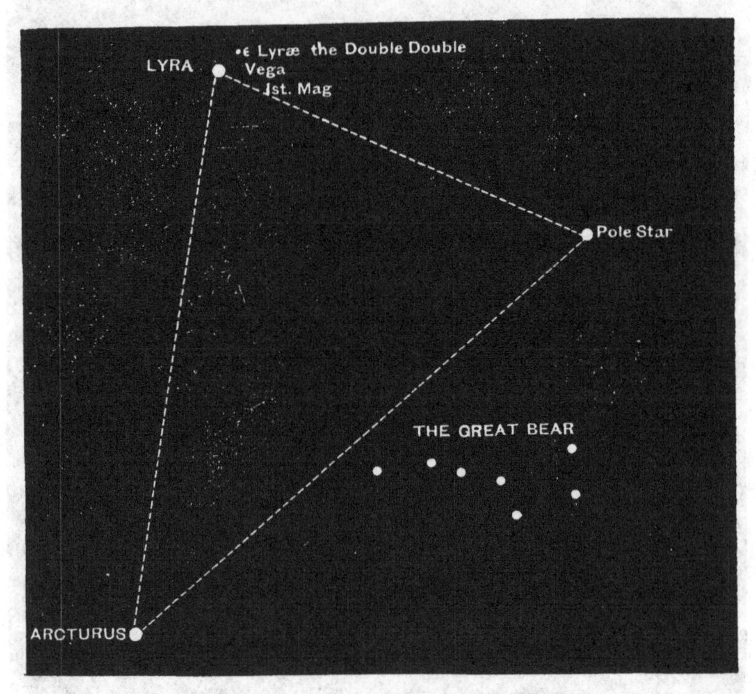

Fig. 77.—The Constellation of Lyra.

southern horizon, encounters the brilliant red star Cor Scorpionis, or the Scorpion's Heart (Antares), which was the first star mentioned as having been seen with the telescope in the daytime.

The first magnitude star, Vega, in the constellation of the Lyre, can be, perhaps, most readily found at the corner of a bold triangle, of which the Pole Star and Arcturus form the base (Fig. 77). The brilliant whiteness of Vega cannot fail to arrest the attention, while the small group of neighbouring stars which form the Lyre produces one of the best defined constellations.

Near Vega is another important constellation, known as the Swan or Cygnus. The brightest star will be identified as the vertex of a right-angled triangle, of which the line from Vega to the Pole Star is the base, as shown in Fig. 78. There are in Cygnus five principal stars, which form a constellation of rather remarkable form.

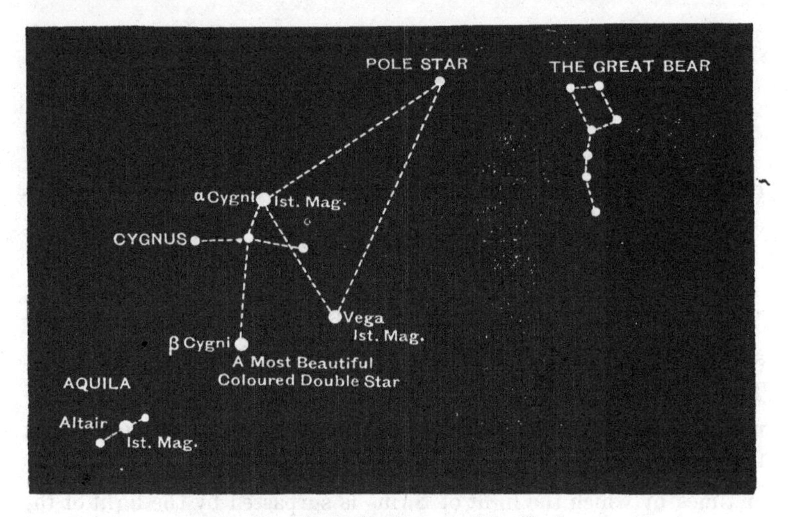

Fig. 78.—Vega, the Swan, and the Eagle.

The last constellation which we shall here describe is that of Aquila or the Eagle, which contains a star of the first magnitude, often known as Altair; this group can be readily found by a line from Vega over β Cygni, which passes near the line of three stars, forming the characteristic part of the Eagle.

We have taken the opportunity to indicate in these sketches of the constellations the positions of some other remarkable telescopic objects, the description of which we must postpone to the following chapters.

CHAPTER XIX.

THE DISTANT SUNS.

Comparison between the Sun and the Stars—Sirius Contrasted with the Sun—Stars can be Weighed, but not Measured—The Companion of Sirius—Determination of the Weights of Sirius and his Companion—Dark Stars—Variable Stars—Enormous Number of Stars.

WIDE indeed is the contrast between the splendour of the noon-day sun and the feeble twinkling of even the brightest of the stars. This contrast, so forcible to our ordinary observation, can be submitted to the test of actual measurement. Let us take the most brilliant star, Sirius. We can determine by experiment the number of times by which the light of Sirius is surpassed by the light of the sun. It is true we cannot make that comparison directly. In the bright daylight Sirius cannot even be seen, much less can his light be measured. But we can take the full moon as an intermediate step between the glory of the sun and the feeble twinkle of Sirius. It has been found that the light from the sun is about 600,000 times as great as the light from the full moon. Sirius and the full moon can be compared together, and by suitable photometers the quantity of light from the two bodies can be measured. It can be shown that the light of about 33,000 stars, equal to Sirius, would produce an illumination equal to that of the moon. We have thus the necessary figures for comparing the brilliancy of the sun with the brilliancy of Sirius, and the result is indeed significant. The light from the sun is about twenty thousand million times as great as the light from Sirius.

But do these figures truly represent the relative importance of the sun and Sirius? Recollect where we are placed. Our earth is near the sun; it is very far from Sirius. Our earth is not properly placed for an impartial comparison between the splendour of the sun and the splendour of Sirius. To make such a comparison, the earth should be placed midway between the two bodies, so that we could look at Sirius on one side and the sun on the other, under precisely similar circumstances. Into such a position our earth never has come, and never will come. How, then, is the real comparison to be accomplished? In this, as in many other cases, it is possible to determine by calculation what it is impossible to ascertain by experiment. It has been found, by observation, that Sirius is about one million times as far from us as the sun. If we take the distance of Sirius from the earth, and sub-divide it into one million equal parts, each of these parts would be long enough to span that great distance of 92,700,000 miles from the earth to the sun. If, therefore, the earth were to be placed half-way between the sun and Sirius, the earth would be five hundred thousand times as far away from the sun as it is at present. What would be the effect of this change upon the light received from the sun? Take the light of a candle at one foot as unity, then at two feet the light is reduced to one-fourth, and at three feet to one-ninth, and so on. The light received from a luminous source varies "inversely as the square of the distance." Transplanted to a distance 500,000 times as great as our present distance from the sun, to what a vast extent must his radiance be diminished! At present our sun is as bright as 20,000,000,000 stars each equal to Sirius; but viewed from this central point, the light is reduced to an almost incredible extent. Our sun would have totally lost his pre-eminence, and would only send to us one-twelfth part of the light which we now receive from Sirius. On the other hand, by shifting our station from the vicinity of the sun to a point half-way to Sirius, the light of Sirius will become intensified in the ratio of four to one. Assuming, then, that the earth was placed in the correct position for testing the comparative splendour of Sirius and the sun, we should find, first, that Sirius was increased four-fold, and that the sun was so enormously reduced in intensity as to look only one-twelfth of the original brightness of Sirius. The consequence is irresistible, and fatal to the pre-eminence of the sun. We may state it thus: Sirius sheds actually forty-eight times as much light around it as does our sun; we are not exactly entitled to say that Sirius is forty-eight times as big as the sun, but we can say that Sirius is forty-eight times as brilliant or as splendid. In making this calculation we have taken a lower determination of the brightness of Sirius relatively to the sun than some other careful observations would have warranted. It will thus be seen that if there be any uncertainty in our result it must only be as to whether Sirius is not really more than forty-eight times as bright as the sun.

Is this surpassing glory of Sirius merely due to excessive brilliancy, or is Sirius really a body which vastly surpasses our sun, both in intrinsic importance as well as in radiance? The question can only be fully answered by measuring the circumference of Sirius, and then placing him in the weighing scales. At first sight it might seem more difficult to achieve the latter than the former. How is it possible that Sirius, at a distance of a hundred billions of miles, should be actually weighed? Yet we can weigh Sirius, and we cannot measure him. We have no idea what the actual bulk of Sirius may be, but as we find that he is certainly heavier than the sun, we may conclude that he is probably larger than the sun, though we can have no certain knowledge on this point.

It is certainly a very remarkable fact, that out of the thousands of stars with which the heavens are adorned, no single star has yet been found which certainly shows an appreciable disc in the telescope. We are aware that some skilful observers have thought that certain small stars do show discs; but we may lay this aside, and appeal only to the ordinary fact that our best telescopes turned on the brightest stars show merely glittering points of light, so hopelessly small as to elude our most delicate micrometers. The ideal astronomical telescope is, indeed, one which will show Sirius or any other bright star as nearly identical as possible with Euclid's definition of a point, being that which has no parts and no magnitude. It will throw some light on this question to consider the telescopic appearance which the sun would present if viewed from Sirius. This is a question which we can answer. We know the diameter of the sun, and we can calculate how large our sun would look if viewed from the standpoint of Sirius. The answer is indeed significant. The size of the sun from this distance would correspond to the size of a halfpenny 1,600 miles away. It is hopeless to expect that a quantity so minute as this could be detected by any telescope. When we have a telescope sufficiently powerful to show animals on the moon, or to show buildings on the planet Mars, then we may hope that we shall see the discs of the stars. But this is utterly beyond our reach. The weakness is not merely in the telescope; it is inherent in every circumstance of the problem. The stability of our earth is not sufficient to afford a secure foundation for a telescope of such capacities, while even the purest skies would form media far too turbid. We ought, therefore, to feel no surprise at not being able to detect any disc of Sirius; were his diameter ten times that of the sun, and his bulk, therefore, one thousand times as great, the task would be equally hopeless.

Abandoning, therefore, all hope of measuring Sirius, let us attempt to weigh him. Here we can be successful, as we shall proceed to demonstrate. The story is, indeed, one of no little interest in the history of astronomical discovery.

The splendid pre-eminence of Sirius had caused it to be observed with minute care from the earliest times in the history of astronomy. Each generation of astronomers devoted time and labour to determine the exact place of the brightest star in the heavens. A vast mass of observations as to the place of Sirius among the stars had thus been accumulated, and it was found that, like many other stars, Sirius had what astronomers call proper motion. Comparing the place of Sirius now in the heavens with the place of Sirius one hundred years ago, there is a difference of about two minutes (131″) in its situation. This is a small quantity: it is so small that the unaided eye could not see it. Were we suddenly to be transplanted back one century ago, we should still see Sirius in its well-known place to the left of Orion. Careful alignment by the eye would hardly detect that Sirius was moving, in two, or even in three or in four centuries. But the accuracy of the meridian circle grasps these minute quantities, and gives to them their true significance. To the eye of the astronomer, Sirius, instead of creeping along with a movement which centuries will not show, is pursuing its majestic course with a velocity appropriate to his dimensions. It is easy to calculate that Sirius must be sweeping along at the rate of a thousand miles a minute.

It is of the utmost importance to observe that though the velocity of Sirius is *about* 1,000 miles a minute, yet it is sometimes a little more and sometimes a little less than its mean value. To the astronomer this fact is pregnant with significance. Were Sirius one isolated star, attended only by planets of comparative insignificance, there could be no irregularity in its motion. If it were once started with a velocity of 1,000 miles a minute, then it must preserve that velocity. Neither the lapse of centuries nor the mighty length of the journey can alter it. The path of Sirius would be inflexible in its direction; and it would be traversed with unalterable velocity.

The fact, then, that Sirius was not moving uniformly was of such interest that it arrested the attention of Bessel when he discovered the irregularities in 1844. Believing, as Bessel did, that there must be some adequate cause for these disturbances, it was hardly possible to doubt what the cause must be. When motion is disturbed there must be force in action, and the only force that we recognise in such cases is that known as gravitation. But gravity can only act from one body to another body; so that when we seek for the derangement of Sirius by gravitation, we are obliged to suppose that there must be some mighty and massive body near Sirius. The question was taken up again by Peters and by Auwers, who were able to discover, from the irregularities of Sirius, the nature of the path of the disturbing body. They were able to show that this disturbing body must revolve around Sirius in a period of about fifty years; and although they could not tell the actual distance of the unknown body from Sirius, yet they were able to point out the direction in which it must lie.

In many respects this was a problem analogous to the ever memorable discovery of the planet Neptune. In each case the unknown body had been manifested by the perturbations it produced, and in each case the unknown object was discovered by the calculations of the mathematician, before it was revealed to the penetration of the astronomer. Nearly twenty years had elapsed after Bessel had predicted the disturber of Sirius, before the telescopic discovery which confirmed it was made. The circumstances under which that discovery was made are not, indeed, so dramatic as those which attended the discovery of Neptune, but yet they have an interest of their own. In February, 1862, Messrs. Alvan Clark and Sons, the celebrated telescope makers, were completing a superb 18-inch object-glass for the Chicago Observatory. Turning the instrument on Sirius, for the purpose of trying it, the practised eye of the younger Clark soon detected something unusual, and he exclaimed, "Why, father, the star has a companion!" The father looked, and there was a faint companion due east from the bright star, and distant about ten seconds. This was exactly the predicted direction of the companion of Sirius, and yet the observers knew nothing of the prediction. As the news of this discovery spread, many great telescopes were pointed on Sirius; and it was found that when observers knew exactly where to look for the object, many instruments would show it. The new companion star to Sirius lay in the true direction, and it was now watched with the keenest interest, to see whether it also was moving in the way it should move, if it were really the body whose existence had been foretold. Four years of observation showed that this was the case, so that hardly any doubt could remain that the telescopic discovery had been made of the star which had caused the inequality in the motion of Sirius. The correspondence between the observed motions and the predicted motions has not since proved quite exact; for the observed companion appears to have moved about half a degree per annum more rapidly than the calculated companion. This difference, though larger than was expected, may be partly due to the inevitable errors of the difficult observations from which the movements of the theoretical companion were computed.

Recent researches have exhibited the movements of Sirius in a very unexpected way. Mr. Huggins demonstrated with his spectroscope in 1868 that Sirius was receding from the sun, while the observers at Greenwich have found that since 1881 Sirius has been approaching the sun .

The discovery of the attendant of Sirius and the measures which have been made thereon, give us an answer to the question—"What is the weight of Sirius?" Let us attempt to illustrate this subject. It must, no doubt, be admitted that the numerical estimates we employ are to be received with a certain degree of caution. The companion of Sirius is a difficult object to observe, and the measurements require great delicacy. We are, therefore, hardly as yet in a position to speak with accuracy as to the periodic time in which the companion completes its revolution. We shall, however, using the best observations available, take this time to be forty-nine years. We also know the distance from Sirius to his companion, and we may take it to be about thirty-seven times the distance from the earth to the sun. It is useful, in the first place, to compare the revolution of the companion around Sirius with the revolution of the outermost planet, Neptune, around the sun. Taking the earth's distance as unity, the radius of the orbit of Neptune is about thirty, and Neptune takes 165 years to accomplish a complete revolution. We have no planet in the solar system at a distance of thirty-seven, but from Kepler's third law it is very easy to calculate that if there were such a planet its periodic time must be about 225 years. We have now the necessary materials for making the comparison between the mass of Sirius and the mass of the sun. A body revolving around Sirius at a certain distance completes its journey in 49 years. A body revolving around the sun at the same distance completes its journey in 225 years. The quicker the body is moving the greater must be the centrifugal force, and consequently the greater must be the attractive power of the central body. It can be easily shown from the principles of dynamics that the attractive power varies inversely as the square of the periodic time. Hence, then, the attractive power of Sirius must bear to the attractive power of the sun, the proportion which the square of 225 has to the square of 49. As the distances are in each case supposed to be equal, the attractive powers will be proportional to the masses, and hence we conclude that the mass of Sirius, together with that of his companion, is to the mass of the sun in the ratio of 20 to 1. We had already learned that Sirius was much brighter than the sun; now we have learned that it is also much more massive.

Before we leave the consideration of Sirius, there is one additional point of considerable interest which it is necessary to consider. It is remarkable to observe the contrast between the brilliancy of Sirius and his companion. Sirius is a star far transcending all other stars of the first magnitude, while his companion is extremely faint. Even if it were completely withdrawn from the dazzling proximity of Sirius, the companion would be only a small star of the eighth or ninth magnitude, far below the limits of visibility of the unaided eye. To put the matter in numerical language, Sirius is 5,000 times as bright as its companion, but only about twice as heavy! Here is a very great contrast; and this point will appear even more forcible if we contrast the companion of Sirius with our sun. The companion is *seven times as heavy as our sun*; seven suns equal to ours in one pan of the scales, would only just turn with the companion in the other pan; but in spite of its inferior bulk, our sun is much more powerful as a light-giver. One hundred of the companions to Sirius would not give as much light as our sun! This is a result of very considerable significance. It teaches us that besides the great bodies in the universe which attract attention by their brilliancy, there are also other bodies of stupendous mass which have but little brilliancy, probably some of them none at all. This suggests a greatly enhanced conception of the majestic scale of the universe. It almost invites us to the belief that the universe which we behold, bears but a small ratio to the far larger part which is invisible in the sombre shades of night. In the wide extent of the material universe we have here or there a star or a mass of gaseous matter sufficiently heated to become luminous, and thus to become visible from the earth; but our observation of these luminous points can tell us little of the remaining contents of the universe.

For the purposes of practical astronomy it has been found convenient to divide the stars into groups, according to their relative degrees of brightness. In this way we denote the brightness of the star by a certain number which is called the magnitude of the star, and the lower the number which expresses the magnitude, the brighter is the star. Of the stars of the first magnitude, which include all the brightest stars in the heavens, there are about twenty. Among this number Sirius is included, though if the classification were to be carried out with logical precision, a distinct class of exceptional brilliancy would have to be created for the reception of Sirius alone.

The stars of the second magnitude are those in which there is one distinct step downwards from the brilliancy of the first magnitude. The brighter stars in the constellation of the Great Bear may be taken as examples. In the entire heavens we have about 65 stars of the second magnitude. Immediately below the second magnitude we have the stars of the third magnitude, to the number of 190. Next comes the fourth, 425; the fifth, 1,100; and so on down to the sixth, 3,200, which completes the stars visible to the unaided eye. In stars of telescopic magnitude we have the seventh, to the number of about 13,000; while the eighth has 40,000, and the ninth 142,000.

It will thus be seen that the number of stars increases when we approach the lower magnitudes, and when we come to the magnitudes still lower than the ninth the numbers speedily swell from thousands to millions. The minutest stars visible in powerful telescopes are usually stated to be of the fourteenth or fifteenth magnitude, while in the very greatest instruments magnitudes two or three steps lower can be observed.

The number of stars visible without a telescope in England

may be estimated at about 3,000. Argelander has given to the world a well-known catalogue of the stars in the northern hemisphere, accompanied by a series of charts on which these stars are depicted. All the stars of the first nine magnitudes are included, as well as a very large number of stars lying between the ninth and the tenth magnitudes. The total number of these stars is 324,188, and yet they are all within reach of a telescope of three inches in aperture!

Amid the hosts of stars, a considerable number specially attract our attention by the peculiar changes they undergo in brilliancy. They are known as variable stars; some of them run through a cycle of changes in a day or two, some take many months. Some appear once, and never appear again. Some are conspicuous to the unaided eye, some are faint telescopic objects. Though the number of variable stars is very large, yet, compared with the ordinary fixed stars, they must be regarded as very infrequent.

The most celebrated of all the variable stars is that known as Algol, whose position in the constellation of Perseus is shown in Fig. 70. This star is very conveniently placed for observation, being visible every night in the northern hemisphere, and its wondrous and regular changes can be observed without any telescopic aid. Every one who desires to become acquainted with the great truths of astronomy should be able to recognise this star, and should have also followed it during one of its periods of change. Algol is usually a star of the second magnitude, but in a period between two or three days, or, more accurately, in an interval of 2 days 20 hours 48 minutes and 55 seconds, its brilliancy goes through a most remarkable cycle of variations. The series commences with a gradual decline of the star's brightness, which in the course of three or four hours falls from the second magnitude down to the fourth. At this lowest stage of brightness Algol remains for about twenty minutes, and then begins to increase, until in three or four hours it regains the second magnitude, at which it continues for about 2 days 13 hours, when the same series commences anew. It seems that the period required by Algol to go through its changes is itself subject to a slow but certain variation.

The claim of our sun to be admitted as a star having been conceded, the question arises as to whether he shall be entitled to the distinction of being admitted to the select class of variable stars, or whether he shall not take rank among the far more numerous class which dispense their beams with uniformity. I do not think we can have much hesitation in answering the question. It is obvious that the light from our sun is for all practical purposes absolutely constant. No doubt the spots and the other objects on the sun are variable, but it would be merely fantastic to speak of the sun as a variable star. It is no doubt variable to the close observation of terrestrial beings who are placed in its immediate vicinity; it is not variable in the sense in which we speak of Algol, or any of the other variable stars which diversify the brilliancy of the sky.

That the sun is no more than a star, and the stars are no less than suns, is the cardinal doctrine of astronomy. The imposing magnificence of this truth is only realised when we attempt to estimate the countless myriads of stars. This is a problem on which our calculations are necessarily vain. Let us, therefore, invoke the aid of the poet to attempt to express the] innumerable, and conclude this chapter with the following lines of Mr. Allingham:—

> "But number every grain of sand,
> Wherever salt wave touches land;
> Number in single drops the sea;
> Number the leaves on every tree,
> Number earth's living creatures, all
> That run, that fly, that swim, that crawl.
> Of sands, drops, leaves, and lives, the count
> Add up into one vast amount,
> And then for every separate one
> Of all those, let a flaming SUN
> Whirl in the boundless skies, with each
> Its massy planets, to outreach
> All sight, all thought: for all we see,
> Encircled with infinity,
> Is but an island."

CHAPTER XX.

DOUBLE STARS.

Interesting Stellar Objects—What is a Double Star?—Stars Optically Double—The great Discovery of the Binary Stars made by Herschel—The Binary Stars describe Elliptic Paths—Why is this so important?—The Law of Gravitation—Special Double Stars—Castor—Mizar—The Pole Star—The Coloured Double Stars—β Cygni, γ Andromedæ.

THE sidereal heavens contain few more interesting objects for the telescope than can be found in the numerous class of double stars. They are to be counted now in thousands; indeed, many thousands can be found in the catalogues devoted to this special branch of astronomy. Many of these objects are, no doubt, small and comparatively uninteresting, but some of them are among the most conspicuous stars in the heavens. We shall in this brief account select for special discussion and illustration a few of the more remarkable double stars. We shall particularly notice some of those that can be readily observed with a small telescope, and we have indicated on the sketches of the constellations in a previous chapter how the positions of these objects in the heavens can be ascertained.

In 1678, about 100 years before Herschel's observations commenced, it had been shown by Cassini that certain stars, which to the unaided eye appeared single points of light, really consisted of two or more stars, so close together that the telescope was required for their separation. The number of these objects was gradually increased by fresh discoveries, until in 1781 (the same year in which Herschel discovered Uranus) a list, containing eighty double stars, was published by the astronomer Bode. These interesting objects claimed the attention of Herschel during his memorable researches. The list of known doubles rapidly swelled. Herschel's discoveries are to be enumerated by hundreds, while he also commenced systematic measurements of the distance by which the stars were separated, and the direction in which the line joining them pointed. It was these measurements which ultimately led to one of the most important and instructive of all Herschel's discoveries. When, in the course of years, his observations were repeated, Herschel found that in some cases the relative position of the stars had changed. He was thus led to the discovery that in many of the double stars the components are so related that they revolve around each other. Mark the importance of this result. We must remember that the stars are suns, comparable, it may be, with our sun in magnitude; so that here we have the pleasing spectacle of a pair of twin suns in revolution. But this is not the chief point of interest in this great discovery. There is nothing very surprising in the fact that movements should be observed, for in all probability every body in the universe is in motion. It is the particular character of the movement which is specially interesting and instructive.

It had been in the first instance supposed that the proximity of the two stars forming a double was really only accidental. It was thought that amid the vast host of stars in the heavens, it not unfrequently happened that one star was so nearly behind another that when the two were viewed in a telescope they produced the effect of a double star. No doubt many of the so-called double stars are produced in this way. Herschel's discovery shows that this explanation will not always answer, but that in many cases we really have two stars close together, and physically connected.

When the measurements of the distances and the positions of double stars made during many years by many astronomers had been accumulated, they were handed over to the mathematician. There is one peculiarity about these observations: they have not—they cannot have—the accuracy which can be obtained in the determination of the places of the planets, or in meridian work generally. This latter work is so accurate that almost any superstructure of mathematical reasoning can be reared on its foundation. But the case is very different when we come to the binary stars. If the distance between the pair of stars forming a binary be four seconds, the orbit we have to scrutinize is only as large as the apparent size of a penny-piece at the distance of one mile. It would require very careful measurement to make out the form of a penny a mile off, even with good telescopes. If the penny were tilted a little, it would appear, not circular, but oval; and it would be possible, by measuring this oval, to determine how much the penny was tilted. All this requires skilful work; the errors, viewed intrinsically, may not be great, but viewed with reference to the whole size of the quantities under consideration, they are very appreciable. We therefore find the errors of observation far more prominent in observations of this class, than is generally the case when the mathematician assumes the task of discussing the labours of the observer.

The interpretation of Herschel's discovery was not accomplished by himself; the light of mathematics was turned on his observations of the binary stars by Savary, and afterwards by other mathematicians. Under their searching inquiry the errors of the observations were disclosed, and they were purified from the grosser part of their inaccuracy. Mathematicians could then apply to their corrected materials the methods of inquiry with which they were familiar; they could deduce with fair precision the actual shape of the orbit of the binary stars, and the position of the plane in which that orbit is contained. The result is not a little remarkable. It has been proved that the motion of each of the stars is performed in an ellipse which contains the centre of gravity of the two stars in its focus. This has been actually shown to be true in many binary stars; it is believed to be true in all. But why is this so important? Is not motion in an ellipse common enough? does not the earth revolve in an ellipse around the sun? and do not the planets also revolve in ellipses?

It is this very fact that elliptic motion is so common in the planets of the solar system which renders its discovery in binary stars of such importance. From what does the elliptic motion in the solar system arise? Is it not due to the law of attraction, discovered by Newton, which states that every mass attracts every other mass with a force which varies inversely as the square of the distance? That law of attraction had been found to pervade the whole solar system, and it explained the movements of the bodies of our system with marvellous fidelity. But the solar system, consisting of the sun, and the planets, with their satellites, the comets, and a host of smaller bodies, formed merely a little island group in the universe. In the economy of this tiny cosmical island the law of gravitation reigns supreme; before Herschel's discovery we never could have known whether that law was not merely a piece of local legislation, specially contrived for the exigencies of our particular system. This discovery gave us the knowledge which we could have gained from no other source. From the binary stars came a whisper across the vast abyss of space. That whisper told us that the law of gravitation was not peculiar to the solar system. It told us the law extended to the distant shores of the abyss in which our island is situated. It gives us grounds for believing that the law of gravitation is obeyed throughout the length, the breadth, the depth, and the height of the entire visible universe.

One of the finest binary stars is that known as Castor, the brighter of the two principal stars in the constellation of Gemini. The position of Castor on the heavens is indicated in Fig. 73, page 382. Viewed by the unaided eye, Castor resembles a single star; but with a moderately good telescope it is found that what seems to be one star is really two separate stars, one of which is of the third magnitude, while the other is somewhat less. The angular distance of these two stars in the heavens is not so great as the angle subtended by a line an inch long viewed at a distance of half-a-mile. Castor is one of the double stars in which the components have been observed to possess a motion of revolution. The movement is, however, extremely slow, and the lapse of centuries will be required before a revolution is completely effected.

A very beautiful double star can be readily identified in the constellation of Ursa Major (see Fig. 75, page 384). It is known as Mizar, and is the middle star (ζ) of the three which form the tail. In the close neighbourhood of Mizar is the small star Alcor, which can be readily seen with the unaided eye; but when we speak of Mizar as a double star, it is not to be understood that Alcor is one of the components of the double. In the magnifying power of the telescope Alcor is seen to be transferred a long way from Mizar, while Mizar itself is split up into two suns close together. These components are of the second and the fourth magnitudes respectively, and as the apparent distance is nearly three times as great as in Castor, they are observed with the greatest facility even in a small telescope. This is, indeed, the best double star in the heavens for the beginner to commence his observations upon. We cannot, however, assert that Mizar is a binary star, inasmuch as observations have not yet established the existence of a motion of revolution. Still less are we able to say whether Alcor is also a member of the same group, or whether it may not merely be a star which happens to fall nearly in the line of vision.

Another object of considerable interest is the remarkable double-double star found in the constellation Lyra, and usually known as ε Lyræ (see Fig. 77, page 386). In this case the unaided eye suffices to show a pair of stars so close that they can only just be distinguished. When a telescope is applied to the object, the stars are, of course, at once widely opened, and then the interesting observation is made that each of these is itself composed of two stars extremely close together. The entire object thus consists of four stars in two fine binary pairs, and in each of these pairs revolution is taking place. It is difficult to resist the conclusion that the two pairs are connected together, and each revolving, in some period of stupendous duration, around the common centre of gravity of the two.

The Pole Star is also an interesting double, but it differs widely from the objects already mentioned, and must be regarded as a representative of that class of doubles in which the two components are of very unequal brilliancy. The Pole Star itself is between the second and third magnitude; its companion is, however, only of the ninth magnitude. This object requires a better telescope than is sufficient for the others we have mentioned; in fact, it is sometimes regarded as a suitable "test" for the performance of a small instrument of from two to three inches in diameter.

A very pleasing class of double stars are those in which we

have the remarkable phenomena of colours, differing in a striking degree from the colours of ordinary stars. Among the latter we find in the great majority of cases, no very characteristic hue; some are, however, more or less tinged with red, some are decidedly ruddy, and some are intensely red. Stars of a bluish or greenish colour are much more rare,* and when a star of this character does occur, it is almost invariably as one of a pair which form a double. The other star of the double is sometimes of the same hue, but more usually it is yellow or ruddy.

One of the loveliest of these objects, and one of those which are fortunately within reach of telescopes of very moderate pretensions, is that found in the constellation of the Swan, and known as β Cygni (Fig. 78, p. 387). This exquisite object is composed of two stars. The larger, about the third magnitude, is of a golden-yellow, or topaz, colour; the smaller, of the fifth magnitude, is of a light blue. These colours are nearly complementary, but still there can be no doubt that the effect is not merely one of contrast. It is indeed the fact that these two stars are both tinged with the hues we have stated, as can be shown by hiding each in succession behind a bar placed in the field of view. It has also been confirmed in a very striking manner by spectroscopic investigation; for Mr. Huggins has shown that the blue star has experienced a special absorption of the red rays, while the more ruddy light of the other star has arisen from the absorption of the blue rays. The contrast of the colours in this object can often be very effectively seen by putting the eye-piece out of focus. The discs thus produced show the contrast of colours better than when the object exhibits merely two stellar points.

Another rather more difficult coloured double star is γ Andromedæ (see Fig. 69, page 376). The larger star of the third magnitude is a deep yellow colour; the other, of the fifth magnitude, has a greenish hue. The interest of this object has been greatly in-

* Perhaps if we could view the stars without the intervention of the atmosphere, blue stars would be more common. The absorption of the atmosphere specially affects the greenish and bluish colours. Professor Langley gives us good reason for believing that the sun itself would be blue if it were not for the effect of the air.

creased by the discovery that the blue star, the smaller of the two, is itself composed of two minute stars, so close together that their separation is quite beyond the reach of ordinary telescopes; instruments of considerable size are required to separate them. γ Andromedæ is, therefore, to be described as a triple star rather than as a double. There are several instances of such complex systems. One of the most interesting triples is found in the constellation of the Crab, where it is known as ζ Cancri. It will be readily identified at one-sixth of the distance from Præsepe to Betelgueze. The brightest star is of the fifth magnitude, while the star near it is between the fifth and the sixth. They have been watched through the course of one complete revolution in a period a little more than sixty years. Their comparatively distant companion of intermediate brightness has only accomplished about one-twelfth of a circuit in that interval. The recent researches of M. Seeliger have added to the complexity of this system by demonstrating with a high degree of probability the existence of a dark attendant circulating around the third star in about eighteen years.

Such are a few of these double and multiple stars. Their numbers are being daily augmented; indeed one observer—Mr. Burnham, of Chicago—has within the last ten or twelve years added by his own researches no fewer than 1,000 new doubles to the list of those previously known.

The interest in this class of objects must necessarily be increased when we reflect that, small as these stars appear to be in our telescopes, they are in reality suns of great size and splendour, in many cases rivalling our own sun, or, perhaps, even surpassing him. Whether these suns have planets attending upon them we cannot tell; the light reflected from the planet would be utterly inadequate to the penetration of the vast extent of space which separates us from the stars. If there be planets surrounding these objects, then, instead of a single sun, such planets will be illuminated by two, or, perhaps, even more suns. What wondrous effects of light and shade must be the result! Sometimes both suns will be above the horizon together, sometimes only one sun, and sometimes both will be absent. Especially remarkable would be the condition of a planet whose suns were of the coloured type. To-day we have a red sun illuminating the heavens, to-morrow it would be a blue sun, and, perhaps, the day after both the red sun and the blue sun will be in the firmament together. What endless variety of scenery such a thought suggests! There are, however, grave dynamical reasons for doubting whether the conditions under which such a planet would exist could be made compatible with life in any degree resembling the life with which we are familiar. The problem of the movement of a planet under the influence of two suns is one of the most difficult that has ever been proposed to mathematicians, and it is, indeed, impossible in the present state of analysis to solve with accuracy all the questions which it implies. It seems not at all unlikely that the disturbances of the planet's orbit would be so great, that it would be exposed to vicissitudes of light and of temperature far transcending those experienced by a planet moving, like the earth, under the supreme control of a single sun.

CHAPTER XXI.

THE DISTANCES OF THE STARS.

Sounding-line for Space—The Labours of W. Herschel—His Reasonings Illustrated by Vega—Suppose this Star to recede 10 times, 100 times, or 1,000 times—Some Stars 1,000 times as far as others—Herschel's Method incomplete—The Labours of Bessel—Meaning of Annual Parallax—Minuteness of the Parallactic Ellipse Illustrated — The case of 61 Cygni — Different Comparison Stars used—Difficulty owing to Refraction—How to be avoided—The Proper Motion of the Star—Bessel's Preparations—The Heliometer—Struve's Investigations—Can they be Reconciled?—Researches at Dunsink—Conclusion obtained—Accuracy which such Observations admit — Examined — How the Results are Discussed — The Proper Motion of 61 Cygni — The Permanence of the Sidereal Heavens — Changes in the Constellation of the Great Bear since the time of Ptolemy — The Star Groombridge 1,830 — Large Proper Motion—Its Parallax—Velocity of 200 Miles a Second—The New Star in Cygnus—Its History—No Appreciable Parallax—A Mighty Outburst of Light—The Movement of the Solar System through Space—Herschel's Discovery—Journey towards Hercules—Probabilities—Conclusion.

WE have long known the dimensions of the solar system with more or less accuracy. Our knowledge includes the distances of the planets and the comets from the sun, as well as their movements. We have also considerable knowledge of the diameters and the masses of many of the different bodies which belong to the solar system. We have long known, in fact, many details of the isolated group nestled together under the protection of the sun. The problem for consideration in the present chapter involves a still grander survey than is required for measures of our solar system. We propose to carry the sounding-line across the vast abyss which separates the group of bodies closely associated about our sun, from the other stars which are scattered through the realms of space. For centuries the great problem of star distance has engaged the attention of those who have studied the heavens. It would be impossible to attempt here even an outline of the various researches which have been made on the subject. In the limited survey which we can make, we must glance first at the remarkable speculative efforts which have been directed to the problem, and then we shall refer to those labours which have introduced the problem into the region of accurate astronomy.

It was Herschel who first attempted, in a scientific manner, to give any accurate conception of the extent and magnitude of the sidereal system. There was in particular one form of speculation which Herschel employed with great success, for it possesses very strong claims on our imagination. It would, indeed, have been idle, or worse than idle, to indulge in speculation whenever the means of certainly ascertaining the truth were available. This Herschel never did. But when a majestic problem was incessantly being proposed which he could find no means of certainly solving, what course was he to adopt? He was entitled to attempt its solution by analogy and by probability, in default of more trustworthy guides.

We now know the distances of a few of the stars, perhaps of twenty or thirty, with more or less accuracy; but of the distances of the great majority we are still ignorant, while of the thousands of nebulæ we have not yet found the distance of even a single one. Our knowledge of the distances of the sidereal objects is therefore scanty in the extreme, but the little that we do know is vastly more than Herschel knew. It was not until many years after Herschel's death that the great problem of finding the distance of a star was first solved by Bessel.

During Herschel's celebrated sweeps of the sidereal heavens, hundreds of thousands of objects—perhaps we might say millions—passed before his view. We do not now refer to the members of the solar system, the comets, or the planets and their satellites: we are speaking solely of the sidereal heavens, whose denizens are vastly more numerous, including the stars, the clusters, and the nebulæ. All this countless host passed before the review of Herschel, yet there was not a single one of all these objects of which he knew the distance.

To so enthusiastic an astronomer as Herschel the questions "How far?" and "How big?" were ever suggested during his nightly watches. He first exhausted all the means he could think of to obtain an answer to these questions, and having failed, he then—but not until then—sought to ascertain by reasoning, an answer to those problems which his observations had failed to determine. Herschel's reasoning on this point is most plausible; and we now make use of it to supplement and to extend the positive knowledge of the distances of the stars which has been acquired from observation. It is by this combination that we seek to obtain an adequate conception of the scale on which the vast fabric of the universe is constructed. It would be useless to attempt to enter into any detail on this point; let it suffice to give a single illustration of the kind of reasoning which Herschel employed in his speculations.

Take, for instance, the beautiful star Vega, or *a* Lyræ, the brightest star in the constellation of Lyra, and, with but few exceptions, the brightest star in the whole sky. To this star let us apply the arguments of Herschel. We shall proceed to sound the depths of the universe to which the telescope can penetrate, and we shall use the distance of Vega as the unit of our measurement. Herschel has shown how we can estimate the distances of the faintest stars, not only of those which are visible to the unaided eye, but even of the very faintest and most remote stars which can be shown in the most powerful telescope.

Imagine Vega to be moved away from the earth; then, as it receded further and further from our view its brilliancy would gradually decline. Speed on the journey, nor let a halt be made until Vega has moved to a distance which is ten times its present distance. The brilliancy has now decreased so much that Vega is no longer one of the brightest stars in the heavens. At ten times its present distance from the earth it would have sunk in brilliancy to a star of the sixth magnitude, and would only be just visible to the unaided eye.

Let Vega again resume its journey. It will now become invisible to the unaided eye, but the telescope can still follow the star as its distance rapidly increases. Let us not make another halt until the distance has again increased tenfold, and has become

100 units. Yet even at this vast distance Vega can still be seen with a telescope of moderate pretensions. Once again let the star start on its voyage. Let it travel tenfold the distance which it had hitherto attained; and now, for the last time, let the wanderer stop when it has plunged into space to a distance 1,000 times as great as it was at first. It has long passed from the ken of the ordinary telescope, but it would still remain within view of the colossal instruments of which there are now so many in the world. Herschel himself said that his telescopes could pursue the star until it had retreated 900 times as far as it was at first. Telescopes have been much improved since his time, so that we shall not exaggerate if we take 1,000 as the present limit of visibility. In all this there is no assumption whatever; we shall have to make an assumption presently, but we have not made it yet. There can be no doubt that if Vega were moved to ten times its distance it would still be faintly visible to the unaided eye. There can be no doubt that if Vega were moved 100 times as far, it would still be visible in small telescopes; while if it were 1,000 times as far, it would even then not have passed beyond the scrutiny of the grand instruments now in use.

In coming to the application of these principles, we here for the first time make an assumption. Herschel at this point adopted a sort of statistical reasoning, in which he deals with the stars, not in units, or even tens, but in thousands and millions. No one will suppose for a moment that the stars are all the same size: in fact, we know perfectly well that there is a vast difference in the actual sizes of some of the stars; but when we deal with the stars in their thousands, we may assume some principles which are, at all events, intelligible. We must, however, apprise the reader that these results are to be received with caution. They have not been demonstrated by observation, they cannot be demonstrated by calculation. They are plausible, but nothing more.

In the absence of all information to the contrary, Herschel thought it would be reasonable to assume that the average sizes of the stars distributed far and wide through space were the same as the average sizes of the stars with which we are familiar. He saw clearly that those stars which are brightest would seem the faintest of all stars if they were moved 1,000 times as far, but there is no reason to suppose that the faint stars which we do see are intrinsically smaller than the bright ones around us. All would be explained perfectly if we suppose that the average size of the stars in different regions of space is the same, and if we made the further supposition that the faintest stars are, as a general rule, about 1,000 times as far off as the brightest.

Such was the principle of Herschel's method of probing the heavens; but it was incomplete. He saw that the relative distances of the faint stars and the bright stars must be as we have described, but what was the absolute value of those distances? This was the problem which baffled Herschel all his life. He never succeeded in finding the actual distance of one of the stars from the earth. He tried it in the early part of his career; it was, in fact, while he was so engaged that he discovered Uranus. He tried it subsequently in a different manner, and again he was unsuccessful in his immediate object, though the observations led to the splendid discovery of the binary stars. The problem has been solved since Herschel's time, and we have been able to understand why he was frustrated. The distances are so great that the observations required are of the utmost delicacy. Herschel's great instruments, though of vast space-penetrating power, were wanting in the nice refinements of measurement necessary for this purpose. But now that the distances of some of the stars have been ascertained, we can make use of Herschel's reasoning to determine the distances of the others with some degree of plausibility.

No attempt to solve the problem of the absolute distances of the stars was successful until many years after Herschel's labours were closed. Fresh generations of astronomers, armed with fresh appliances, have for many years pursued the object with unremitting diligence, but for a long time the effort seemed hopeless. The distances of the stars were so great that they could not be ascertained until the utmost refinements of mechanical skill and the most elaborate methods of mathematical calculation were brought to converge on the difficulty. At last it was found that the problem was beginning to yield. A few stars have been induced to disclose the secret of their distance. We are able to give some answer to the question—How far are the stars? though it must be confessed that our reply up to the present moment is both hesitating and imperfect. Even the little knowledge which has been gained, possesses interest and importance. As often happens in similar cases, the discovery of the distance of a star was made independently about the same time by two or three astronomers. The name of Bessel stands out conspicuously in this memorable chapter of astronomy. Bessel proved (1840) that the distance of the star known as 61 Cygni was a measurable quantity. His demonstration possessed such unanswerable logic that universal assent could not be withheld. Almost simultaneously with the classical labours of Bessel, we have Strûve's measurement of the distance of Vega, and Henderson's determination of the distance of the southern star α Centauri. Great interest was excited in the astronomical world by these discoveries, and the Royal Astronomical Society awarded its gold medal to Bessel. It appropriately devolved on Sir John Herschel to deliver the address on the occasion of the presentation of the medal: that address is a most eloquent tribute to the labours of the three astronomers. We cannot resist quoting the few lines in which Sir John said:—

"Gentlemen of the Royal Astronomical Society,—I congratulate you and myself that we have lived to see the great and hitherto impassable barrier to our excursion into the sidereal universe, that barrier against which we have chafed so long and so vainly—*æstuantes angusto limite mundi*—almost simultaneously overleaped at three different points. It is the greatest and most glorious triumph which practical astronomy has ever witnessed. Perhaps I ought not to speak so strongly; perhaps I should hold some reserve in favour of the bare possibility that it may be all an illusion, and that future researches, as they have repeatedly before, so may now fail to substantiate this noble result. But I confess myself unequal to such prudence under such excitement. Let us rather accept the joyful omens of the time, and trust that, as the barrier has begun to yield, it will speedily be effectually prostrated."

Before proceeding further, it will be convenient to explain briefly how it is possible for the distance of a star to be measured. The problem has to be approached in a very different manner to that of the sun's distance, which we have already discussed in these pages. The observations for the determination of stellar parallax are founded on the familiar truth that the earth revolves around the sun. We may for our present purpose assume that

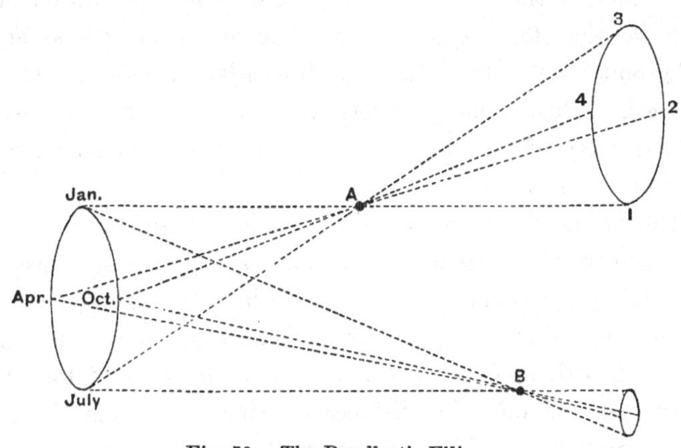

Fig. 79.—The Parallactic Ellipse.

the earth revolves in a circular path. The centre of that path is at the centre of the sun, and the radius of the path is 92,700,000 miles. Owing to our position on the earth, we observe the stars from a point of view which is constantly changing. In summer the earth is 185,400,000 miles distant from the position which it had in winter. It follows that the apparent positions of the stars, as projected on the background of the sky, must present corresponding changes. We do not now mean that the actual positions of the stars are really displaced. The changes are only apparent, and while oblivious of our own motion, which produces the displacements, we attribute the changes to the stars.

On the diagram in Fig. 79 is an ellipse with certain months —viz., January, April, July, October—marked upon its circumference. This ellipse may be regarded as a miniature picture of the earth's orbit around the sun. In January the earth is at the spot so marked; in April it has moved a quarter of the whole journey; and so on round the whole circle, returning to its original position in the course of one year. When we look from the position of the earth in January, we see the star A projected against the point of the sky marked 1. Three months later, the observer with his telescope is carried round to April; but he now sees the star projected to the position marked 2. Thus, as the observer moves around the whole orbit in the annual revolution of the earth, so the star appears to move round in an ellipse on the background of the sky. In the technical language of astronomers, we speak of this as the parallactic ellipse, and it is by measuring the major axis of this ellipse that we determine the distance of the star from the sun.

The figure shows another star, B, more distant from the earth and the solar system generally than the star previously considered. This star also describes an elliptic path. We cannot, however, fail to notice that the parallactic ellipse belonging to B is much smaller than that of A. The difference in the sizes of the ellipses arises from the different distances of the stars from the earth. The nearer the star is to the earth the greater is the ellipse, so that the nearest star in the heavens will describe the largest ellipse, while the most distant star will describe the smallest ellipse. We thus see that the distance of the star is inversely proportional to the size of the ellipse, and if we measure the angular value of the major axis of the ellipse, then, by a manipulation which every mathematician understands, the distance of the star can be expressed as a multiple of a radius of the earth's orbit. Assuming that radius to be 92,700,000 miles, the distance of the star is obtained by simple arithmetic. The difficulty in the process arises from the fact, that these ellipses are so small that our micrometers often fail to detect them. Take, for example, the Great Bear. Each of the stars in the Great Bear describes an ellipse in the course of a year. It would, however, be quite an exaggeration to suppose that the actual shape and position of the Great Bear is appreciably distorted by parallax. Indeed, the dimensions of the ellipses have an almost immeasurably small ratio to the dimensions of the constellation.

How shall we adequately describe the extreme minuteness of the parallactic ellipses in the case of even the nearest stars? In the technical language of astronomers, we may state that the longest diameter of the ellipse can never subtend an angle of more than two seconds. In a somewhat more popular manner, we would say that one thousand times the major axis of the very largest parallactic ellipse would not greatly exceed the diameter of the full moon. For a still more simple illustration, let us endeavour to think of a penny-piece placed at a distance of two miles. If looked at edgeways it will be linear, if tilted a little it would be elliptic; but the ellipse would, even at that distance, be greater than the greatest parallactic ellipse of any star in the sky. Suppose a sphere described around an observer with a radius of two miles. If a penny-piece were placed on this sphere, in front of each of the stars, every parallactic ellipse would be totally concealed.

The star in the Swan known as 61 Cygni is not remarkable either for its size or for its brightness. It is barely visible to the unaided eye, and there are some thousands of stars which are apparently larger and brighter. It is, however, a very interesting example of that remarkable class of objects known as double stars. It consists of two nearly equal stars close together, and evidently connected by a bond of mutual attraction. The attention of astronomers is also specially directed towards the star by its large proper motion. In virtue of that proper motion, the two components are carried together over the sky at the rate of five seconds annually. A proper motion of this magnitude is extremely rare, yet we do not say it is unparalleled, for there are some few stars which have a proper motion even more rapid; but the remarkable duplex character of 61 Cygni, combined with the large proper motion, render it an unique object, at all events, in the northern hemisphere.

When Bessel proposed to undertake the great research with

which his name will be for ever connected, he determined to devote one, or two, or three years to the continuous observations of one star, with the view of measuring carefully its parallactic ellipse. How was he to select the object on which so much labour was to be expended? It was all important to choose a star which should prove sufficiently near to reward his efforts by exhibiting a measurable parallax. Yet he could have but little more than surmise

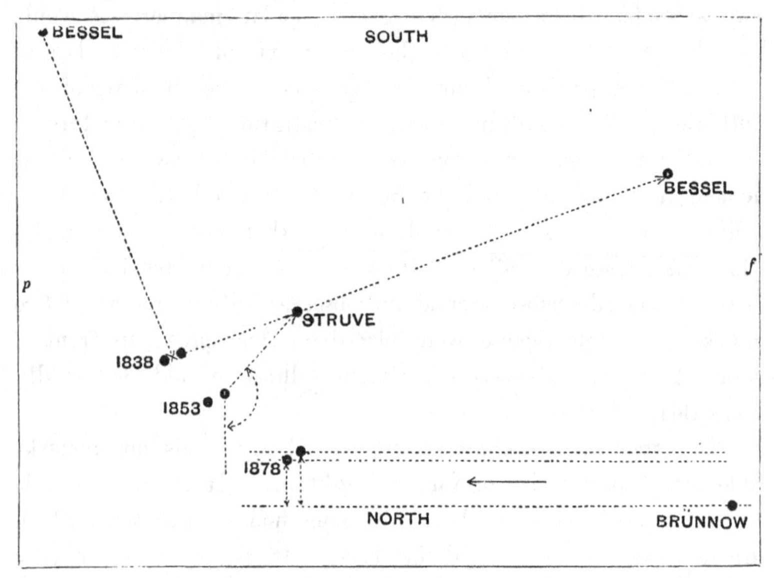

Fig. 80.—61 Cygni and the Comparison Stars.

and analogy as a guide. It occurred to him that the exceptional features of 61 Cygni afforded the necessary presumption, and he determined to apply the process of observation to this star. He devoted the greater part of three years to the work, and succeeded in discovering its distance from the earth.

Since the date of Sir John Herschel's address, 61 Cygni has received the devoted and scarcely remitted attention of astronomers. In fact, we might say that each succeeding generation undertakes a new discussion of the distance of this star, with the view of confirming or of criticising the original discovery of Bessel. The diagram here given (Fig. 80) is intended to illustrate the recent history of 61 Cygni.

When Bessel engaged in his labours, the pair of stars forming the double were at the point indicated on the diagram by the date 1838. The next epoch occurred fifteen years later, when the younger Struve undertook his researches, and the pair of stars had by that time moved to the position marked 1853. Finally, when the same object was more recently observed at Dunsink Observatory, the pair had made still another advance to the position indicated by the date 1878. Thus, in forty years this double star had moved over an arc of the heavens upwards of three minutes in length. The actual path is, indeed, more complicated than a simple rectilinear movement. The two stars which form the double have a certain relative velocity, in consequence of their mutual attraction. It will not, however, be necessary to take this into account, as the displacement thus arising in the lapse of a single year is far too minute to produce any inconvenient effect on the parallactic ellipse.

There is an ever present enemy with which the parallax seeker has to contend. That enemy is the atmosphere, which, even when free from clouds and fogs, is a source of uncertainty and of error. On the clearest and finest night, in the most superb climate, the refraction of the atmosphere distorts each star from its true place. The direction in which the telescope must be pointed to see the star is not the true direction; it is not the direction in which the telescope would be pointed were the atmosphere absent. The grosser part of this distortion is, no doubt, amenable to calculation; it can be allowed for, and the true place of the star can be approximately ascertained. The amount of refraction under specified circumstances, as to the elevation of the object above the horizon, the temperature of the air and the height of the barometer, admits of being calculated with an accuracy sufficient for many purposes in astronomy; but the calculated result is still only an approximation to the truth. There are minute irregularities in the refraction which render its amount so variable that we cannot entirely eliminate its effects, and the corrected places of the stars mav, from this cause alone, be occasionally erroneous to the extent of a second, or even more. For the present purpose errors of this magnitude are fatal. The parallactic ellipse which we want to measure is, perhaps, itself not a second in diameter. If, therefore, our observations are wrong by a large fraction of a second, the errors would bear such a large proportion to the total quantity that success would be hopeless.

There is, however, one way, and probably only one way, of evading the difficulty. It is founded on the fact which we have endeavoured to illustrate by the two stars shown in Fig. 79. A star which is very distant has a small parallactic ellipse; a star which is near has a comparatively large ellipse. If, therefore, two stars can be seen in the same field of view of the telescope, and if one is much more remote than the other, the sizes of the parallactic ellipses will be very different. On the diagram, Fig. 80, four stars are indicated besides 61 Cygni; two of these bear the name of Bessel, one bears the name of Struve, and the fourth has the name of Brünnow. These stars are small telescopic objects: they are merely units of the vast host with which the whole heavens teem. They are much inferior in brightness to 61 Cygni; they have not its remarkable duplicity, they have not its remarkable proper motion. They are more remote from the earth than 61 Cygni, which we are to suppose standing out boldly in the foreground, while the others are the distant objects in the picture. The parallactic ellipse described by 61 Cygni is therefore larger than that described by these other stars. Bessel proposed to measure the apparent angular distance from 61 Cygni to the stars marked with his name at every available opportunity. This distance must be constantly changing by the effect of parallax. If the reader have been imbued with geometrical instinct, he will understand how from these measures the form and the position of the parallactic ellipse can be certainly ascertained.

It may be urged that the refraction must surely derange these measures also. No doubt each of the stars is greatly displaced by refraction, but fortunately, they are both displaced by almost the same amount. The distance between the two stars is thus nearly the same as it would be if refraction were entirely absent, and the minute outstanding difference can be accurately allowed for. By this happy contrivance for eliminating the effect of refraction, the research for annual parallax is brought within the grasp of practical astronomy. The measures of the angular distance, though thus purged from refraction, are, however, still complicated by the proper motion of 61 Cygni. The movement of the star in its parallactic ellipse becomes amalgamated with the rectilinear movement due to proper motion. This circumstance will not embarrass the experienced computer. He is able to disentangle the composite movement, and determine with accuracy the elliptic and the rectilinear components.

Bessel commenced the preparations for his great work in a somewhat startling manner. He ordered the opticians to cut the object-glass of his telescope in half. One-half was rigidly fixed in the tube in its proper position, the other half was mounted on a slide, so that its movement was rectilinear, and at right angles to the axis of the telescope. An instrument so treated constitutes a heliometer. When directed towards a star, each half of the object-glass forms a distinct image of the star. If the halves be in their ordinary position the two images of the star are coincident, and the telescope performs like one of the usual type. As, however, the movable part of the object-glass changes its position the two images separate, and the distance between the images is equal to the distance through which the half object-glass has been moved. The movement is effected by a screw of delicate workmanship, so that the distance admits of being determined with accuracy.

The application of the heliometer to the present problem will be easily understood. The telescope is directed to 61 Cygni, so that the comparison star shall be also in the field, and then as the two halves of the object-glass separate so do the two images of the comparison star. Gradually the screw is turned until one of these two images is brought exactly to the central point between the two components of 61 Cygni. The distance through which the half object-glass has been displaced is then to be read off, and it will represent the apparent distance from the centre of 61 Cygni to the comparison star. For three years Bessel devoted himself to these measurements; and the result of his observations was the discovery and the measurement of the parallactic ellipse. It was, indeed, but a small object; it only looked as large as a penny-piece six or seven miles off. But as he found the same ellipse from each of the two comparison stars independently, and as he had exhausted every refinement which consummate skill could suggest, his labours elicited the glowing language of Herschel which has been already cited.

Fifteen years later (1853) the celebrated Russian astronomer, Otto Struve, undertook the labour of a new determination of the distance of 61 Cygni. He determined to vary the method of observation employed by Bessel in every possible detail. Bessel had measured the distance of the central point between the components of 61 Cygni from his comparison stars. Struve chose a single component of 61 Cygni and a comparison star marked on the figure by his name, which is different from either of those used by Bessel. Struve also rejected the divided object-glass, and used instead the instrument previously described in this work and known as the parallel wire micrometer. With this he measured the distance and the position angle of the comparison star from the upper component of 61 Cygni for a period of one year. Struve's labours were also rewarded with success, and he was enabled to obtain the parallactic ellipse from his measurements of the distance, while he obtained an independent determination from his measures of the position angle. The two investigations led to identical results.

Struve and Bessel had thus each shown that the parallactic ellipse—and therefore the distance of 61 Cygni—was a measurable quantity, but there was a substantial difference between their results. Struve's determination indicated that 61 Cygni was much nearer to us than Bessel's work would have led us to suppose. Bessel's distance is, indeed, half as much again as Struve's. Bessel concluded that the distance was about sixty billions of miles; Struve thought it could not be more than forty billions of miles. There was thus a discrepancy between the two astronomers of twenty billions of miles. Is a discrepancy of twenty billions of miles so very serious? No doubt, viewed by our terrestrial standards, the amount is enormous. It would take about 300,000 years even to count twenty billions. But the discrepancy is not so very significant when it has occurred in these colossal magnitudes, the stellar distances. We do not know the parallax of even a dozen of stars well enough to specify their distance accurately to a distance of within twenty billions of miles. A parallax-seeker would feel contented and satisfied with his work if he could be sure that his error was well within this limit.

The case of 61 Cygni is, however, exceptional. It is one of our nearest neighbours in the heavens. We can never find its distance accurately to one or two billions of miles; but still we have a consciousness that an uncertainty amounting to twenty billions is too large. We shall presently show that we believe Struve was right, yet it does not necessarily follow that Bessel was wrong. The apparent paradox can be easily explained. It would not be easily explained if Struve had used the *same comparison star* as Bessel had done; but Struve's comparison star was different from either of Bessel's, and this is probably the cause of the discrepancy. It will be recollected that the whole essence of the process consists of the comparison of the small ellipse made by the distant star, with the larger ellipse made by the nearer star. If the two stars were at the same distance, the process would be wholly inapplicable. In such a case, no matter how near the stars were to the earth, no parallax could be detected. For the method to be completely successful, the comparison star should be at least eight times as far as the principal star. Bearing this in mind, it is quite possible to reconcile the measures of Bessel with those of Struve. We need only assume that Bessel's comparison stars are about three times as far as 61 Cygni, while Struve's comparison star is at least eight or ten times as far. We may add that, as the comparison stars used by Bessel are brighter than that of Struve, there really is a presumption that the latter is the most distant of the three.

We have here a characteristic feature of this method of determining parallax. Even if all the observations and the reductions of a parallax series were mathematically correct, we could not with strict propriety describe the final result as the parallax of one star.

It is only the *difference* between the parallax of the star and that of the comparison star. We can therefore only assert that the parallax sought cannot be less than the quantity determined. Viewed in this manner, the discrepancy between Struve and Bessel entirely vanishes. Bessel asserted that the distance of 61 Cygni could not be *more* than sixty billions of miles. Struve did not contradict this—nay, he certainly confirmed it—when he showed that the distance could not be more than forty billions.

A quarter of a century has elapsed since Struve made his observations. Those observations have not been challenged; they have been, on the whole, confirmed by other investigations. In a critical review of the whole subject several years ago, Auwers had shown that Struve's determination is more worthy of confidence than is that of any other astronomer. Yet, notwithstanding this authoritative dictum, the question has been recently opened afresh. Dr. Brünnow, the recent Astronomer-Royal of Ireland, had successfully measured the parallax of several stars; the instrument he used being the South Equatorial at Dunsink of twelve inches aperture, which we have described in a previous chapter. He commenced a series of observations on the parallax of 61 Cygni which were continued and completed by his successor. Brünnow chose a fourth comparison star (marked on the diagram), different from any of those which had been used by the earlier observers. The method of observing which Brünnow employed was quite different to that of Struve, though the filar micrometer was used in both cases. Brünnow sought to determine the parallactic ellipse by measuring the difference in declination between 61 Cygni and the comparison star. In the course of a year it is found that the difference in declination undergoes a periodic change, and from that change the parallactic ellipse can be computed. In the first series of observations we measured the difference of declination between the preceding star of 61 Cygni and the comparison star; in the second series we took the other component of 61 Cygni and the same comparison star. We had thus two completely independent determinations of the parallax resulting from two years' work. The first of these makes the distance forty billions of miles, and the second makes it almost exactly the same. There can be no doubt that this work supports Struve's determination in correction of Bessel's, and therefore we may perhaps sum up the present state of our knowledge of this question by saying that the distance of 61 Cygni is much nearer to the forty billions of miles which Struve found, than to the sixty billions which Bessel found.*

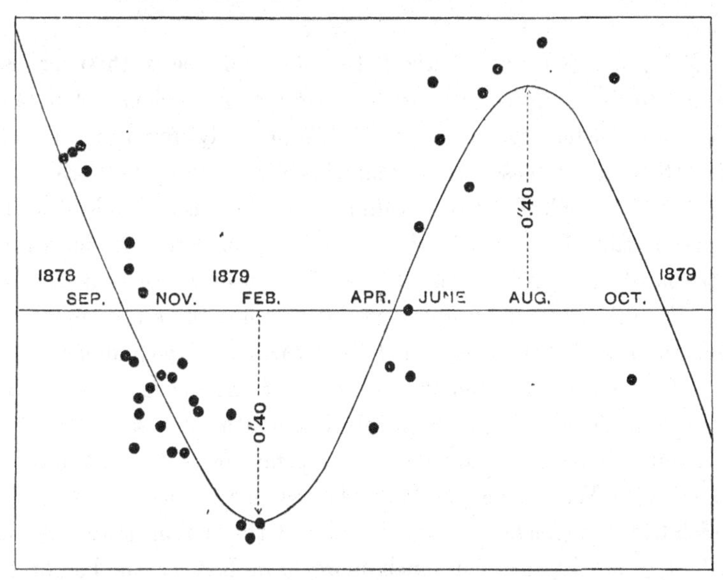

Fig. 81.—Parallax in Declination of 61 Cygni.

It is desirable to give the reader the means of forming his own opinion as to the quality of the evidence which is available in such researches. The diagram in Fig. 81 here shown has been constructed with this object. It is intended to illustrate the second series of observations of difference of declination which we have made at Dunsink. Each of the dots represents one night's observations. The height of the dot is the observed difference of declination between 61 (B) Cygni and

* The distance of 61 Cygni has again been recently investigated by Prof. Asaph Hall, of Washington, who has obtained a result practically coincident with that found at Dunsink.

the comparison star. The distance along the horizontal line—or the abscissa, as a mathematician would call it—represents the date. These observations are grouped more or less regularly in the vicinity of a certain curve. That curve expresses where the observations should have been, had they been absolutely perfect. The distances between the dots and the curve may be regarded as the errors which have been committed in making the observations.

Perhaps it will be thought that in many cases these errors appear to have attained very undesirable dimensions. Let us, therefore, hasten to say that it was precisely for the purpose of setting forth these errors that this diagram has been shown; we have to exhibit the weakness of the case no less than its strength. The errors of the observations are not, however, intrinsically so great as might at first sight be imagined. To perceive this, it is only necessary to interpret the scale on which this diagram has been drawn, by comparison with familiar standards. The distance from the very top of the curve to the horizontal line denotes an angle of only four-tenths of a second. This is about the apparent diameter of a penny-piece at a distance of *ten miles!* We can now appraise the true magnitude of the errors which have been made. It will be noticed that no one of the dots is distant from the curve by much more than half of the height of the curve. It thus appears that the greatest error in the whole series of observations amounts to but two or three tenths of a second. This is equivalent to our having pointed the telescope to the upper edge of a penny-piece fifteen or twenty miles off, instead of to the lower edge. This is not a great blunder. A rifle team whose errors in pointing were more than a hundred times as great, might still easily win every prize at Wimbledon. The parallactic ellipse is, however, so small, that the errors, minute as they are, bear a large proportion to the total quantity under consideration. This it is which constitutes the weakness of parallax observations, but notwithstanding these difficulties the results are seemingly entitled to confidence. It is in the very nature of such work that by increasing the number of observations the errors tend to counteract each other. To ensure the neutralization of errors, the star is observed on as many nights as possible, and a considerable number of observations are usually secured on each night. As the number of observations is increased, so does the error of the final result diminish.

The observations being all completed, the astronomer finds before him a vast mass of, perhaps, fifteen thousand figures. The parallax, of which he is in search, can be sufficiently expressed by two digits—indeed, almost by a single one. With the collection of these observations the functions of the *observer* are at an end. It now becomes the duty of the *mathematician* to elaborate from this mass of detail the one or two figures which embody the final result. He is not only able to elicit the most probable value of the parallax, but he can appraise the value of his result, and express numerically the degree of confidence to which it is entitled; in mathematical language, he determines the probable error of the parallax, as well as its most probable value.

We have entered into the history of 61 Cygni with some detail, because it is the star whose distance has been most studied. We do not say that 61 Cygni is the nearest of all the stars; it would, indeed, be very rash to assert that any particular star was the nearest of all the countless millions in the heavenly host. We certainly know one star which seems nearer than 61 Cygni; it lies in one of the southern constellations, and its name is α Centauri. This star is, indeed, of memorable interest in the history of the subject. Its parallax was first determined at the Cape of Good Hope by Henderson; subsequent researches have confirmed his observations, and the elaborate investigations of Mr. Gill have proved that the parallax of this star is about three-quarters of a second, so that it is only two-thirds of the distance of 61 Cygni. By what sagacious intuition was Henderson guided to a comparatively near star in the southern hemisphere, and Bessel to one in the northern?

61 Cygni arrested our attention, in the first instance, by the circumstance that it had the large proper motion of five seconds annually. We have also ascertained that the annual parallax is

about half a second. The combination of these two statements leads to a result of considerable interest. It teaches us that 61 Cygni must each year traverse a distance of not less than ten times the radius of the earth's orbit. Translating this into ordinary figures, we learn that this star must travel nine hundred and twenty million miles per annum. It must move between two and three million miles each day, but this can only be accomplished by maintaining the prodigious velocity of thirty miles per second. There seems to be no escape from this conclusion. The facts which we have described, and which are now sufficiently well established, are inconsistent with the supposition that the velocity of 61 Cygni is less than thirty miles per second; the velocity may be greater, but less it cannot be.

For forty years, for one hundred years, we know that 61 Cygni has been moving in the same direction and with the same velocity. Prior to the existence of the telescope we have no observation to guide us; we cannot, therefore, be absolutely certain as to the earlier history of this star, yet it is only reasonable to suppose that 61 Cygni has been moving from remote antiquity with a velocity comparable with that it has at present. If disturbing influences were entirely absent, there could be no trace of doubt about the matter. *Some* disturbing influence, however, there must be; the only question is whether that disturbing influence is sufficient to seriously modify the assumption we have made. A powerful disturbing influence might greatly alter the velocity of the star; it might deflect the star from its rectilinear course; it might even force the star to move around a closed orbit. We do not, however, believe that any disturbing influence of this magnitude need be contemplated, and there can be no reasonable doubt that 61 Cygni moves at present in a path very nearly straight, and with a velocity very nearly uniform.

As the distance of 61 Cygni from the sun is forty billions of miles, and as its velocity is thirty miles a second, it is easy to find how long the star would take to accomplish a journey equal to its distance from the sun. The time required will be about 40,000 years. In the last 400,000 years, 61 Cygni will have moved over a distance ten times as great as its present distance from the sun, whatever be the direction of motion. This star must therefore have been about ten times as far from the earth 400,000 years ago as it is at present. Though this epoch is incredibly more remote than any historical record, it is perhaps not incomparable with the duration of the human race; while compared with the vast lapse of geological time, such periods seem trivial and insignificant. Geologists have long ago repudiated mere thousands of years; they now claim millions, and many millions of years, for the performance of geological phenomena. If the earth has existed for the millions of years which geologists assert, it becomes reasonable for astronomers to speculate on the phenomena which have transpired in the heavens in the lapse of similar ages. By the aid of our knowledge of star distances, combined with an assumed velocity of thirty miles per second, we can make the attempt to peer back into the remote past, and show how great are the changes which our universe seems to have undergone.

In a million years 61 Cygni will, apparently, have moved through a distance which is twenty-five times as great as its present distance from the sun. Whatever be the direction in which 61 Cygni is moving—whether it be towards the earth or from the earth, to the right or to the left, it must have been about twenty-five times as far off a million years ago as it is at present; but even at its present distance 61 Cygni is a small star; were it ten times as far it could only be seen with a good telescope; were it twenty-five times as far it could barely be a visible point in our greatest telescopes.

The conclusions arrived at with regard to 61 Cygni may be applied with varying degrees of emphasis to other stars. We are thus led to the conclusion that many of the stars with which the heavens are strewn are apparently in slow motion. But this motion though apparently slow may really be very rapid. When standing on the sea-shore, and looking at a steamer on the distant horizon, we can hardly notice that the steamer is moving. It is true that by looking again in a few minutes we can detect a change in the place; but the motion of the steamer seems slow.

Yet if we were near the steamer we would find that it was rushing along at the rate of many miles an hour. It is the distance which causes the illusion. So it is with the stars: they seem to move slowly because they are very distant, but were we near them, we could see that in the majority of cases their motions are enormously rapid, and probably a thousand times as fast as the swiftest steamer that ever ploughed the ocean.

It thus appears that the permanence of the sidereal heavens, and the fixity of the constellations in their relative positions, are only ephemeral. When we rise to the contemplation of such vast periods of time as the researches of geology disclose, the durability of the constellations vanishes! In the lapse of those stupendous ages stars and constellations gradually dissolve from view, to be replaced by others of no greater permanence.

In order to convey some idea of the magnitudes of the quantities with which we have to deal when we are discussing the proper motion of the stars, let us study the effects of proper motion on the principal stars of the Great Bear. Night after night we see these stars in the same relative positions, and the general outline of the constellation has remained unaltered for centuries. Homer spoke of it thousands of years ago as the bear which feared to dip into the sea, alluding to the fact that in these latitudes this constellation never sets.

But it is very interesting to inquire whether some changes may not have occurred in the stars of the Great Bear since the earliest times. Fortunately, we have the means of answering the question with a considerable degree of accuracy. Nearly two thousand years ago Ptolemy constructed a catalogue, in which he recorded the position of a large number of fixed stars. Ptolemy had no telescope, but he had constructed certain instruments for measuring the positions of the stars in the heavens. No doubt we should now consider these contrivances excessively rude and inadequate, and vastly inferior to those used every day by modern astronomers. Yet the antiquity of Ptolemy's observations confers on them a positive value, which more than compensates for their inaccuracy. Among the stars which Ptolemy observed were those of the Great

Bear; and from his observations we are able to construct the map of the Great Bear as it appeared to him, and compare it with the present appearance of the same constellation. The comparison shows that the shape of the Great Bear has not changed much for the last 2,000 years. There is, however, *some* difference between the past and present places of the stars. A part of the differences can be explained by the inevitable errors of the crude methods of measuring used by Ptolemy; but making every allowance for these errors, there can still be no doubt that the positions of the stars have perceptibly changed in 2,000 years.

We next proceed to the history of a very remarkable star, which has been the subject of much parallax research and of much ingenious speculation. The star will be found in the Great Bear, but is only of the sixth or the seventh magnitude, and therefore invisible to the unaided eye. It is one of that myriad host of stars still remaining unchristened, so that astronomers refer to it either by its place on the heavens, or by citing Groombridge's well known star catalogue, in which the star is No. 1,830. Yet, though small and inconspicuous, this object is in one respect the most remarkable star known among the thousands in the northern hemisphere. It is distinguished by the exceptionally large amount of its proper motion. In virtue of this proper motion, Groombridge 1,830 moves over no less than seven seconds of arc per annum. It should also be remembered that our telescopes only give us that component of the proper motion which is projected on the surface of the heavens. The resultant proper motion must of course be generally greater than this single component. By Mr. Huggins' spectroscopic method, to be described in a future chapter, it might be possible to measure the velocity of the star along the line of sight. If this could be ascertained, then by the composition of velocities we could determine the numerical value of the true proper motion. In the absence of such measures, we can only assert that the proper motion of this star may be greater, but cannot be less, than seven seconds per annum. We have calculated the most probable value of the proper motion from the theory of probabilities, and have found that it is just as likely to exceed nine seconds per annum as to fall short thereof; we shall, however, only assume the velocity to be seven seconds, as that will be amply sufficient for our purpose.

We first translate the conception of the proper motion into other forms in which it may be more readily intelligible. Imagine a star placed at the Pole, and moving with the velocity of seven seconds per annum across the heavens to the Belt of Orion. This is a distance of nearly ninety degrees, and the entire journey would occupy about forty-six thousand years. Even in a few months the movement would be large enough to be readily detected by telescopic observation, while in a period of two hundred and fifty years this star would traverse a distance on the surface of the heavens equal to the apparent diameter of the full moon.

When we speak of a *large* proper motion, it is to be understood that we refer for comparison to the ordinary proper motions of ordinary stars. Compared with the proper motions of objects belonging to the solar system, the proper motions of the stars are extremely small. The slowest and most remote planet, Neptune, moves over eight hundred times as much arc per annum as the most rapidly moving star. We must be prepared to find that although Groombridge 1,830 only changes its place at the rate of seven seconds annually, yet, that as its distance is enormously great, the intrinsic velocity of the star may attain portentous dimensions.

Several attempts have been made to measure the distance of Groombridge 1,830. Astronomers were specially attracted to the research by the exceptional proper motion, which seemed to suggest that a measurable distance might be expected. Strange to say, however, this anticipation has not been realised; the star is not among the sun's nearer neighbours. Perhaps the most successful determinations are those obtained by Struve and by Brünnow. The latter has, indeed, produced a memoir upon the parallax of Groombridge 1,830, which is a classical work in this branch of astronomical literature. Both these observers have shown that this star is at least four or five times as far from the earth as the star 61 Cygni, which engaged our attention in the earlier part of this chapter. By the combination of the measured distance and the known proper motion, we arrive at a result which is very startling. An observer situated on the star and looking at the solar system, would find the radius of the earth's orbit to subtend an angle of one-tenth of a second; we, on the other hand, see that the star moves over seven seconds in a year. From this it follows that the annual journey of Groombridge 1,830 must be no less than seventy times the radius of the earth's orbit; in other words, the velocity of this star is so enormous that it would travel from the earth to the sun in about five days. It does not seem possible that there can be any substantial error in this conclusion; indeed, the known facts would be quite consistent with the supposition that the star travelled over a distance equal to one hundred radii of the earth's orbit per annum. We have taken the lowest estimate that can be reconciled with the observations, in stating the velocity to be seventy radii per annum. A very simple calculation will transform this result into more familiar language.

We thus obtain, finally, the extraordinary result that the actual velocity of this star is not less than two hundred miles per second! In ten minutes Groombridge 1,830 has travelled 120,000 miles. In one minute this wonderful star would perform the journey from London to Pekin. If our earth moved equally fast, the journey around the sun would be accomplished in about a month, instead of the year which is now required. We do not assert that so great a velocity is absolutely unparalleled in the universe. In the full flush of its perihelion swoop, a comet, as it passes close to the sun, may attain this velocity, or may even exceed it. This is, however, a mere explosive outburst of vigour on the part of the comet; the great velocity is soon abated, a reaction sets in, and in the remote parts of its orbit the motion is very slow. The velocity of Groombridge 1,830 is no mere spasmodic effort; with a stately uniformity, worthy of the dignity of a majestic sun, it sweeps along, alike inflexible in the direction of its motion and in the velocity with which its journey is performed.

But it not unfrequently happens that a parallax research proves abortive. The labour is accomplished, the observations are reduced and discussed, and yet no value of the parallax can be obtained. The distance of the star is so vast that our base line, although it is nearly two hundred millions of miles long, is too short to bear any appreciable ratio to the distance of the star. Even from such failures, however, information may often be drawn.

Let me illustrate this by an account derived from our experience at Dunsink. On the 24th November, 1876, a well-known astronomer—Dr. Schmidt, of Athens—noticed a new bright star of the third magnitude in the constellation Cygnus, in a place where he was sure that no corresponding star had existed four days previously. The star which thus suddenly burst into splendour seems never to have been observed before. The charts and catalogues were searched, and they showed no star in that position; so that if Nova Cygni were really only an outbreak of brilliancy in a small star, that star must have been so small as to have escaped the attention of all the surveying astronomers who had previously examined that region. The suddenly acquired splendour of Nova Cygni was not of long duration. In a single week it had ceased to be a conspicuous object; in a fortnight it was no longer visible to the unaided eye; and at the last it dwindled down to a minute telescopic point.

This is, in any case, a very remarkable career. We do not say that it is unparalleled, as there have been other similar cases recorded; but it may be regarded as certain that there was never a new star which was honoured by so much attention as was shown to Nova Cygni. From the outbreak of splendour to its decline, this wonderful star has been relentlessly pursued by equatorials, meridian circles, micrometers, and the paraphernalia of the modern observatory. By the spectroscope especially many interesting facts have been ascertained. It has been shown that the constitution of this star is totally different from that of the ordinary stars. The spectroscope proves that the brilliant splendour of Nova Cygni was due to the presence of glowing and incandescent gases, while the same subtle instrument showed that when the star was declining to invisibility, the last rays which reached the earth were emitted not from a solid but from some gaseous body. It is impossible to resist the conclusion, that the ephemeral splendours of the star were connected with some prodigious outbreak of luminous vapour, on the cause of which it is hard to resist hazarding a speculation. Most astronomers are of opinion that in all probability the sudden outburst of splendour was due to a collision between two bodies, which dashed together with a high relative velocity. It is known that such a collision would be an adequate cause of brilliant incandescence. If, for example, two bodies, each equal to the earth, and moving, as the earth does, with a velocity of eighteen miles a second, were to collide, a prodigious flash of light would be the consequence, while the heat developed would be adequate to dissipate the entire mass of both into glowing vapour.

On the 20th of November Nova Cygni was invisible. Whether it first burst forth on the 21st, 22nd, or 23rd, no one can tell; but on the 24th it was discovered. Its brilliancy even then seemed to be waning; so, presumably, it was brightest at some moment between the 20th and 24th of November. The outbreak must thus have been comparatively sudden, and we know of no cause which would account for such a phenomenon more simply than a gigantic collision. The decline in the brilliancy was much more tardy than its growth, and more than a fortnight passed before the star relapsed into insignificance—two or three days for the rise, two or three weeks for the fall. Yet even two or three weeks was a short time in which to extinguish so mighty a conflagration. It is comparatively easy to suggest an explanation of the sudden outbreak; it is not equally easy to understand how it can have been subdued in a few weeks. A good-sized iron casting in one of our foundries takes nearly as much time to cool as sufficed to abate the celestial fires in Nova Cygni!

On this ground it seemed not unreasonable to suppose that perhaps Nova Cygni was not really a very extensive conflagration. But, if such were the case, the star must have been comparatively *near* to the earth, since it presented so brilliant a spectacle and attracted so much attention. It therefore appeared a plausible object for a parallax research; and consequently a series of observations were recently made at Dunsink. We were at the time too much engaged with other work to devote very much labour to a research which might, after all, only prove illusory. We simply made a sufficient number of micrometric measurements to test whether a *large* parallax existed. It has been already pointed out how each star appears to describe a minute parallactic ellipse, in consequence of the annual motion of the earth and by measurement of this ellipse the parallax—and therefore the distance—of the star can be determined. Under ordinary circumstances, when the parallax of a star is being investigated, it is necessary to measure the position of the star in its ellipse on many different occasions, distributed over a period of at least an entire year. The method we adopted was much less laborious. It was sufficiently accurate to test whether or not Nova Cygni had a *large* parallax, though it might not have been delicate enough to disclose a small parallax. At a certain date, which can be readily computed, the star is at one end of the parallactic ellipse, and six months later the star is at the other end. By choosing suitable times in the year for our observations, we can measure the star in those two positions when it is most deranged by parallax. It was by observations of this kind that we sought to detect the parallax of Nova Cygni. Its distance from a neighbouring star was carefully measured by the micrometer at the two seasons when, if parallax existed, those distances should show their greatest discrepancy ; but no certain difference between these distances could be detected. The observations, therefore, failed to reveal the existence of a parallactic ellipse—or, in other words, the distance of Nova Cygni was too great to be measured by observations of this kind.

I feel certain that if Nova Cygni had been one of the nearest stars these observations would not have been abortive. We are therefore entitled to believe that Nova Cygni must be at least 20,000,000,000,000 miles from the solar system ; and the suggestion that the brilliant outburst was of small dimensions must be at once abandoned. The intrinsic brightness of Nova Cygni, when at its best, cannot have been greatly if at all inferior to the brilliancy of our sun himself. If the sun were withdrawn from us to the distance of Nova Cygni, it would seemingly have dwindled down to an object not more brilliant than the variable star. How the lustre of such a stupendous object declined so rapidly, remains, therefore, a mystery not easy to explain.

Have we not said that the outbreak of brilliancy in this star occurred between the 20th and the 24th of November, 1876? It would be more correct to say that the tidings of that outbreak reached our system at the time referred to. The real outbreak must have taken place years previously—three years, at all events, and possibly much more. At the time that the star excited such commotion in the astronomical world it had really relapsed again into insignificance, for the rays of light which start from the star take years before they traverse the mighty distance to the solar system.

In connection with the subject of the present chapter we have to consider a great problem which was proposed by Sir William Herschel. He saw that the stars were animated by proper motion; he saw also that the sun is a star, one of the countless host of heaven, and he was therefore led to propound the stupendous question as to whether the sun, like the other stars which are its peers, was also in motion.

Consider all that this great question involves. The sun has around it a retinue of planets and their attendant satellites, the comets, and a host of smaller bodies. The question is, whether all this superb system is revolving around the sun *at rest* in the middle, or whether the whole system—sun, planets, and all—is not moving on bodily through space.

Herschel was the first to solve this noble problem; he discovered that our sun and all the splendid retinue by which it is attended are moving bodily in space. He not only discovered this, but he ascertained the direction in which the system was moving, as well as the velocity with which that movement was performed. He has shown that the sun and his system is now rapidly hastening towards a certain point of the constellation Hercules. The velocity with which the motion is performed corresponds to the magnitude of the system: quicker than the swiftest rifle-bullet that was ever fired, the sun, bearing with it the earth and all the other planets, is now sweeping onwards. We on the earth participate in that motion. Every half-hour we are about ten thousand miles nearer to the constellation of Hercules than we would have been if the solar system was not animated by this motion. As we are proceeding at this stupendous rate towards Hercules, it might at first be supposed that we ought soon to get there; but the distances of the stars in Hercules are enormously great, as are those of the stars elsewhere, and we may be certain that the sun and his system must travel at the present rate for far more than a million years before we have crossed the abyss between our present position and the frontiers of Hercules.

It remains to explain the method of reasoning which Herschel adopted, by which he was able to make this great discovery. It may sound strange to hear that the detection of the motion of the sun was not made by looking at the sun himself; all the observations of the sun with all the telescopes in the world would never tell us of that motion, for the simple reason that the earth, from whence our observations must be made, participates in it. A passenger in the cabin of a ship usually becomes aware that the ship is moving by the roughness of the sea; but if the sea be perfectly calm, then, though the tables and chairs in the cabin are moving as rapidly as the ship, yet we do not see them moving, because we are also travelling with the ship. If we could not go out of the cabin, nor look out through the windows, we would never know whether the ship was moving or at rest; nor could we have any idea as to the direction in which the ship was going, or as to the velocity with which that motion was performed.

The sun, with his attendant host of planets and satellites, may be likened to the ship. The planets may revolve around the sun just as the passengers may move about in the cabin, but as the passengers, by looking at objects on board, can never tell where the ship is going to, so we, by merely looking at the sun, or the other planets or members of the solar system, can never tell whether our system as a whole is in motion.

The conditions of a perfectly uniform movement along a perfectly calm sea are not often fulfilled on the waters with which we are acquainted, but the course of the sun and his system is untroubled by any disturbance, so that the majestic progress is conducted with the most absolute uniformity. We feel neither the rolling nor the pitching, neither the jolts nor the rattle of machinery, which apprise us of motion in the ship; and as all the planets are travelling with us, we can get no information from them as to the common motion by which the whole system is animated.

The passengers are, however, at once apprised of the ship's motion when they go on deck, and when they look at the sea surrounding them. Let us suppose that their voyage is nearly accomplished, that the distant land appears in sight, and, as evening approaches, the harbour is discerned into which the ship is to enter. Let us suppose that the harbour has, as is often the case, a narrow entrance, and that its mouth is indicated by a lighthouse on each side. When the harbour is still a long way off, near the horizon, the two lights are seen close together, and now that the evening has closed in, and the night has become quite dark, these two lights are all that is visible. While the ship is still some miles from its destination the two lights look close together, but as the distance decreases the two lights seem to open out; gradually the ship gets nearer while the lights are still opening, till finally, when the ship enters the harbour, instead of the two lights being directly in front, as at the commencement, one of the lights is passed by on the right hand, while the other is similarly found on the left. If, then, we are to discover the motion of the solar system, we must, like the passenger, look at objects unconnected with our system, and learn our own motion by their apparent movements. But are there any objects in the heavens unconnected with our system? If all the stars were like the earth, merely the appendages of our sun, then we never could discover whether we were at rest or whether we were in motion: our system might be in a condition of absolute rest, or it might be hurrying on with an inconceivably great velocity, for anything we could tell to the contrary. But the stars do not belong to the system of our sun; they are, rather, suns themselves, and do not recognise the sway of our sun, as this earth is obliged to do. The stars will, therefore, act as the external objects by which we can test whether our system is voyaging through space.

With the stars as our beacons, what ought we to expect if our system be really in motion? Remember that when the ship was approaching the harbour the lights gradually opened out to the right and left. But the astronomer has also lights by which he can observe the navigation of that vast craft, our solar system, and these lights will indicate the path along which he is borne. If our solar system be in motion, we should expect to find that the stars were gradually spreading away from that point in the heavens towards which our motion tends. This is precisely what we do find. The stars in the constellations are gradually spreading away from a central point in the constellation of Hercules, and hence we infer that it is towards Hercules that the motion of the solar system is directed.

There is one great difficulty in the discussion of this question. Have we not had occasion to observe that the stars themselves are in actual motion? It seems certain that all the stars, including the sun himself as a star, have each an individual motion of their own. The motions of the stars as we see them are partly apparent as well as partly real; they partly arise from the actual motion of each star and partly from the motion of the sun, in which we partake, and which produces an apparent motion of the star. How are these to be discriminated? Our telescopes and our observations can never effect this decomposition directly. To accomplish the analysis, Herschel resorted to certain geometrical methods. His materials at that time were but scanty, but in his hands they proved adequate, and he boldly announced his discovery of the movement of the solar system.

So majestic an announcement demanded the severest test which the most refined astronomical resources could suggest. There is a certain powerful and subtle method which astronomers use in the effort to interpret nature. Bishop Butler has said that probability is the guide of life. If we wish to see this maxim applied with logical perfection, we must visit the study of a mathematician when he is discussing a problem like that now before us. The proper motion of a star has to be decomposed into two parts, one real and the other apparent. When several stars are taken, we may conceive an infinite number of ways into which the movements of each star can be so decomposed. Each one of these conceivable divisions will have a certain element of probability in its favour. It is the business of the mathematician to determine the amount of that probability. The case, then, is as follows :—Among all the various systems one must be true. We cannot lay our finger for certain on the true one, but we can take that which has the highest degree of probability in its favour, and thus follow the precept of Butler to which we have already referred. A mathematician would describe his process by calling it the method of least squares. We do not say that Herschel employed these refinements. Indeed, Sir William Herschel was not a mathematician in the extended sense of the word, and he employed methods which would not be sufficiently capacious to deal with the ample materials now accumulated. Since Herschel's discovery, one hundred years ago, many a mathematician using observations of thousands of stars has attacked the same problem. The stars in the northern hemisphere have been discussed with far greater completeness; the stars in the southern hemisphere, which Herschel knew not, have been made to bear their testimony. The stars which move quickly have been interrogated, and so have the stars which hardly move at all. Mathematicians have exhausted every refinement which the theory of probabilities can afford, but only to establish conclusively the truth of that splendid theory which seems merely to have been one of the flashes of Herschel's genius.

We thus learn that our whole system, comprising the sun in the centre, with the planets which circulate around the sun, the comets, and the incredible host of minute bodies, are all together bound on a stupendous voyage through the realms of space. The progress of our sun is marked by the dignified solemnity worthy of so majestic a body. The sun requires almost two days to move through a length equal to his own diameter. Every two days the solar system accomplishes a stage of about a million miles in its journey towards the constellation of Hercules.

CHAPTER XXII.

THE SPECTROSCOPE.

A New Department of Science—The Materials of the Heavenly Bodies—Meaning of Elementary Bodies—Chemical Analysis and Spectroscopic Analysis—The Composite Nature of Light—Whence Colours?—The Rainbow—The Prism—Passage of Light through a Prism—Identification of Metals by the Rays they emit when Incandescent—The great Discovery of the Identity of the D-Lines with Sodium—The Dark Lines in the Solar Spectrum Interpreted—Metals present in the Sun—Examination of Light from the Moon or the Planets—The Prominences surrounding the Sun—Photographs of Spectra—Measurement of the Motion of the Stars along the Line of Sight.

THE revelations of the spectroscope form an important chapter in modern science. By its aid a mighty stride has been taken in our attempt to comprehend that elaborate system of suns and other bodies which adorn the skies.

The subject is widely dissimilar to those branches of astronomy which had been previously cultivated. The discoveries of the early astronomers had taught us how the earth revolves around the sun. We have been told how the distances of the different bodies were to be measured, and how their sizes were found. More wonderful still, the older astronomy narrated how we could weigh one body in the heavens against another. We knew that the substance of which the sun was made glowed with intense heat; but we could not tell, we had no means of telling, what was the nature of that substance. To this the old astronomy could render no answer whatever; and, indeed, half a century ago it would have been thought incredible that such a question could be answered. If this be the case with regard to the sun, what are we to say of the other and still less known bodies of the heavens? What is the physical nature of the other planets? What are the substances present in the stars, and in those dim nebulæ which lie on the confines of the visible universe? It is the triumph of modern astronomy to have afforded an answer to these questions: not, indeed, with all the fulness that might be desired, and that, perhaps, may yet be expected; but still they have been answered in some degree, and the results are of great interest.

Let us first enunciate the problem in a definite shape; and here we must call in the aid of chemistry to enable us to obtain distinct views on the subject. What are we to understand by the substances of which the sun is composed? What do we even mean by the substances of which the earth is composed? At a first glance we would say that our globe is composed of rocks and clay, of air and water, but the chemist will strive to render our ideas more accurate. He shows us how rocks are composed of certain other substances, which he calls elements, and the elements are substances which he cannot decompose into anything else. He will also tell us how the air is composed of two elementary substances, oxygen and nitrogen, mingled together; and how water is composed of oxygen and hydrogen in a state of combination. The chemist pursues his analysis through every solid, liquid, and gas on the globe; he decomposes animal or vegetable substances into their elements, and the outcome of his labours is to demonstrate that every particle of matter cn our globe, or in the atmosphere surrounding it, is composed of one or more elementary bodies, and that the whole number of such elementary bodies is about sixty or rather more. The elements may be solids, like iron, or gold, or carbon; or they may be gases, like oxygen, or hydrogen, or nitrogen. Many of the elements are extremely rare. About twenty of the more ordinary ones are all that need concern us at present.

We are now enabled to state a special problem of modern astronomy. Are these elements of which the earth is composed peculiar to the earth, or are they found on the other bodies which teem throughout space? Take, for example, the most abundant of all metals, iron, which exists in rich profusion near the surface of the earth, and which exists perhaps in no less abundance in the interior of the earth. Is iron a product limited to this, our globe? or does it enter into the composition of other bodies in the universe? This is one of the questions which modern science has answered. How that answer has been given it is the object of this chapter to unfold.

Quite a new phase of astronomy is here opened up. Great telescopes revealed faint objects, and showed the features clearly which small telescopes only showed dimly, but all the telescopes in the world would not answer the question as to whether iron was found in the sun. A totally distinct branch of inquiry was necessary, which is known as spectrum analysis. We could not expect to actually *see* iron in the sun. The sun is itself a mighty glowing globe, infinitely hotter than a Bessemer converter or a Siemens furnace; if iron is in the sun, that iron is not only white-hot and molten, but actually driven off into vapour; but vapour of iron is not distinguishable. How would you recognise it? How would you know it if commingled with the vapour of many other metals or other substances? It is, in truth, a delicate piece of analysis to discriminate iron in the glowing atmosphere of the sun. But the spectroscope is adequate to the task, and it renders its analysis with an evidence that is absolutely convincing.

Wide indeed is the gulf which separates this branch of analysis from those to which chemists were formerly restricted. To analyse a body in the old fashion, the chemist must have a sample of that body, and then by his re-agents and his test tubes, he could determine what the body contained. But how is the chemist to procure a sample of the sun? He can never procure a sample of any body external to the earth. No doubt, in the case where a body, which we call a meteorite, falls to the earth, we certainly can submit to chemical analysis an actual fragment derived from external space. It is a matter of significance to observe that when a chemist analyses a meteorite, he finds in it no element with which he is not already familiar; nay, rather, he is struck by the remarkable fact that while nearly all meteorites contain iron, some are almost entirely composed of that widely distributed metal. We cannot, however, for the reasons set forth in Chapter XVII., draw any reliable conclusions from meteorites as to the constituents of bodies exterior to the earth. In our attempts to analyse the sun, the moon, the planets, or the stars, we are therefore deprived of the ordinary resources of chemistry. To deal with the problem a new branch of science has been created.

What we receive from the sun is warmth and light. The intensely heated mass of the sun radiates forth its beams in all directions with boundless prodigality. Each beam we feel to be warm, and we see to be brilliantly white, but a more subtle analysis than mere feeling or mere vision is required. Each sunbeam really bears indelible marks of its origin. These marks are not visible until a special process has been applied, but then the sunbeam can be made to tell its story, and it will disclose to us the nature of the sun's constitution.

We speak of the sun's light as colourless, just as we speak of water as tasteless, but both of these expressions relate rather to our own feelings than to anything really characteristic of water or of sunlight. We regard the sunlight as colourless because it forms, as it were, the background on which all other colours are depicted. The fact is, that white is so far from being colourless, that it contains every hue known to us blended together in certain proportions. The sun's light is really extremely composite; Nature herself tells us this if we will but give her the slightest attention. Whence come the beautiful hues with which we are all familiar? Look at the lovely tints of a garden; the red of the rose is not in the rose itself. All the rose does is to grasp the sunbeams which fall upon it, extract from these beams the red which is in them, and radiate that red light to our eyes. Were there not red rays commingled with the other rays in the sunbeam, there could be no red rose to be seen by sunlight. This point is, in fact, well known in ordinary conversation; a lady will often say that a dress which looks very well in the daylight does not answer in the evening. The reason is that the dress is intended to show certain colours which exist in the sunlight; but these colours do not exist to the same degree in gaslight, and consequently the dress has a different hue. The fault is not in the dress, the fault lies in the gas; and when the electric light is used, it sends forth beams more nearly resembling those from the sun, and the colours appear again with all their intended beauty.

The most glorious natural indication of the nature of the sunlight is seen in the rainbow. Here the sunbeams are refracted and reflected from the tiny globes of water falling as rain-drops from the clouds; these convey to us the sunlight, and in doing so decompose the white beams into the seven primary colours—red, orange, yellow, green, blue, indigo, and violet.

The bow set in the cloud is typical of that great department of modern science which we discuss in this chapter. The globes of water decompose the solar beams; and we follow the course suggested by the rainbow, and analyze the sunlight into its

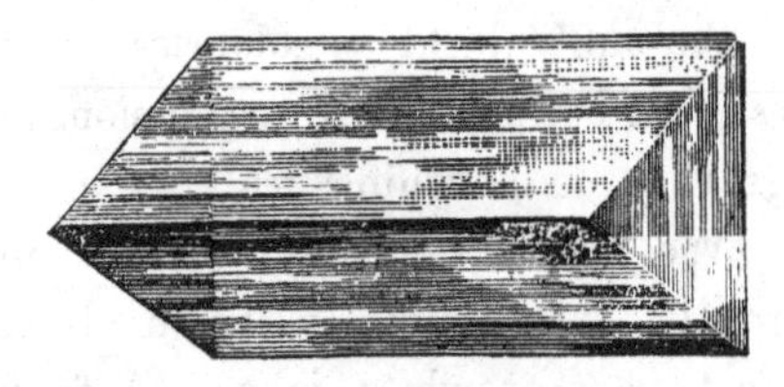

Fig. 82.—The Prism.

constituents. We are enabled to do this with scientific accuracy when we employ that admirable invention, the spectroscope. The beams of white sunlight really consist of innumerable beams of every hue all commingled together. Every shade of red, of yellow, of blue, and of green can be found in a sunbeam. The magician's wand, with which we can strike the sunbeam and instantly sort out into perfect order the tangled skein, is the simple instrument known as the glass prism. We have represented a prism in its simplest form in the adjoining figure (Fig. 82). It is a piece of pure and homogeneous glass, ground and polished to the shape of a wedge. When a ray of light from the sun or from any source falls upon the prism, it passes through the transparent glass and emerges on the other side; a remarkable change is, however, impressed on the ray by the influence of the glass. It is bent by refraction from the path it originally pursued, and it is compelled to follow a different path. If, however, the prism bent all rays of light equally, then it would be of no service in the analysis of light; but it fortunately happens that the prism acts with varying efficiency on the rays of different hues. A red ray is not refracted so much as a yellow ray; a yellow ray is not refracted so much as a blue one. It consequently happens that when the composite beam of sunlight in which all the different rays are blended, passes through the prism, they emerge in the manner shown in the annexed figure. Here then we have the source of the analysing power of the prism: it bends the different hues unequally, and consequently the beam of composite sunlight, after passing through the prism, no longer shows mere white light, but

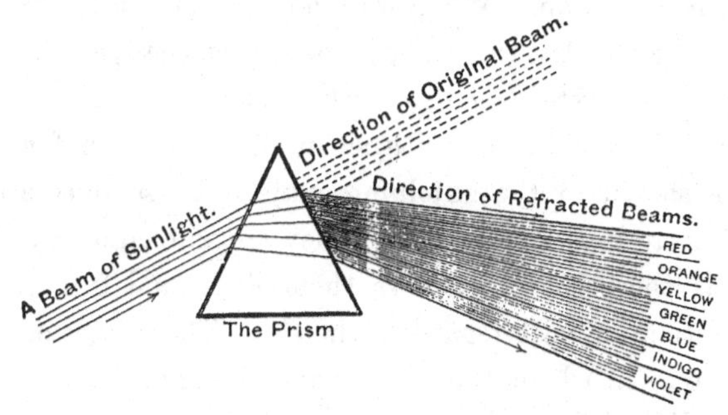

Fig. 83.—Dispersion of Light by the Prism.

is expanded into a coloured band of light, with hues like the rainbow, passing from deep red at one end through every intermediate grade to the violet.

We have in the prism the means of decomposing the light from the sun, or the light from any other source, into its component parts. The examination of the quality of the light when analysed will often enable us to learn something of the constitution of the body from which this light has emanated. Indeed, in some simple cases the mere colour of a light will be sufficient to indicate the source from which the light has come. There is, for instance, a splendid red light sometimes seen in displays of fireworks, due to the metal strontium. The eye can at once identify the element by the mere colour of the flame. There is also a very characteristic yellow light produced by the flame of common salt burned with spirits of wine. Sodium is the important constituent of salt, so here we can recognise another substance merely by the colour it emits when burning. We may also mention a third substance, magnesium, which burns with a brilliant white light, eminently characteristic of the metal.

The three metals strontium, sodium, and magnesium are thus to be identified by the colours they produce when incandescent. In this simple observation lies the germ of the modern method of research known as spectrum analysis. We examine with the prism the colours of the sun and the colours of the stars, and from this examination we can learn something of the materials which enter into their composition. When we study the sunlight through the prism, it is found that the coloured band—or the spectrum, as it is called—does not extend continuously from one end to the other, but is shaded over by a multitude of dark lines, a few only of which are shown in the adjoining plate. These lines are a permanent feature in the spectrum. They are as characteristic of the sunlight as the prismatic colours themselves, and full of interest and of information with regard to the sun. These lines are, indeed, the characters in which the history and the nature of the sun are written. Viewed through an instrument of adequate power, dark lines are to be found crossing the solar spectrum in hundreds and in thousands. They are of every variety of strength and faintness; their distribution seems guided by no simple law. At some parts of the spectrum there are but few lines; in other regions they are crowded so closely together that it is difficult to separate them. They are in some places so exquisitely fine and delicate, that they never fail to excite the admiration of every one who looks at this interesting spectacle in a good instrument.

There can be no better method of expounding the rather difficult subject of spectrum analysis than by actually following the steps of the original discovery which first gave a clear demonstration of the significance of the lines. Let us concentrate our attention specially upon that line of the solar spectrum marked D. This line, when seen in the spectroscope, is found to consist of two lines, very delicately separated by a minute interval, one of these lines being slightly darker than the other. Suppose that while the attention is concentrated on these lines, the flame of an ordinary spirit-lamp coloured by common salt be held in front of the instrument, so that the ray of direct and brilliant solar light passes through the flame before entering the instrument. The observer sees at once the two lines known as D flash out with a greatly increased blackness and vividness, while there is no other perceptible effect on the spectrum. A few trials show that this intensifying of the D lines is due to the vapour of sodium, arising from the salt burning in the lamp through which the sunlight has passed.

It is quite impossible that this marvellous connection between sodium and the D lines of the spectrum can be a merely casual matter. Even if there were only a single line concerned, it would be in the highest degree unlikely that the coincidence should arise by accident; but when we find the sodium affecting both of the two close lines which form D, our conviction that there must be some profound connection between these lines and sodium rises to absolute certainty. Suppose, now, that the sunlight be cut off, and that all other light be excluded save that emanating from the glowing vapour of sodium in the spirit flame. We shall then find, on looking through the spectroscope, that we have no longer all the colours of the rainbow; the light from the sodium is now concentrated into two bright yellow lines, filling precisely the position which the dark D lines occupied in the solar spectrum, and the darkness of which the sodium flame seemed to intensify. We must here remove an apparent paradox. How is it, that though the sodium flame produces two *bright* lines when viewed in the absence of other light, yet it actually appears to intensify the two *dark* lines in the sun's spectrum? The explanation of this conducts us at once to the cardinal doctrine of spectrum analysis. The so-called dark lines in the solar spectrum are only dark *by contrast* with the brilliant illumination of the rest of the spectrum. A good deal of solar light really lies in the dark lines, though not enough to be seen when the eye is dazzled by the brilliancy around. When the flame of the spirit-lamp charged with sodium intervenes, it sends out a certain amount of light, which is entirely localised in these two lines. So far, it would seem that the influence of the sodium flame ought to be manifested in diminishing the darkness of the lines and rendering them less conspicuous. As a matter of fact, they are far more conspicuous with the sodium flame than without it. This arises from the fact that the sodium flame possesses the remarkable property of cutting off the sunlight which was on its way to those particular lines; and thus, though the sodium contributes some light to the lines on the one hand, yet on the other it intercepts a far greater quantity of the light that would otherwise have illuminated those lines, and hence they became darker with the sodium flame than without it.

Here, then, we are conducted to a remarkable principle, which has led to the interpretation of the dark lines in the spectrum of the sun. We find that sodium vapour, when heated, gives out light of a very particular type, which, viewed through the prism, is concentrated in two lines. But the sodium vapour possesses also this property, that light from the sun can pass through it without any perceptible absorption, save and except of those particular rays which are of the same refrangibilities as the two lines in question. In other words, we say that if the heated vapour of a substance gives a spectrum of bright lines, corresponding to lights of various refrangibilities, this same vapour will act as an opaque screen to lights of those special refrangibilities, while remaining transparent to light of every other kind. This is a matter of some little complexity, but it is of such momentous importance in the theory of spectrum analysis that we cannot avoid giving a further example. Let us take the element iron, which in a very striking degree illustrates the law in question. In the solar spectrum some hundreds of the dark lines are known to correspond with the spectrum of iron. This correspondence is exhibited in a vivid manner when, by suitable contrivance, the light of an electric spark from poles of iron is examined in the spectroscope simultaneously with the solar spectrum. It can be shown that hundreds of the lines in the sun are identical in position with the lines in the spectrum of iron. But the spectrum of iron, as here described, consists of bright lines, while those with which it is compared in the sun consists of dark lines. These dark lines can be completely understood if we suppose that the vapour arising from intensely heated iron be present in the atmosphere which surrounds the luminous strata on the sun. This vapour would then, by the law in question, stop precisely the same rays as it emits when incandescent, and hence we learn the important fact that iron, no less than sodium, must, in one form or another, be a constituent of the sun.

Such is, in brief outline, the celebrated discovery of modern times which has given an interpretation to the dark lines of the solar spectrum. A large number of terrestrial substances have been examined, in conjunction with the solar spectrum, and thus it has been established that several of the elements known on the earth are present in the sun. These elements are as follows:—hydrogen, sodium, barium, calcium, magnesium, aluminium, iron, manganese, chromium, cobalt, nickel, zinc, copper, titanium, cadmium, strontium, cerium, uranium, lead, potassium.

It would, indeed, lead us far beyond the prescribed limits of this volume if we were to attempt to give any detailed account of the beautiful researches in the application of spectrum analysis to astronomy. It is, however, impossible that we could entirely pass over discoveries of such interest; and, therefore, we shall give a brief—though it must be a very brief—account of some of the principal points.

When we apply the spectroscope to the moon or to the planets, we recognise in the light they send us the features characteristic of the solar light. The reflected sunlight, even from the moon, is not bright enough to show all the lines that can be seen when the sun is examined directly. Venus, when examined with care, will probably show more of the dark lines in the solar spectrum than the moon. In the case of the planets, also, we often find certain additional lines, or bands of lines, added to the spectrum. These lines have arisen from the further special absorption of sunlight which takes place in the planet's atmosphere. It should also be mentioned that some of the lines in the solar spectrum are to be referred to the absorption in our own atmosphere. This fact is clearly brought out when the sun is observed near sunset, for then the rays have a vast depth of air to traverse, and, accordingly, we find the atmospheric bands to be defined with increased vividness.

We have hitherto spoken of the light from the sun, without attempting to distinguish the light from the different parts of the sun. The spectroscope can, however, be applied to each part of the sun separately, and thus it has been possible to establish remarkable differences between the spectra of the several parts. The most interesting discovery made in connection with this subject was that accomplished almost simultaneously by Janssen and Lockyer, which has given a wide extension to the possibilities of solar observation. In our chapter on the sun we have dwelt on the very remarkable objects known as the solar prominences, which become visible on the rare occasions of a total eclipse of the sun. The remarkable discovery just referred to disclosed a method by which the solar prominences could be observed at any time with the aid of the spectroscope.

The prominences are composed of glowing gas, principally hydrogen; and the light, instead of being spread all over the spectrum, like ordinary solar light, is mainly concentrated into the few lines which constitute the spectrum of hydrogen. When the spectroscope is directed around the edge of the sun the rays of ordinary diffused sunlight are distributed over the whole length of the spectrum, and are, of course, correspondingly weakened. The light from the prominences, being all concentrated into the lines, is, therefore, relatively increased in intensity to such an extent that they become visible.

The first line of Plate XIII. represents the solar spectrum, on which a few only of the principal lines are introduced for comparison. It is, however, to be understood that the solar spectrum really exhibits thousands of lines when viewed with a spectroscope of adequate power. Among the varied spectra of the stars different types have been recognised, the characteristic features of which are also depicted. The first type of stars are those which are intensely white, of which we may take Sirius and Vega as examples. In the

PLATE XIII.

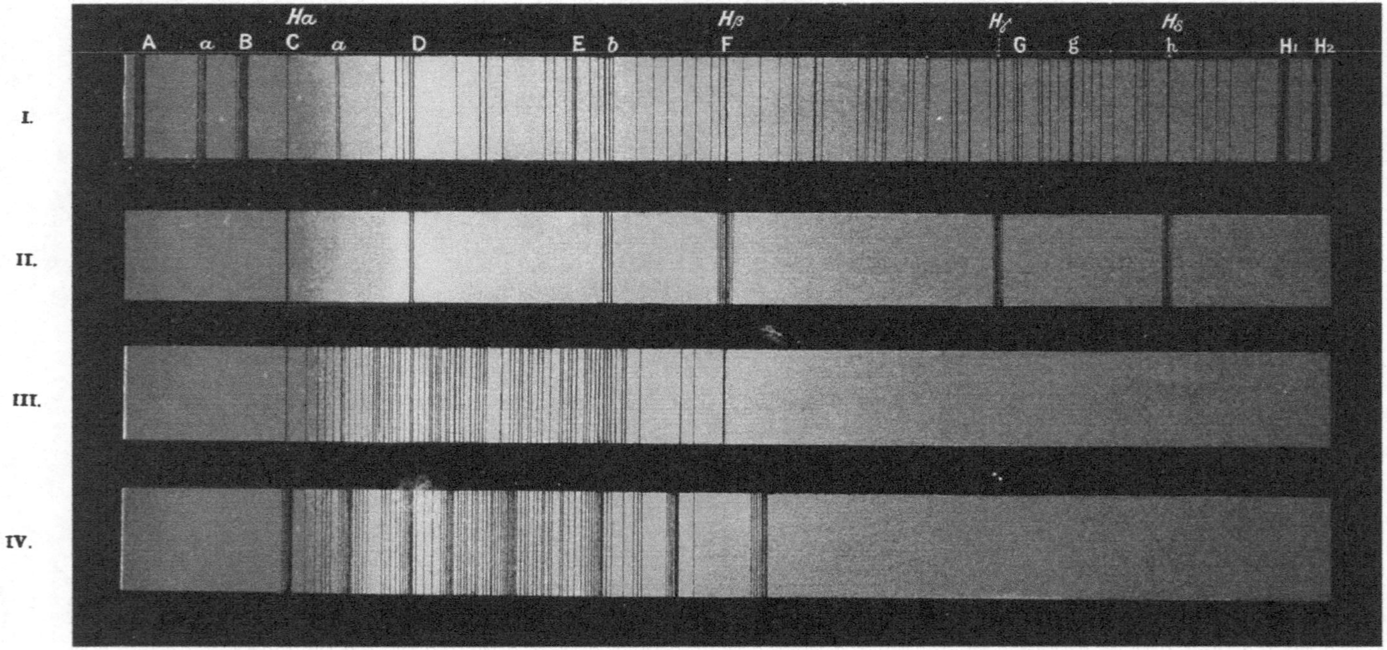

$H\alpha$ $H\beta$ $H\gamma$ $H\delta$

A a B C a D E b F G g h H_1 H_2

I.

II.

III.

IV.

SPECTRA OF THE SUN AND STARS.

I. SUN. III. ALDEBARAN.
II. SIRIUS. IV. BETELGEUZE.

spectra of these stars there are comparatively few lines. The lines of hydrogen are however very strong, but the metallic lines of sodium and magnesium are barely perceptible. The presence of hydrogen in the atmosphere of these stars has been confirmed in a very remarkable manner by the photographs of the spectra of these stars which have been obtained by Dr. Huggins. In these photographs, lines can be seen in that part of the spectrum invisible to the eye, and these lines also have been found to coincide with lines in the photographic spectrum of hydrogen.

The second type of stars exhibit spectra closely resembling that of the sun; a good illustration of this type is found in the star Capella. Aldebaran has also a spectrum belonging to the second type, though verging on the third. The chief hydrogen lines are still conspicuous, but many metallic lines are coming strongly out.

The spectrum of *a* Orionis, or Betelgueze, exhibits a typical spectrum of the third class, which contains many metallic lines and shaded bands; but the most characteristic feature of this type is the absence of the hydrogen lines.

Another most remarkable achievement in spectrum analysis has been accomplished by Dr. Huggins in his very beautiful discovery of the spectroscopic method of measuring the motion of stars along the line of sight. The place of a line of hydrogen, for instance, will be slightly moved to one side or the other, according as the star be approaching the earth or receding from the earth. The amount of displacement is very small; but in the case of several stars large enough to enable an approximation to be made to the velocity with which the star is moving towards or moving from the earth.

The theory of this method is beautifully verified by observations on the sun. As the eastern edge of the sun is approaching and the western is receding, there is a corresponding difference in the spectra of the two edges, and the observed amount gives a velocity of rotation practically coincident with that otherwise known.

CHAPTER XXIII.

STAR CLUSTERS AND NEBULÆ.

Interesting Sidereal Objects—Stars not Scattered uniformly—Star Clusters—Their Varieties—The Cluster in Perseus—The Globular Cluster in Hercules—The Milky Way—A Cluster of Minute Stars—Nebulæ distinct from Clouds—Number of known Nebulæ—The Constellation of Orion—The Position of the Great Nebula—The Wonderful Star θ Orionis—The Drawing of the Great Nebula in Lord Rosse's Telescope—Photographs of this Wonderful Object—The Importance of Accurate Drawings—The Great Survey—Photographs of the Heavens—Magnitude of the Nebula—The Question as to the Nature of a Nebula—Is it composed of Stars or of Gas?—How Gas can be made to Glow—Spectroscopic Examination of the Nebula—The Great Nebula in Andromeda—Its Examination by the Spectroscope—The Annular Nebula in Lyra—Resemblance to Vortex Rings—Planetary Nebulæ—Drawings of Several Remarkable Nebulæ—Distance of Nebulæ—Conclusion.

WE have already mentioned Saturn as one of the most glorious telescopic spectacles in the heavens. Setting aside the obvious claims of the sun and of the moon, there are, perhaps, two other objects visible from these latitudes which rival Saturn in the splendour and the interest of their telescopic picture. One of these objects is the star cluster in Hercules; the other is the great nebula in Orion. We may take these objects as typical of the two great classes of bodies to be discussed in this chapter, under the head of Star Clusters and Nebulæ.

The stars that, to the number of several millions, bespangle the sky, are not scattered uniformly. We see that while some regions are comparatively bare of stars, others contain stars in profusion. Sometimes we have a small group, like the Pleiades; sometimes we have a stupendous region of the heavens strewn over with stars, as in the Milky Way. Such objects are called star clusters. We find every variety in the clusters; sometimes the stars are remarkable for their brilliancy, sometimes for their enormous numbers, and sometimes for the remarkable form in which they are grouped.

Sometimes a star cluster is adorned with brilliantly-coloured stars; sometimes the stars are so close together that their separate rays cannot be disentangled; sometimes the stars are so minute or so distant that the cluster is barely distinguishable from the nebula.

Of the clusters remarkable at once both for richness and brilliancy of the individual stars, we may mention the cluster in the Sword-handle of Perseus. The position of this object is marked on Fig. 70, page 378. To the unaided eye a dull spot is visible, which in the telescope expands into two clusters separated by a short distance. In each of them we have innumerable stars, crowded together so as to fill the field of view of the telescope. The splendour of this object is appreciated when we reflect that each one of these stars is itself a brilliant sun, perhaps rivalling our own sun in lustre. There are, however, regions in the heavens near the Southern Cross, of course invisible from northern latitudes, in which parts of the Milky Way present a richer appearance even than the cluster in Perseus.

The most striking type of star cluster is well exhibited in the constellation of Hercules. In this case we have a group of minute stars apparently in a roughly globular form. Fig. 84 represents this object as seen in Lord Rosse's great telescope, and it shows three radiating streaks, in which the stars seem less numerous than elsewhere. It is estimated that this cluster must contain from 1,000 to 2,000 stars, all concentrated into an extremely small part of the heavens. Viewed in a very small telescope, this object resembles a nebula. The position of the cluster in Hercules is shown in a diagram previously given (Fig. 75, page 384). We have already referred to this glorious aggregation of stars as one of the three especially interesting objects in the heavens.

The Milky Way forms a girdle which, with more or less regularity, sweeps completely around the heavens; and when viewed with the telescope, is seen to consist of myriads of minute stars. In some places the stars are much more numerous than elsewhere. All these stars are incomparably more distant than the sun, which they entirely surround, so it is evident that our sun, and of course the system which attends him, lies actually inside the Milky Way. It seems tempting to pursue the thought here suggested, and to reflect that the whole Milky Way may, after all, be merely a star cluster, comparable in size with some of the other star clusters which we see, and that viewed from a remote point in space, the Milky Way would seem to be but one of the countless clusters of stars containing our sun as an indistinguishable unit.

Fig. 84.—The Globular Cluster in Hercules.

When we direct a good telescope to the heavens, we shall occasionally meet with one of the remarkable celestial objects which are known as nebulæ. They are faint cloudy spots, or stains of light on the black background of the sky. They are nearly all quite invisible to the naked eye. These celestial objects must not for a moment be confounded with clouds, in the ordinary meaning of the word. The latter exist only suspended in our atmosphere, while nebulæ are immersed in the depths of space. Clouds shine only by the light of the sun, which they reflect to us; nebulæ shine with no borrowed light: they are self-luminous. Clouds change from hour to hour; nebulæ do not change even from year to year. Clouds are far smaller than the earth; while the smallest nebula known to us is incomparably greater than the sun. Clouds are within a few miles of the earth; nebulæ are not within many millions of miles.

The systematic study of the nebulæ may be said to have commenced with the colossal labours of William Herschel at Slough. The scheme which Herschel proposed for this task was indeed a comprehensive one. He determined to make a survey of the entire heavens with a powerful telescope, and to note all the features of interest which he could detect. But though we can thus summarily describe the undertaking, yet a little reflection will show the gigantic amount of labour which it entailed. Considering how rapidly we can sweep our eyes over the heavens, it might seem a very easy matter to turn a telescope to one spot after another, until the whole sky has been reviewed. The two cases are, however, as different as possible; a glance of the eyes takes in an enormous region of the heavens, while the field of a large telescope only includes a very small region. This is a point which very often surprises those who for the first time look through a large telescope. People sometimes expect to see the whole northern hemisphere, and perhaps the signs of the zodiac, all at once. It is even unreasonable to ask to be shown the Great Bear in a large telescope; the telescope can be pointed to special parts of the constellation; but if we want a comprehensive view of the whole, we must take an opera-glass, or something of that description, not a great and powerful instrument. A large telescope will hardly show even so much as the whole of the moon at once. When we look through the eye-piece, we find the entire field filled with the brilliant body of the moon, and it will be necessary to move the telescope a little up and down, and a little to the right and to the left, in order to bring the whole surface of the moon under review. The moon only occupies a very small portion of the sky; but small as that portion is, the field of view in a large telescope is not so great. Suppose the whole surface of the heavens to be covered over with moons quite close together, the apparent size of each being the same as that of our moon, the heavens would then form a mosaic of about 200,000 pieces. Assuming that a great telescope can show about half of the moon at once, the whole surface of the heavens would form about 400,000 fields of view. In such an instrument a complete survey of the heavens must therefore involve no less than 400,000 examinations of what the telescope can reveal in one view. This estimate includes the southern heavens; if we confine our thoughts to that portion of the heavens which is visible from these latitudes, we may perhaps say that about a quarter of a million fields of view have to be carefully examined. The scheme, then, which Herschel sketched out for his labours at Slough involved bringing a quarter of a million fields under review; it will readily be believed that considerable organisation was necessary to enable so vast an undertaking to be accomplished. It was necessary to provide that none of the fields should be allowed to escape, and that none should be observed oftener than was necessary.

But there is another way in which we can obtain an adequate conception of the enormous labour which Herschel contemplated, and which he lived in great part to complete. Instead of counting the number of fields of view which he would have to examine, let us attempt to form an estimate of the number of objects that would be likely to come under his notice. We need hardly say that such an estimate must be only an approximate one, but for the purpose of conveying to the reader some idea of the extent of Herschel's labours, it will be quite sufficient. There are several different classes of objects in the heavens, but the objects which are most numerous and most characteristic are the fixed stars. The constellations, with which every one is familiar, are formed of fixed stars of every conceivable brightness, from the splendour of Sirius down to the merest point of light that can be discerned in the most powerful telescope. Stars can be found of every tint, from the red at one end of the spectrum to the blue at the other, and they are scattered in boundless profusion over the whole extent of the heavens.

Suppose that a great telescope like that which Herschel used in his researches be employed for the purpose of counting the stars, each one of the 250,000 fields of view would have to be regularly inventoried. In almost every one of those fields of view some stars would be seen; in many fields there would be a large number of stars; while in others the stars would be found in countless multitudes.

Immediately after Herschel and his sister had settled at Slough he commenced his immortal review of the heavens in a systematic manner. For observations of this kind it is essential that the sky be free from cloud, while even the light of the moon is sufficient to obliterate the fainter and more interesting objects. It was in the long and fine winter nights, when the stars were shining brilliantly and the pale path of the Milky Way extended across the heavens, that the labour was to be done. The great telescope being directed to the heavens, the ordinary diurnal motion by which the sun and stars appear to rise and set carries the stars across the field of view in a majestic panorama. The stars enter slowly into the field of view, slowly move across it, and slowly leave it, to be again replaced by others. Thus the observer, by merely remaining passive at the eye-piece, sees one field after another pass before him, and is enabled to examine their contents. It follows, that even without moving the telescope a long narrow strip of the heavens is brought under review, and by moving the telescope slightly up and down the width of this strip can be suitably increased. On another night the telescope is brought into a different position, and another strip of the sky is examined; so that in the course of time the whole heavens can be carefully scrutinised.

Herschel stands at the eye-piece to watch the glorious procession of celestial objects. Close by, his sister Caroline sit at her desk, pen in hand, to take down the observations as they fall from her brother's lips. In front of her is a chronometer from which she can note the time, and a contrivance which indicates the altitude of the telescope, so that she can record the exact position of the object in connection with the description which her brother dictated. Such was the splendid scheme which this brother and sister had arranged as the object of their life-long devotion. The discoveries which Herschel was destined to make were to be reckoned not by tens or by hundreds, but by thousands. The records of these discoveries are to be found in the "Philosophical Transactions of the Royal Society," and they are among the richest treasures of those volumes. It was left to Sir John Herschel, the

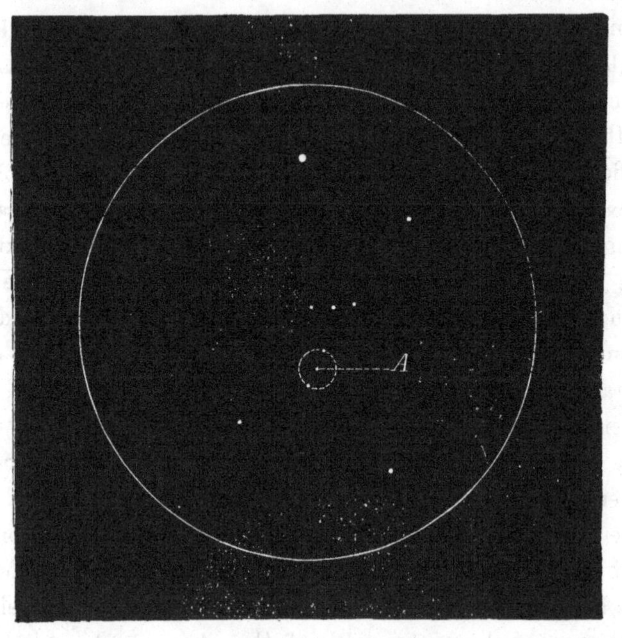

Fig. 85.—The Constellation of Orion, showing the position of the Great Nebula.

only son of Sir William, to complete his father's labour by extending the survey to the southern heavens. He undertook with this object a journey to the Cape of Good Hope, and sojourned there for the years necessary to complete the great work.

As the result of the gigantic labours thus inaugurated, there are now three or four thousand nebulæ known to us, and with every improvement of the telescope fresh additions are made to the list. They are scattered over both hemispheres, and some are to be found in every constellation. They differ from one another as four thousand pebbles selected at random on a sea-beach might differ—

namely, in form, size, colour, and material—but yet have, like the pebbles, a certain generic resemblance to each other. To describe this class of bodies in any detail would altogether exceed the limits of this chapter; we shall merely select a few of the nebulæ, choosing naturally those of the most remarkable character, and also those which are representatives of the different groups into which nebulæ may be divided.

We have already alluded to the great nebula in the constellation of Orion as being one of the most interesting objects in the heavens. It is alike remarkable whether we look at its size or its brilliancy, the care with which it has been studied, or the success which has attended the efforts to learn something of its character. To find this object, we refer to Fig. 85 for the sketch of the chief stars in this constellation, where the letter A indicates the middle one of the three stars which form the sword-handle of Orion. Above the handle are the three stars which form the well-known belt so conspicuous in the wintry sky. The star A, when viewed attentively with the unaided eye, presents a somewhat misty appearance. In the year 1618 Cysat directed a telescope to this star, and saw surrounding it a curious luminous haze, which was the great nebula. Ever since his time this object has been diligently studied by many astronomers, so that innumerable observations have been made of it, and even whole volumes have been written which treat of nothing else. Any ordinary telescope will show the object to some extent, but the more powerful the telescope the more are the curious details revealed.

In the first place, the star A (θ Orionis) is in itself the most striking multiple star in the whole heavens. It consists really of six stars, represented in the next diagram (Fig. 86); these points are so close together that their commingled rays cannot be distinguished without a telescope. Four of them are, however, easily seen in quite small instruments, but the two smaller stars require telescopes of considerable power. And yet these stars are suns, comparable, it may be, with our sun in magnitude.

It is not a little remarkable that this unrivalled group of six suns should be surrounded by the renowned nebula; the nebula or the multiple star would, either of them alone, be of exceptional interest, and here we have a combination of the two. It seems impossible to resist drawing the conclusion that the multiple star really lies in the nebula, and not merely along the same line of vision. It would, indeed, seem to be at variance with all probability to suppose that the presentation of these two exceptional objects in the same field of view was merely accidental. If the multiple star be really in the nebula, as seems most likely, then this object affords evidence that in one case at all events the distance of a nebula is a quantity of the same magnitude as the distance of a

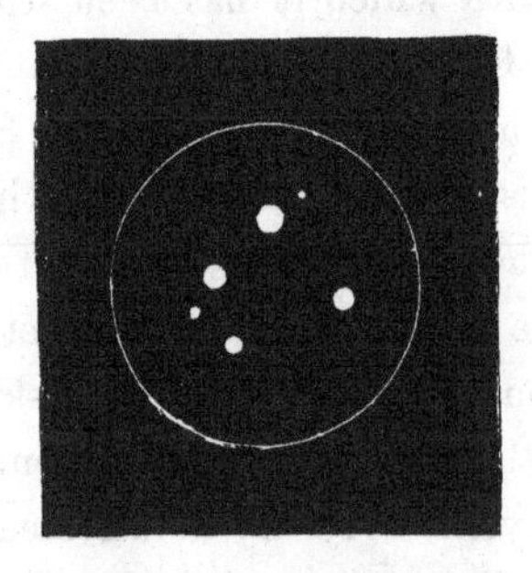

Fig. 86.—The Multiple Star (θ Orionis) in the Great Nebula of Orion.

star. This is unhappily almost the entire extent of our knowledge of the distances of the nebulæ from the earth.

The great nebula of Orion surrounds the multiple star, and extends out to a vast distance into the neighbouring space. The dotted circle drawn around the star marked A in figure 85 represents approximately the extent of the nebula, as seen in a moderately good telescope. The nebula is of a faint bluish colour, impossible to represent in a drawing. Its brightness is much greater in some places than in others; the central parts are, generally speaking, the most brilliant, and the luminosity gradually fades away as the edge of the nebula is approached. In fact, we can hardly say that the nebula has any definite boundary, for with each increase of telescopic power faint new branches can be seen. There seems to be an empty space in the nebula immediately surrounding the multiple star, but it is not

PLATE XIV.

unlikely that this is merely an illusion, produced by the contrast of the brilliant light of the stars. At all events, the spectroscopic examination of the nebula seems to show that the nebulous matter is continuous over the stars.

The plate of the great nebula in Orion which is here shown represents, in a reduced form, the elaborate drawing of this object which has been made by the Earl of Rosse with the great reflecting telescope at Parsonstown.* A telescopic view of the nebula shows a large number of stars, two hundred or more, scattered over its surface. It is not necessary to suppose that these stars are all immersed in the substance of the nebula as the multiple star appears to be; they may be either in front of it, or, less probably, behind it, so as to be projected on the same part of the sky. In the production of this drawing, the first task was to construct an accurate chart of these stars, to serve as a skeleton on which the nebula could be delineated. The execution of the elaborate drawing taxed the powers of the great telescope for a period of some years. It must, however, be remembered that it is only during a part of the year that the nebula can be seen. Very fine nights are necessary for observing these objects. There must be no clouds, and the moon must be absent, for its bright light would quite extinguish the fainter portions of a nebula. The Milky Way should be seen clearly if the night is to be employed in drawing such objects. In fact, taking the weather alone into consideration, there are in the British Islands only about one hundred hours in the whole year which are really suited for the very best astronomical observations. When it is further remembered that the great telescope at Parsonstown is a meridian instrument, with only a limited range of lateral motion, it is plain that the nebula in Orion will only be within its reach for an hour, or rather less, each evening. Unless, then, the sky be clear at that particular hour, the whole night is lost so far as this drawing is concerned, and hence it cannot be a matter of surprise that much time has been expended in the completion of a work requiring so much careful comparison and examination. There is, indeed, considerable difficulty in seeing the fainter portions of this

* I am indebted for this drawing to the kindness of Messrs. De la Rue.

nebula. The long projecting arms, which seem so plain in the drawing, can only be seen by the most careful scrutiny with a practised eye. A little will be seen one night: this will be verified on a subsequent evening, and the faint object will be traced a little further. But this faintness is only to be understood as applying to the outlying portions. The central part of the nebula is bright enough to be seen in any telescope, and in the great instrument presents a scene of surpassing glory.

A considerable number of drawings of this unique object have been made by other astronomers. Among these, we must mention that executed by Professor Bond, in Cambridge, Mass., which possesses a degree of faithfulness in detail that every student of this object is bound to acknowledge. Of late years also successful attempts have been made to photograph the great nebula. The late Professor Draper was fortunate enough to obtain some admirable photographs. In England, too, Mr. Common has taken a most excellent photograph of the nebula, which has deservedly gained for its author the high honour of the Royal Astronomical Society's gold medal.

It may be asked, What is the object of devoting such elaborate care to the examination and drawing of this single object? The answer to this question is really two-fold. As a geographer would survey a new country for the sake of ascertaining the details of its mountains, its plains, and its rivers; as he would sketch the most important features in its scenery, and fill his note-books with measurements and with descriptions; so does the astronomer consider a simple survey of the heavens to be one of the numerous duties which it is incumbent on him to discharge. The contents of the heavens are to be carefully measured, they are to be accurately described, they are to be artistically drawn; and the result ultimately aimed at is a complete and exhaustive delineation of the magnitude, weight, position, and peculiar features of every comet or planet, every star and every nebula, which the utmost powers of our greatest instruments can disclose to us. We can hardly hope that this majestic scheme will be ever fully realised. The task is too great for human labour to accomplish completely, but important contributions to the undertaking are made every year. In pursuance of this object, the lives of some astronomers are devoted to the preparation of catalogues in which the positions and magnitudes of the fixed stars are recorded. Others are engaged in the delineation of charts of portions of the sky, in which the minutest stars are depicted.* Others, again, expend their energies in making drawings of the features visible on the moon, or on drawings and measures of the planets, or in the measurement of double stars, or in tracing the faint outlines of the nebulæ. These are the results of special surveys of different provinces in the vast kingdom. Now, of all the objects in the heavens beyond the confines of our solar system, one of the most splendid and the most remarkable is that now before us; and hence, in the great celestial survey it is of paramount importance that we should possess faithful drawings and measurements of every detail that can be seized of the great nebula in Orion.

The laborious work of surveying is the first and the simplest duty of the astronomer. Far more difficult, and perhaps far more interesting problems are to follow. One of these problems is to determine the movements of the heavenly bodies, and another—the most difficult of all—is to determine their nature. The drawings of the great nebula in Orion are designed to aid in the solution of both

* The tedious process of forming star charts by measurements or estimations of each separate star, is apparently to be superseded by the more comprehensive methods of photography. It has been found that very minute telescopic stars can be made to produce a sensible picture when an exposure of suitable length has been given. Stars of the 14th and 15th magnitudes are thus shown, and it is even possible for stars to be recorded on the photographic plate, which are unable to excite vision in the most powerful telescope. A plate taken by Mr. J. Roberts contained an area of the sky about two degrees square. On comparison of this photograph with the well-known map of Argelander, Father Perry states that he found not only all the fifty-eight stars of the map, but at least thirty-two new stars, all probably below the 9th or 10th magnitude. At the Paris Observatory an exposure of an hour gave a negative of about five square degrees of the heavens, on which 2,790 stars from the 5th to the 14th magnitude were depicted, while traces of stars of the 15th magnitude were visible. It is hoped, therefore, that in some six or eight years a complete survey of the heavens on this splendid scale may be accomplished. Many important results may be anticipated from this research: for instance, a minor planet is at once revealed by its motion, which gives a streak instead of a point.

these great problems; to show whether the nebula has any appreciable movements, and to elucidate its actual structure and physical condition. It is a most interesting question to decide, whether this great nebula is really in a process of change. Looked at from night to night, it does not alter; even from year to year it still seems the same. . But is this permanence real? Do not those arms of light or the bright central regions in any degree wax or wane—become absorbed or extended, dissipated or dispersed—like the ordinary clouds whose evanescence is proverbial? This is a question of the very deepest interest; and how is it to be answered? The life of an astronomer is not long enough, even if his memory were good enough, to pronounce finally upon this question. But by making an elaborate drawing with the utmost care and precision, by measuring the relative positions of the stars in the nebula with the last degree of exactness, and by careful comparison ascertaining that each minute part of the drawing represents the corresponding part of the nebula as closely as it is possible for art to represent nature; then those drawings can at a future time be compared with the nebula, and the question as to its permanence can thus be investigated. This has already been done, and the labour expended upon these drawings has been amply rewarded. The entire subject has been critically examined, and it has been demonstrated that in parts of the mighty nebula certain changes *are* in progress, though probably much more time must elapse before the full import of these changes can be understood.

The magnitude of this gigantic object can perhaps never become fully known to us. In fact, the light fades away so gradually that we cannot tell how large a portion of sky it really occupies. What can be seen certainly occupies an area of the sky two or three times as great as that covered by the full moon. But what shall we say of the real magnitude of this object? As we do not know its distance, we cannot, of course, attempt any estimate of its bulk, though it is possible to enunciate a minor limit to its dimensions. The magnitude of our earth, vast as it is, will not serve as a means of comparison; we must have a far larger unit of measurement for a bulk so stupendous. The earth sweeps around the sun in a mighty path, whose diameter is not less than 185,400,000 miles. Let us imagine a sphere so mighty that this circle would just form a girdle around its equator, and let this gigantic globe be the measure wherewith to enunciate the bulk of the vast nebula of Orion. It can be demonstrated that a million of these mighty globes rolled into one would not equal the great nebula in bulk; though how much greater than this the nebula may really be we have no means of ascertaining.

The actual nature of the nebula in Orion offers a problem of the greatest interest, especially when we consider the question as to the materials of which this object is composed? This is a problem which for a long time presented a matter of controversy, but happily the discoveries of Dr. Huggins have, in a great measure, decided the matter.

We have already given a brief description of some of those objects known as star clusters, and we have, in particular, referred to the great cluster in Hercules. This is a superb object when seen in an instrument of adequate power, but in a small instrument it is not seen exactly as a cluster of stars; it is, rather, a dull hazy point of light, not by any means unlike the appearance of some of the nebulæ seen in a great instrument. As it was found that many apparently nebulous objects were shown to be really star clusters when adequate telescopic power was applied, it seemed perhaps plausible to contend that all nebulæ might be really only clusters of stars; sunk, however, to so great a depth in space, that even the largest telescopes failed to disentangle the several rays of the different stars, and thus produced merely the hazy appearance of a nebula.

How far can this reasoning be applied to the great nebula in Orion? We have, fortunately, one or two very interesting observations bearing on this point. On a particularly fine night, when the speculum of the great six-foot telescope at Parsonstown was in its finest order, the skilled eye of the late Earl of Rosse and of his then assistant, Mr. Stoney, detected in the densest part of the nebula myriads of minute stars, which had never before been recognised by human eye. Unquestionably, the commingled rays of these stars contributes not a little to the brilliancy of the nebula. But there still remains the question as to whether the entire luminosity of the great nebula can be so explained, or whether the light thereof may not partly arise from some other source. This question is one which must necessarily be forced on the attention of any observer who has ever enjoyed the privilege of viewing the great nebula through a telescope of power really adequate to render justice to its beauty. It seems impossible to believe that the bluish light of such delicately graduated shades has really arisen merely from stellar points. The object is so soft and so continuous—might we not almost say so ghost-like?—that it is impossible not to believe that we are really looking at some gaseous matter.

But here a difficulty may be suggested. The nebula is a luminous body, but ordinary gas is invisible. We do not see the gases which surround us and form the atmosphere in which we live. How, then, if the nebula consisted merely of gaseous matter, would we see it shining on the far distant heavens? A well-known experiment will at once explain this difficulty. We take a tube containing a very small quantity of some gas: for example, hydrogen; this gas is usually invisible: no one could tell that there was any gas in the tube, or still less could the kind of gas be known; but pour a stream of electricity through the tube, and instantly the gas begins to glow with a violet light. What has the electricity done for us in this experiment? Its sole effect has been to heat the gas. It is, indeed, merely a convenient means of heating the gas and making it glow. It is not the electricity which we see, it is rather the gas heated by the electricity. We infer, then, that if the gas be heated it becomes luminous. The gas does not burn in the ordinary sense of the word; no chemical change has taken place. The tube contains exactly the same amount of hydrogen after the experiment that it did before. It glows with the heat just as red-hot iron glows. If, then, we could believe that in the great nebula in Orion there were vast volumes of rarefied gas in the same physical condition as the gas in the tube while the electricity was passing, then we should expect to find that this gas would actually glow.

PLATE XV.

THE GREAT NEBULA IN ANDROMEDA.

To settle the question as to the real nature of the nebulæ, we must call in the aid of that refined method of investigation known as spectrum analysis. We have explained in a previous chapter the principle of this method in its application to the stars; let us now see how it can be applied to the nebulæ. The spectrum of a gaseous nebula is a truly remarkable sight. Instead of the continuous band of colours, crossed by dark lines, which is characteristic of the spectrum of a star, the nebular spectrum consists of but three or four bright lines. Two of these lines correspond to the spectrum of hydrogen, another belongs to nitrogen, while the fourth seems to belong to some element not hitherto identified with any terrestrial substance.

This remarkable discovery made by Dr. Huggins has been corroborated and extended in a very interesting manner by the more recent labours of the same astronomer, who has actually succeeded in obtaining a photograph of the spectrum of the great nebula in Orion. The photograph shows a strong line in that invisible part of the spectrum extending beyond the violet rays, and this line has been proved to be in all probability identical with one due to hydrogen.

The labours of Dr. Huggins have thus solved to a considerable extent the important problem as to the constitution of the nebula in Orion. We see that it consists in part of stars, making up, perhaps, in number for what they want in size. These stars are bathed in and surrounded by a stupendous mass of glowing gas, partly consisting of that gas which enters so largely into the composition of our ocean, namely, hydrogen, and partly of that which is so important an ingredient in our atmosphere, namely, nitrogen; and further, that these are mingled with some other gaseous substance of a nature at present unknown.

Space will only admit of a brief reference to a few other nebulæ, and among these, a chief place must be given to the great nebula in Andromeda. This is visible to the unaided eye, and has, indeed, not unfrequently been mistaken for a comet. Its telescopic appearance is shown in Plate XV., which has been copied from one of Mr. Trouvelot's beautiful drawings made at

Harvard College Observatory. Two dark channels in the nebula cannot fail to be noticed, and the number of faint stars scattered over its surface is also a point to which attention may be drawn. To find this object, we must look out for Cassiopeia and the Great Square of Pegasus, and then the nebula will be easily perceived in the position shown on p. 376. It differs in a very remarkable degree from the gaseous nebulæ of which that in Orion is a type. When we examine the nebula in Andromeda with the spectroscope, we find a faint continuous band of light, of a totally different character to that shown in the four bright lines of Orion. In fact, the spectroscopic evidence would seem to indicate that there was no glowing gas in the nebula in Andromeda, and that it may really be due to the existence of a vast cluster of minute stellar points. Those who are experienced in the observation of nebulæ, will be aware of a peculiar bluish hue observable in the gaseous nebulæ, which is not present in those of the Andromeda type. In fact, the difference is so marked, that it is quite possible to say to which of the two classes the nebula really belongs, even before the crucial test of the spectroscope has been applied.

Among the various other classes of nebulæ, perhaps the most striking are those known as the Annular, or the Ring Nebulæ. The most celebrated of these objects is found in the constellation of Lyra, and its position can be readily determined from the annexed sketch of the principal stars in the constellation (Fig. 87). This annular nebula belongs to the gaseous class of objects, though the lines in its spectrum are not so numerous as those in the nebula of Orion. We here give two attempts to delineate this curious object (Fig. 88). The drawing will, at all events, serve the purpose of enabling the student to identify the nebula when he meets with it in the heavens. The smaller picture shows it as seen in an instrument of moderate size; the larger one represents it as seen in one of our more powerful telescopes. The latter view discloses a number of stellar points; it also shows the remarkable fringe surrounding the ring, while the interior of the ring is seen not to be absolutely dark.

The nebula in Lyra is the most conspicuous ring nebula in the heavens, but it is not to be supposed that it is the only member of this class. Altogether, there are about a dozen of these objects. It seems difficult to form any adequate conception of the nature of

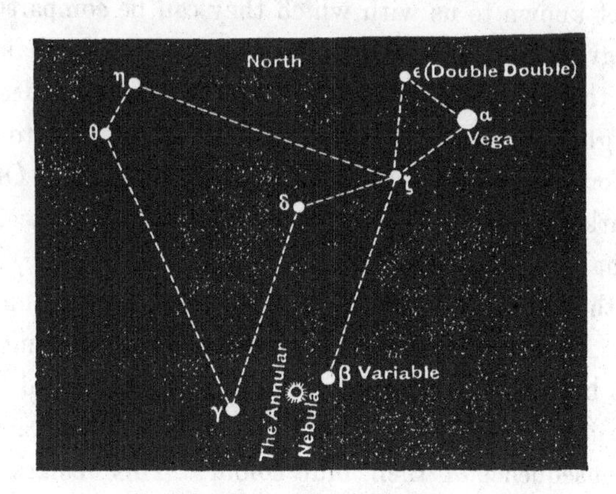

Fig. 87.—Lyra, with the Annular Nebula.

such a body. It is, however, impossible to view the annular nebulæ without being, at all events, reminded of those elegant objects known as vortex rings. Who has not noticed a graceful

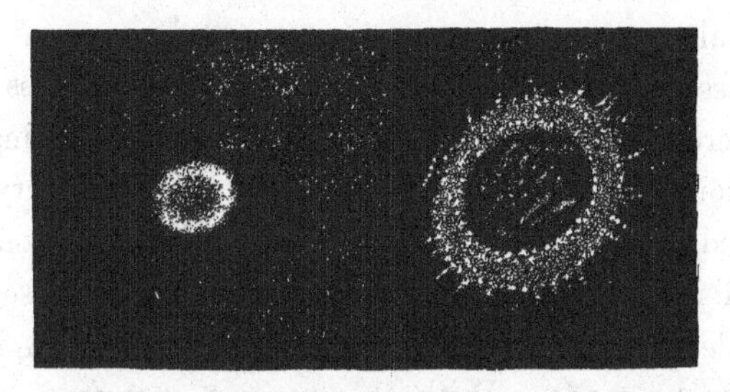

Fig. 88.—The Annular Nebula in Lyra.

ring of steam which occasionally escapes from the funnel of a locomotive, and ascends high into the air, only dissolving some time after the steam not so specialised has disappeared? Such vortex rings can be produced artificially by a cubical box, one open side of which is covered with canvas, while on the opposite side of the box is a circular hole. A tap on the canvas will cause a vortex ring to start from the hole; and if the box be filled with smoke, this ring will be visible for many feet of its path. It would certainly be far too much to assert that the annular nebulæ have any real analogy to vortex rings; but if they have not, there is, at all events, no other object known to us with which they can be compared.

The heavens contain a number of minute but brilliant objects known as the planetary nebulæ. They can only be described as globes of glowing bluish-coloured gas, small enough to be often mistaken for a star when viewed through a telescope. One of the most remarkable of them lies in the constellation Draco, and can be found half-way between the Pole Star and the star γ Draconis. Some of the more recently discovered planetary nebulæ are extremely small, and they have indeed only been distinguished from small stars by the spectroscope. It is also to be noticed that such objects are a little out of the stellar focus in the refracting telescope in consequence of their blue colour. This remark does not apply to a reflecting telescope, as this instrument conducts all the rays to a common focus.

There are many other forms of nebulæ: there are long nebulous rays; there are the wondrous spiral nebulæ which have been disclosed in Lord Rosse's great reflector; there are the double nebulæ; there are some very mysterious variable nebulæ. But all these various objects we must merely dismiss with this passing reference. There is a great difficulty in making pictorial representations of such nebulæ. Most of them are very faint—so faint, indeed, that they can only be seen with close attention even in powerful instruments. In making drawings of these objects, it is impossible to avoid intensifying the features if an intelligible picture is to be made. With this caution, however, we present Plate XVI., which exhibits several of the more remarkable nebulæ as seen through Lord Rosse's great telescope.

There is one problem of the very greatest interest with regard to the nebulæ, which astronomers often turn over in their thoughts, and which they as often despair of seeing satisfactorily settled. That problem is this—how to find the distance of a nebula. The difficulties of finding the distance of a star are so great, that it is only by the most lavish expenditure of time and of patience that

PLATE XVI.

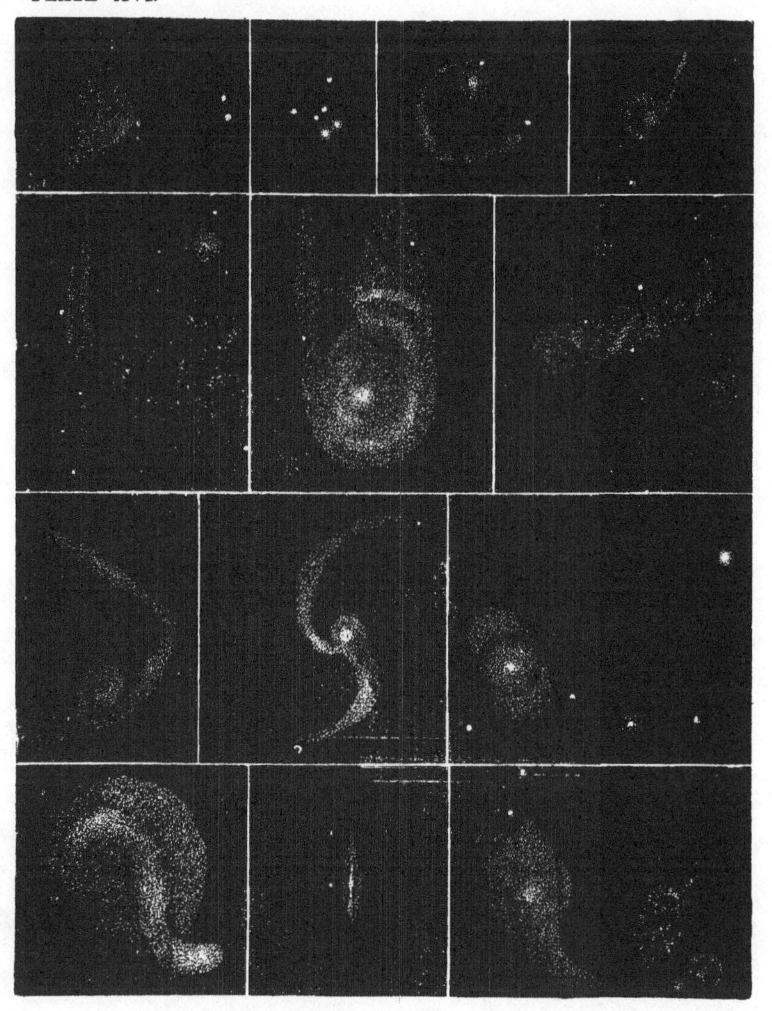

NEBULÆ

OBSERVED WITH LORD ROSSE'S GREAT TELESCOPE.

it can be accomplished; but the difficulties are very much greater and apparently insurmountable in the case of the nebulæ, and no method has yet been devised which will enable us to solve this mighty problem. Our knowledge on the subject is merely of a negative character. We cannot tell how great the distance may be, but we are able in some cases to assign a minor limit to that distance. Our ordinary measures of miles are quite unsuitable for such distances as we shall have to encounter. We must call in a far longer measuring tape, and, fortunately, we have one well suited for the purpose. The most appropriate unit for such magnitudes is the velocity of light, which sweeps along with the prodigious speed of 180,000 miles a second. Moving at this rate, how long will the journey take from the nebula to the earth? It is believed that some of these nebulæ are sunk in space to such an appalling distance, that the light takes centuries before it reaches the earth. We see these nebulæ, not as they are now, but as they were centuries ago. At this climax we bid farewell to the nebulæ. We have reached a point where man's intellect begins to fail to yield him any more light, and where his imagination has succumbed in the endeavour to realise what he has gained.

CHAPTER XXIV.

THE PRECESSION AND NUTATION OF THE EARTH'S AXIS.

The Pole is not a Fixed Point—Its Effect on the Apparent Places of the Stars—The Illustration of the Peg-top—The Disturbing Force which acts on the Earth—Attraction of the Sun on a Globe—The Protuberance at the Equator—The Attraction of the Protuberance by the Sun and by the Moon produces Precession—The Efficiency of the Precessional Agent varies inversely as the Cube of the Distance—The Relative Efficiency of the Sun and the Moon—How the Pole of the Earth's Axis Revolves round the Pole of the Ecliptic.

THE position of the pole of the heavens is most conveniently indicated by the bright star known as the Pole Star, which lies in its immediate vicinity. Around this pole the whole heavens appear to rotate once in a sidereal day; and we have hitherto always referred to the pole as though it were a fixed point in the heavens. This language is sufficiently correct when we only embrace a moderate period of time in our review. It is no doubt true that the pole lies near the Pole Star at the present time. It did so during the lives of the last generation, and it will do so during the lives of the next generation. All this time, however, the pole is steadily moving in the heavens, so that the time will at length come when the pole will have departed a long way from the present Pole Star. This movement of the pole is incessant. It can be easily detected and measured by the instruments in our observatories, and astronomers are familiar with the fact that in all their calculations it is necessary to hold special account of this movement of the pole. It produces an apparent change in the place of a star, which is known by the term "precession."

The movement of the pole is very clearly shown in the accompanying figure, for which I am indebted to the kindness of Professor C. Piazzi Smyth. The circle shows the track along which the pole moves among the stars. The centre of the circle in the constellation of Draco is the pole of the ecliptic. A complete journey of the pole occupies the considerable period of about 25,867 years. The drawing shows the position of the pole at the

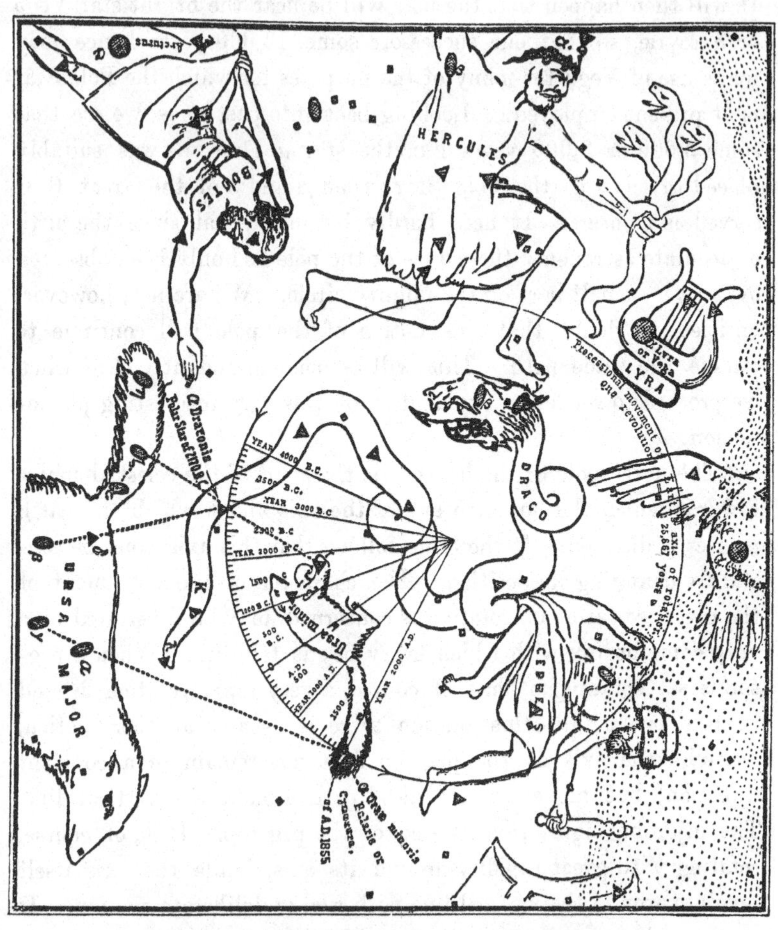

Star-map, representing the precessional movement of the Cœlestial Pole of rotation and especially marking it from the year 4000 B.C. to the year 2000 A.D.

Symbols adopted to represent the magnitudes or brightnesses of the stars, 1st ●; 2nd ●; 3rd ▲; 4th ■.

Fig. 89.

several dates from 4000 B.C. to 2000 A.D. A glance at this map brings prominently before us how casual is the proximity of the pole to the Pole Star. At present, indeed, the distance of the two is actually lessening, but afterwards the distance will increase, until when half of the revolution has been accomplished, the pole will be at a distance of twice the radius of the circle from the Pole Star. It will then happen that the pole will be near the bright star Vega or α Lyræ, so that our successors some 12,000 years hence may make use of Vega for many of the purposes for which the Pole Star is at present employed! Looking back into past ages, we see that some 2,000 or 3,000 years B.C. the star α Draconis was suitably placed to serve as the Pole Star, when β and δ of the Great Bear served as pointers. It need hardly be added, that since the birth of accurate astronomy the course of the pole has only been observed over a very small part of the mighty circle. We are not, however, entitled to doubt that the motion of the pole will continue to pursue the same path. This will be made abundantly clear when we proceed to render an explanation of this very interesting phenomenon.

The north pole of the heavens is the point of the celestial sphere towards which the northern end of the axis about which the earth rotates is directed. It therefore follows that this axis must be constantly changing its position. The character of the movement of the earth, so far as its rotation is concerned, may be illustrated by a very common toy with which every boy is familiar. When a peg-top is set spinning, it has, of course, a very rapid rotation around its axis, but besides this rotation there is usually another motion, whereby the axis of the peg-top does not remain in a constant direction, but moves in a conical path around the vertical line. The adjoining figure gives a view of the peg-top. It is, of course, rotating with great rapidity around its axis, while the axis itself revolves around the vertical line with a very deliberate motion. If we could imagine a vast peg-top which rotated on its axis once a day, and if that axis were inclined at an angle of twenty-three and a half degrees to the vertical, and if the slow conical motion of the axis were such that the revolution of the axis were completed in about 26,000 years, then the movements would resemble those actually made by the earth. The illustration of the peg-top comes, indeed, very close to the actual phenomenon of precession. In each case the rotation about the axis is incomparably more rapid than that of the revolution of the axis itself; in each case also the slow movement is due to an external interference. Looking at the figure of the peg-top we may ask the question, Why does it not fall down? The obvious effect of gravity would seem to say that it is impossible for the peg-top to be in the position shown in the figure. Yet everybody knows that this is possible so long as the top is spinning. If the top were not spinning, it would, of course, fall. It therefore follows that the effect of the rapid rotation of the top so modifies the effect of gravitation, that the latter, instead of producing its apparently obvious consequence, causes the

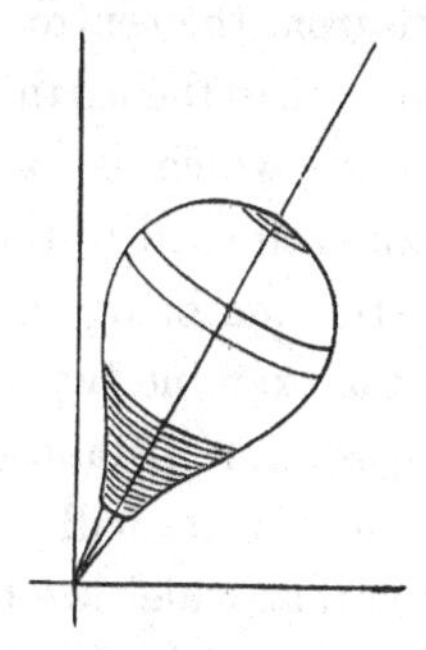

Fig. 90.—Illustration of the motion of Precession.

slow conical motion of the axis of rotation. This is, no doubt, a dynamical question of some difficulty, but it is easy to verify experimentally that it is the case. If a top be constructed so that the point about which it is spinning, and which supports it, shall coincide with the centre of gravity, then there is no effect of gravitation on the top, and there is no conical motion perceived.

If the earth were subject to no external interference, then the direction of the axis about which it rotates must remain for ever constant; but as the direction of the axis does not remain constant, it is necessary to seek for a disturbing force adequate to the production of the phenomena which are observed. We have invariably found that the dynamical phenomena of astronomy can be accounted for by the law of universal gravitation. It is therefore natural to inquire how far gravitation will render an account of the phenomenon of precession; and to put the matter in its simplest form, let us consider the effect which a distant attracting body can have upon the rotation of the earth.

To answer this question, it becomes necessary to define precisely what we mean by the earth; and as for most purposes of astronomy we regard the earth as a spherical globe, we shall commence with this assumption. It seems also certain that the interior of the earth is, on the whole, heavier than the outer portions. It is therefore reasonable to assume that the density increases as we descend; nor is there any sufficient ground for thinking that the earth is much heavier in one part than at any other part equally remote from the centre. It is therefore usual in such calculations to assume that the earth is formed of concentric spherical shells, each one of which is of uniform density; while the density decreases from each shell to the one exterior thereto.

A globe of this constitution being submitted to the attraction of some external body, let us examine the effects which that external body can produce. Suppose, for instance, the sun attracts a globe of this character, what movements will be the result? The first and most obvious result is that which we have already so frequently discussed, and which is expressed by Kepler's laws: the attraction will compel the earth to revolve around the sun in an elliptic path, of which the sun is in the focus. With this movement we are, however, not at this moment concerned. We must inquire how far the sun's attraction will modify the earth's rotation around its axis. It can be demonstrated that the attraction of the sun would be powerless to derange the rotation of the earth so constituted. This is a result which can be formally proved by mathematical calculation. It is, however, sufficiently obvious that the force of attraction of any distant point on a symmetrical globe must pass through the centre of that globe; and as the sun is only an enormous aggregate of attracting points, it can only produce a corresponding multitude of attractive forces; each of these forces passes through the centre of the earth, and consequently the resultant force which expresses the joint result of all the individual forces must also be directed through the centre of the earth. A force of this character, whatever other potent influence it may have, will, at all events, be powerless to affect the rotation of the earth. If the earth be rotating on an axis, the direction of that axis would be invariably preserved; so that as the earth revolves around the sun, it would still continue to rotate around an axis which always remained parallel to itself. Nor would the attraction of the earth by any other body prove more efficacious than that of the sun. If the earth really were the symmetrical globe we have supposed, then the attraction of the sun and moon, and even the influence of all the planets as well, would never be competent to make the earth's axis of rotation swerve for a single second from its original direction.

We have thus narrowed very closely the search for the cause of the "precession." If the earth were a perfect sphere, precession would be inexplicable. We are therefore forced to seek for an explanation of precession in the fact that the earth is not a perfect sphere. This we have already demonstrated to be the case. We have shown that the equatorial axis of the earth is longer than the polar axis, so that there is a protuberant zone girdling the equator. The attraction of external bodies is able to grasp this protuberance, and thereby force the earth's axis of rotation to change its direction.

There are only two bodies in the universe which sensibly contribute to the precessional movement of the earth's axis: these bodies are the sun and the moon. The shares in which the labour is borne by the sun and the moon are not what might have been expected from a hasty view of the subject. This is a point on which it will be desirable to dwell, as it illustrates a point in the theory of gravitation which is of very considerable importance.

The law of gravitation asserts that the intensity of the attraction which a body can exercise is directly proportional to the mass of that body, and inversely proportional to the square of the distance of the attracted point. We can thus compare the attraction exerted upon the earth by the sun and by the moon. The mass of the sun exceeds the mass of the moon in the proportion of about 26,000,000 to 1. On the other hand, the moon is at a distance which, on an average, is about one-386th part of that of the sun. It is thus an easy calculation to show that the efficiency of the sun's attraction on the earth is about 175 times as great as the attraction of the moon. Hence it is, of course, that the earth obeys the supremely important attraction of the sun, and pursues an elliptic path around the sun, bearing the moon as an appendage.

But when we come to that particular effect of attraction which is competent to produce precession, we find that the law by which the efficiency of the attracting body is computed assumes a different form. The measure of efficiency is, in this case, to be found by taking the mass of the body and dividing it by the *cube* of the distance. The complete demonstration of this statement must be sought in the formulæ of mathematics, and cannot be introduced into these pages: we may, however, adduce one consideration which will enable the reader in some degree to understand the principle, though without pretending to be a demonstration of its accuracy. It will be obvious that the nearer the disturbing body approaches to the earth the greater is the *leverage* (if we may use the expression) which is afforded by the protuberance at the equator. The efficiency of a given force will, therefore, on this account alone, increase in the inverse proportion of the distance. The actual intensity of the force itself augments in the inverse square of the distance, and hence the capacity of the attracting body for producing precession will, for a double reason, increase when the distance decreases. Suppose, for example, that the disturbing body is brought to half its original distance from the disturbed body, the leverage is by this means doubled, while the actual intensity of the force is at the same time quadrupled according to the law of gravitation. It will follow that the effect produced in the latter case must be eight times as great as in the former case. And this is merely equivalent to the statement that the precession-producing capacity of a body varies inversely as the cube of its distance.

It is this consideration which gives to the moon an importance as a precession-producing agent which its mere attractive capacity would not have entitled it to. Even though the mass of the sun be 26,000,000 times as great as the mass of the moon, yet when this number is divided by the cube of the relative value of the distances of the bodies (386), it is seen that the efficiency of the moon is more than twice as great as that of the sun. In other words, we may say that one-third of the movement of precession is due to the sun, and two-thirds to the moon.

For the study of the joint precessional effect due to the sun and the moon acting simultaneously, it will be advantageous to consider the effect produced by the two bodies separately; and as the case of the sun is the simpler of the two, we shall take it first. As the earth rotates in its annual path around the sun, the axis of the earth is directed to a point in the heavens which is always $23\frac{1}{2}°$ from the pole of the ecliptic. The precessional effect of the sun is to cause the pole of the earth to revolve, always preserving the same angular distance; and thus we have a motion of the type represented in the diagram. As the ecliptic occupies a position which for our present purpose we may regard as fixed in space, it follows that the pole of the ecliptic is a fixed point on the surface of the heavens; so that the path of the pole of the earth must be a small circle in the heavens, fixed in its position relatively to the surrounding stars. In this we find a motion strictly analogous to that of the peg-top. It is the gravitation of the earth acting upon the peg-top which forces it into the conical motion. The immediate effect of the gravitation is so modified by the rapid rotation of the top, that, in obedience to a profound dynamical principle, the axis of the top revolves in a cone rather than fall down, as it would do were the top not spinning. In a similar manner the immediate effect of the sun's attraction on the protuberance at the equator would be to bring the pole of the earth's axis towards the pole of the ecliptic, but the rapid rotation of the earth modifies this into the conical rotation of precession.

The circumstances with regard to the moon are much more complicated. The moon describes a certain orbit around the earth; that orbit lies in a certain plane, and that plane has, of course, a certain pole on the celestial sphere. The precessional effect of the moon would accordingly tend to make the pole of the earth's axis describe a circle around that point in the heavens which is the pole of the moon's orbit. This point is about 5° from the pole of the ecliptic. The pole of the earth is therefore solicited by two different movements—one a revolution around the pole of the ecliptic, the other a revolution about another point 5° distant, which is the pole of the moon's orbit. It would thus seem that the earth's pole should make a certain composite movement due to the two separate movements. This is really the case, but there is a point to be very carefully attended to which at first seems almost paradoxical. We have shown how the potency of the moon as a precessional agent exceeds that of the sun, and therefore it might be thought that the composite movement of the earth's pole would conform more nearly to a rotation around the pole of the plane of the moon's orbit than to a rotation around the pole of the ecliptic; but this is not the case. The precessional movement is found to be almost absolutely represented by a revolution around the pole of the ecliptic, as is shown in the figure. Here is a point which merits our careful attention, for in it lies the germ of one of the most exquisite of astronomical discoveries.

The plane in which the moon revolves does not occupy a constant position. We are not here specially concerned with the causes of this change in the plane of the moon, but the character of the movement must be enunciated. The inclination of this plane to the ecliptic is about 5°, and this inclination does not vary; but the line of intersection of the two planes does vary, and, in fact, varies so quickly that it completes a revolution in about $18\frac{2}{3}$ years. This movement of the plane of the moon's orbit necessitates a corresponding change in the position of its pole. We thus see that the pole of the moon's orbit must be actually revolving around the pole of the ecliptic, always remaining at the same distance of 5°, and completing its revolution in $18\frac{2}{3}$ years. It will, therefore, be obvious that there is a profound difference between the precessional effect of the sun and of the moon in their action on the earth. The sun invites the earth's pole to describe a circle around a fixed centre; the moon invites the earth's pole to describe a circle around a centre which is itself in constant motion. It fortunately happens that the circumstances of the case are such as to considerably reduce the complexity of the problem. The movement of the moon's plane, only occupying about 18⅔ years, is a very rapid motion compared with the whole precessional movement, which occupies about 26,000 years. It follows that by the time the earth's axis has completed one circuit of its majestic cone, the pole of the moon's plane will have gone round about 1,400 times. Now, as this pole really only describes a comparatively small cone of 5° in radius, we may for a first approximation take the average position which it occupies; but this average position is, of course, the centre of the circle which it describes—that is, the pole of the ecliptic.

We thus see that the average precessional effect of the moon simply conspires with that of the sun to produce a revolution around the pole of the ecliptic. The grosser phenomena of the movements of the earth's axis are to be explained by the uniform revolution of the pole in a circular path; but if we make a minute examination of the track of the earth's axis, we shall find that though it, on the whole, conforms with the circle, yet that it really traces out a delicate sinuous line, sometimes on the inside and sometimes on the outside of the circle. This delicate movement arises from the continuous change in the place of the pole of the moon's orbit. The period of these undulations is 18⅔ years, agreeing exactly with the period of the revolution of the moon's nodes. The amount by which the pole departs from the circle on either side is only about 9·2 seconds—a quantity rather less than the twenty-thousandth part of the radius of the circle. This phenomenon, known as "nutation," was discovered by the beautiful telescopic researches of Bradley in 1747. Whether we look at the theoretical interest of the subject or at the refinement of the observations involved, this achievement of the "Vir incomparabilis," as Bradley has been called by Bessel, is one of the masterpieces of astronomical genius.

CHAPTER XXV.

THE ABERRATION OF LIGHT.

The Real and Apparent Movements of the Stars—How they can be Discriminated—Aberration produces Effects dependent on the Position of the Stars—The Pole of the Ecliptic—Aberration make Stars seems to Move in a Circle—An Ellipse or a Straight Line according to Position—All the Ellipses have Equal Major Axes—How is this Movement to be Explained?—How to be Distinguished from Annual Parallax—The Apex of the Earth's Way—How this is to be Explained by the Velocity of Light—How the Scale of the Solar System can be Measured by the Aberration of Light.

WE have in this chapter to narrate a discovery of a recondite character, which illustrates in a very forcible manner some of the great fundamental truths of Astronomy. Our discussion of it will naturally be divided into two parts. In the first part we must describe the nature of the phenomenon, and then we must give the extremely elegant explanation afforded by the properties of light. The telescopic discovery of aberration, as well as its explanation, are both due to the illustrious Bradley.

The expression *fixed* star, so often used in astronomy, is to be received in a very qualified sense. The stars are, no doubt, well fixed in their places, so far as coarse observation is concerned. The lineaments of the constellations remain unchanged for centuries, and in contrast with the ceaseless movements of the planets, the stars are not inappropriately called fixed. We have, however, had more than one occasion to show throughout the course of this work that the expression "fixed star" is not an accurate one when minute quantities are held in estimation. With the exact measures of modern instruments, many of these quantities are so perceptible, that they have to be always reckoned with in astronomical inquiry. We can divide the movements of the stars into two great classes: the real movements and the apparent movements. The proper motion of the stars and the movements of revolution of the binary stars constitute the real movements of these bodies. These movements are special to each star, so that two stars, although close together in the heavens, may differ in the widest degree as to the real movements which they possess. It may, indeed, sometimes happen, as Mr. Proctor has pointed out, that stars in a certain region are animated with a common movement. In this phenomenon, which has been called star-drift by its discoverer, we have traces of a real movement shared in by a number of stars in a certain group. With this exception, however, the real movements of the stars are governed by no systematic law, and the rapidly rotating binary stars and the other rapidly moving stars are scattered here and there indiscriminately over the heavens.

The apparent movements of the stars have a different character, inasmuch as we find the movement of each star determined by the place which it occupies in the heavens and not by the individual features of the particular object. It is by this means that we can readily discriminate the real movements of the star from its apparent movements, and examine the character of both.

In the present chapter we are concerned with the apparent movements only, and of these there are three, due respectively to precession, to nutation, and to aberration. Each of these apparent movements obeys laws peculiar to itself, and thus it becomes possible to analyse the total apparent motion, and to discriminate the proportions in which the precession, the nutation, and the aberration have severally contributed. We are thus enabled to isolate the effect of aberration as completely as if it were the sole agent of apparent displacement, so that, by an alliance between mathematical calculation and astronomical observation, we can study the effects of aberration as clearly as if the stars were affected by no other motions.

Concentrating our attention solely on the phenomena of aberration, we shall describe its particular effect upon stars in different regions of the sky, and thus ascertain the laws according to which the effects of aberration are exhibited. When this step has been taken, we shall be in a position to give the beautiful explanation of those laws, dependent upon the velocity of light.

At one particular region of the heavens the effect of aberration has a degree of simplicity which is not manifested anywhere else. This region lies in the constellation Draco, at the pole of the ecliptic. At this pole, or in its immediate neighbourhood, each star, in virtue of aberration, describes a circle in the heavens. This circle is very minute: it would take something like 2,000 of these circles together to form an area equal to the area of the moon. Expressed in the usual astronomical language, we would say that the diameter of this small circle is about 40·9 seconds of arc. This is a quantity which, though small to the unaided observation, is really of great relative magnitude in the present state of telescopic research. It is not only large enough to be perceived, but it can be measured with an accuracy which actually does not admit of a doubt, to the hundredth part of the whole. It is also observed that each star describes its little circle in precisely the same period of time; and that period is one year, or, in other words, the time of the revolution of the earth around the sun. It is found that for all stars in this region, be they large stars or small, single or double, white or coloured, the circles appropriate to each have all the same size, and are all described in the same time. Even from this alone it would be manifest that the cause of the phenomenon cannot lie in the star itself. This unanimity in stars of every magnitude and distance requires some simpler explanation.

Further examination of stars in different regions sheds new light on the subject. As we proceed from the pole of the ecliptic, we still find that each star exhibits an annual movement of the same character as the stars just considered. In one respect, however, there is a difference. The apparent path of the star is no longer a circle; it has become an ellipse. It is, however, soon perceived that the shape and the position of this ellipse is governed by the simple law, that the further the star is from the pole of the ecliptic the greater is the eccentricity of the ellipse. The stars at the same distance from the pole have equal eccentricity, and of the axes of the ellipse, the shorter is always directed to the pole, the longer being of course perpendicular thereto. It is, however, found that no matter how great the eccentricity may become, the major axis always retains its original length. It is always equal to about 40·9 seconds, that is, to the diameter of the circle of aberration at the pole itself. As we proceed further and further from the pole of the ecliptic, we find that each star describes a path more and more eccentric, until at length, when we examine a star on the ecliptic, the ellipse has become so attenuated that it has flattened into a line. Each star which happens to lie on the ecliptic oscillates to and fro along the ecliptic through an amplitude of 40·9 seconds. Half-a-year accomplishes the journey one way, and the other half of the year restores the star to its original position. When we come to stars on the other side of the ecliptic, we see the same series of changes proceed in an inverse order. The ellipse, from being actually linear, gradually grows in width, though still preserving the same length of major axis, until at length the stars near the southern pole of the ecliptic are each found to describe a circle equal to the paths pursued by the stars at the north pole of the ecliptic.

The circumstance that the major axes of all these ellipses are of equal length suggests a still further simplification. Let us suppose that every star, either at the pole of the ecliptic or elsewhere, pursues an absolutely circular path, and that all these circles agree, not only in magnitude, but also in being all parallel to the plane of the ecliptic: it is easy to see that this simple supposition will account for the observed facts. The stars at the pole of the ecliptic will, of course, show their circles turned fairly towards us, and we shall see that they pursue circular paths. The circular paths of the stars remote from the pole of the ecliptic will, however, be only seen edgewise, and thus the apparent paths will be elliptical, as we actually find them. We can even calculate the degree of ellipticity which this surmise would require, and we find that it coincides with the observed ellipticity. Finally, when we observe stars actually moving in the ecliptic, the circles they follow would be seen edgewise, and thus the stars would have merely the linear movement which they are seen to possess. All the observed phenomena are thus found to be completely consistent with the supposition that every star of all the millions in the heavens describes once each year a circular path; and that, whether the star be far or near, this circle has always the same apparent diameter, and lies in a plane always parallel to the plane of the ecliptic.

We have now wrought the facts of observation into a form which enables us to examine into the cause of a movement so systematic. Why is it that each star should seem to describe a small circular path? Why should that path be parallel to the ecliptic? Why should it be completed exactly in a twelvemonth? We are at once referred to the motion of the earth around the sun. That movement takes place in the ecliptic. It is completed in a year. The coincidences are so obvious that we feel almost necessarily compelled to connect in some way this apparent movement of the stars with the annual movement of the earth around the sun. If there were no such connection, it would be in the highest degree improbable that the planes of the circles should be all parallel to the ecliptic, or that the time of revolution of each star in its circle should equal that of the revolution of the earth around the sun. As both these conditions are fulfilled, the probability of the connection rises to a value almost infinite.

The important question has then arisen as to why the movement of the earth around the sun should be associated in so remarkable a manner with this universal star movement. There is here one obvious point to be noticed and to be dismissed. We have in a previous chapter discussed the important question of the annual parallax of stars, and we have shown how, in virtue of annual parallax, each star describes an ellipse. It can further be demonstrated that these ellipses are really circles parallel to the ecliptic; so that here we might hastily assume that annual parallax was the cause of the phenomenon discovered by Bradley. A single circumstance will dispose of this suggestion. The circle described by a star in virtue of annual parallax has a magnitude dependent on the distance of the star, so that the circles described by various stars are all of various dimensions, corresponding to the varied distances of different stars. The phenomena of aberration, however, distinctly assert that the circular path of each star is of the same size, quite independently of what its distance may be, and hence annual parallax will not afford an adequate explanation. It should also be noticed that the movements of a star produced by annual parallax are very much smaller than those due to aberration. There is not any known star whose circular path due to annual parallax has a diameter one-twentieth part of that of the circle due to aberration; indeed, in the great majority of cases the parallax of the star is a quantity absolutely insensible.

There is, however, a still graver and quite insuperable distinction between the parallactic path and the aberrational path. Let us, for simplicity, think of a star situated near the pole of the ecliptic, and thus appearing to revolve annually in a circle, whether we regard either the phenomenon of parallax or of aberration. As the earth revolves, so does the star revolve; and thus to each place of the earth in its orbit corresponds a certain place of the star in its circle. If the movement arise from annual parallax, it is easy to see where the place of the star will be for any position of the earth. It is, however, found that in the movement discovered by Bradley the star never has the position which parallax assigns to it, but is, in fact, a quarter of the circumference distant therefrom.

A simple rule will find the position of the star due to aberration. Draw from the centre of the circle a radius parallel to the direction in which the earth is moving at the moment in question, then the extremity of this radius gives the point on the circle where the star is to be found. Tested at all seasons, and with all stars, this law is found to be always verified, and by its means we are conducted to the true explanation of the phenomenon.

We can enunciate the effects of aberration in a somewhat different manner, which will show even more forcibly how the phenomenon is connected with the motion of the earth in its orbit. As the earth pursues its annual course around the sun, its movement at any moment may be regarded as directed towards a certain point of the ecliptic. From day to day, and even from hour to hour, the point gradually moves along the ecliptic, so as to complete the circuit in a year. At each moment, however, there is always a certain point in the heavens towards which the earth's motion is directed. It is, in fact, the point on the celestial sphere towards which the earth would travel continuously, if, at the moment, the attraction of the sun could be annihilated. It is found that this point is intimately connected with the phenomenon of aberration. In fact, the aberration is really equivalent to drawing each star from its mean place towards the apex of the earth's way, as the point is sometimes termed. It can also be shown by observation that the amount of aberration depends upon the distance from the apex. A star which happened to lie on the ecliptic will not be at all deranged by aberration from its mean place when it happens that the apex coincides with the star. All the stars 10° from the apex will be displaced each by the same amount, and all directly in towards the apex. A star 20° from the apex will undergo a larger degree of displacement, though still in the same direction, exactly towards the apex; and all stars at the same distance will be displaced by the same amount. Proceeding thus from the apex, we come to stars at a distance of 90° therefrom. Here the amount of displacement will be a maximum. Each one will be about twenty seconds from its average place; but in every case the imperative law will be obeyed, that the displacement of the star from its mean place lies towards the apex of the earth's way. We have thus given two distinct descriptions of the phenomenon of aberration. In the first we found it convenient to speak of a star as describing a minute circular path; in the other we have regarded aberration as merely amounting to a derangement of the star from its mean place in accordance with specified laws. These descriptions are not inconsistent: they are, in fact, geometrically equivalent; but the latter is rather the more perfect, inasmuch as it assigns completely the direction and extent of the derangement caused by aberration in any particular star at any particular moment.

The question has now been narrowed to a very definite form. What is it which makes each star seem to close in towards the point towards which the earth is travelling? The answer will be found when we make a minute inquiry into the circumstances under which we view a star in the telescope.

The beam of rays from a star falls on the object-glass of a telescope; those rays are parallel, and after they pass through the object-glass they converge to a focus near the eye end of the instrument. Let us first suppose that the telescope is at rest; then if the telescope be pointed directly towards the star, the rays will converge to a point at the centre of the field of view where a pair of cross wires are placed, whose intersection defines the axis of the telescope. The case will, however, be altered if the telescope be moved after the light has passed through the objective; the rays of light in the interior of the tube will pursue a direct path, as before, and will proceed to a focus at the same precise point as before. As, however, the telescope has moved, it will, of course, have carried with it the pair of cross wires; they will no longer be at the same point as at first, and consequently the image of the star will not now coincide with their intersection.

The movement of the telescope arises from its connection with the earth; for as the earth hurries along at a speed of eighteen miles a second, the telescope is necessarily displaced with this velocity. It might at first be thought, that in the incredibly small fraction of time necessary for light to pass from the object-glass to the eye-piece, the change in the position of the telescope must be too minute to be appreciable. Let us suppose, for instance, that the star is situated near the pole of the ecliptic, then the telescope will be conveyed by the earth's motion in a direction perpendicular to its length. If the tube of the instrument be about twenty feet long, it can be readily demonstrated that during the time the light travels down the tube the movement of the earth will convey the telescope through a distance of about one-fortieth of an inch.* This is a quantity very distinctly measurable with the magnifying power of the eye-piece, and hence this derangement of the star's place is very appreciable. It therefore follows that if we wish the star to be shown at the centre

* As the earth carries on the telescope at the rate of 18 miles a second, and as light moves with the velocity of 180,000 miles a second very nearly, it follows that the velocity of the telescope is about one ten-thousandth part of that of light. While the light moves down the tube 20 feet long, the telescope will therefore have moved the ten-thousandth part of 20 feet—*i.e.*, the fortieth of an inch.

of the instrument, the telescope is not to be pointed directly at the star, as it would have to be were the earth at rest, but the telescope must be pointed a little in advance of the star's true position; and as we determine the apparent place of the star by the direction in which the telescope is pointed, it follows that the apparent place of the star is altered by the motion of the earth.

Every circumstance of the change in the star's place admits of complete explanation in this manner. Take, for instance, the small circular path which each star appears to describe. We shall, for simplicity, refer only to the star at the pole of the ecliptic. Suppose that the telescope is pointed truly to the place of the star, then, as we have shown, the image of the star will be at a distance of one-fortieth of an inch from the cross wires. This distance will remain constant, but each night the direction of the star from the cross wires will change, so that in the course of the year it completes a circle, and returns to its original position. We shall not pursue the calculations relative to other stars; suffice it here to say that the movement of the earth has been found adequate to account for the phenomena, and thus the doctrine of the aberration of light is demonstrated.

It remains to allude to one point of the utmost interest and importance. We have seen that the magnitude of the aberration can be measured by astronomical observation. This aberration depends upon the velocity of light, and on the velocity with which the earth's motion is performed. We can measure the velocity of light by independent measurements, in the manner already explained in Chapter XII. We are thus enabled to calculate what the velocity of the earth must be, for there is only one particular velocity for the earth which, when combined with the measured velocity of light, will give the measured value of aberration. The velocity of the earth being thus ascertained, and the length of the year being known, it is easy to find the circumference of the earth's path, and therefore its radius; that is, the distance from the earth to the sun.

Here is indeed a singular result, and one which shows how

profoundly the various phenomena of science are interwoven. We make experiments in our laboratory, and find the velocity of light. We observe the fixed stars, and measure the aberration. We combine these results, and deduce therefrom the distance from the earth to the sun! Although this method of finding the sun's distance is one of very great elegance, and admits of a certain amount of precision, yet it cannot be relied upon as a perfectly unimpeachable method of deducing the great constant. A perfect method must be based on the operations of mere surveying, and ought not to involve recondite physical considerations. We cannot, however, fail to regard the discovery of aberration as a most pleasing and beautiful achievement, for it not only greatly improves the calculations of practical astronomy, but links together several physical phenomena of the greatest interest.

CHAPTER XXVI.

THE ASTRONOMICAL SIGNIFICANCE OF HEAT.

Heat and Astronomy—Distribution of Heat—The Presence of Heat in the Earth—Heat in other Celestial Bodies—Varieties of Temperature—The Law of Cooling—The Heat of the Sun—Can its Temperature be Measured?—Radiation connected with the Sun's Bulk—Can the Sun be Exhausting his Resources?—No marked Change has Occurred—Geological Evidence as to the Changes of the Sun's Heat Doubtful—The Cooling of the Sun—The Sun cannot be merely an Incandescent Solid Cooling—Combustion will not Explain the matter—Some Heat is obtained from Meteoric Matter, but this is not Adequate to the Maintenance of the Sun's Heat—The Contraction of a Heated Globe of Gas—An Apparent Paradox—The Doctrine of Energy—The Nebular Theory—Evidence in Support of this Theory—Sidereal Evidence of the Nebular Theory—Herschel's View of Sidereal Aggregation—The Nebulæ do not Exhibit the Changes within the Limits of our Observation.

THAT a portion of a work on astronomy should bear the title placed at the head of this chapter, will perhaps strike some of our readers as unusual, if not actually inappropriate. Is not heat, it may be said, a question merely of experimental physics? and how can it be legitimately introduced into a discussion of the heavenly bodies and their movements? Whatever weight such objections might have once had, need not now be considered. The recent researches on heat have shown not only that heat has important bearings on astronomy, but that it has really been one of the chief agents by which the universe has been moulded into its actual form. At the present time no work on astronomy could be complete without some account of the remarkable connection between the laws of heat, and the astronomical consequences which follow from obedience to those laws.

In discussing the planetary motions and the laws of Kepler, or in discussing the movements of the moon, the proper motions of the stars, or the revolutions of the binary stars, we proceed on the supposition that the bodies we are dealing with are rigid particles, and the question as to whether these particles are hot or cold does not seem to have any especial bearing. No doubt the ordinary periodic phenomena of our system, such as the revolution of the planets in conformity with Kepler's laws, will be observed for countless ages, whether the planets be hot or cold, or whatever may be the heat of the sun. It must, however, be admitted that the laws of heat introduce certain modifications into the statement of these laws. The effects of heat may not be immediately perceptible, but they exist, they are constantly acting, and in the progress of time they are adequate to effecting the mightiest changes throughout the universe.

Let us briefly recapitulate the circumstances of our system which give to heat its potency. Look first at our earth, which at present seems—on its surface, at all events—to be a body devoid of heat; but a closer examination will at once dispel this idea. Have we not the phenomena of volcanoes, of geysers, and of hot springs, which show that in the interior of the earth heat must exist in far greater intensity than we find on the surface? These phenomena are found in widely different regions of the earth. Their origin is, no doubt, involved in a good deal of obscurity, but yet no one can deny that they indicate vast reservoirs of heat. It would indeed seem that heat is to be found everywhere in the deep inner regions of the earth. If we take a thermometer down a deep mine, we find it records a temperature higher than at the surface. The deeper we descend the higher is the temperature; and if the same rate of progress should be maintained through those depths of the earth which we are not able to penetrate, it can be demonstrated that at twenty or thirty miles below the surface the temperature must be as great as that of red-hot iron.

We find in the other celestial bodies abundant evidence of the present or the past existence of heat. Our moon, as we have already mentioned, affords a very striking instance of a body which must once have been very highly heated. The extraordinary volcanoes on its surface render this beyond any doubt. It is equally certain that those volcanoes have been silent for ages, so that, whatever may be the interior condition of the moon, the surface has now cooled down. Extending our view further, we see in the great planets Jupiter and Saturn evidence that they are still endowed with a temperature far in excess of that which the earth has retained; while when we look at our sun, we see a body in a state of brilliant incandescence, and glowing with a fervour to which we cannot approximate in our mightiest furnaces. The various fixed stars are bodies which glow with heat, like our sun; while we have in the nebulæ objects whose existence is hardly intelligible to us unless we admit that they are possessed of a vast store of heat.

From this rapid survey of the different bodies in our universe, one conclusion is obvious. We may have great doubts as to the actual temperature of any individual body of the system; but it cannot be doubted that there is a wide range of temperature among the different bodies. Some are hotter than others. The stars and suns are perhaps the hottest of all, but it is not improbable that they may be immeasurably outnumbered by the cold and dark bodies of the universe, which are to us invisible, and only manifest their existence in an indirect and casual manner.

The law of cooling tells us that every body radiates heat, and that the quantity of heat which it radiates increases when the temperature of the body increases relatively to the surrounding medium. This law appears to be universal. It is obeyed on the earth, and it would seem that it must be equally obeyed by every other body in space. We thus see that each of the planets and each of the stars is continuously pouring forth in all directions a never-ceasing stream of heat.

This radiation of heat is productive of very momentous consequences. Let us study them, for instance, in the case of the sun. Our great luminary pours forth a mighty flood of radiant heat in all directions. A minute fraction of that heat is intercepted by our earth, and is directly or indirectly the source of all life, and of nearly all movement, on our earth. To pour forth heat as the sun does, it is necessary that his temperature be enormously high. And there are some facts which permit us to form an estimate of what that temperature must actually be.

Every one is acquainted with the use of a burning-glass, by which we can condense the sun's rays to a focus, and produce incandescence or set objects on fire. Large burning-glasses have been constructed, in the focus of which an extraordinary temperature has been obtained. It can, however, be proved that the temperature at the focus cannot be greater, cannot be even equal, to the temperature at the source of heat itself. The effect of a burning-glass is merely equivalent to making a closer approach towards the sun. The rule is indeed a simple one. The temperature at the focus of the burning-glass is the same as that of a point placed at such a distance from the sun that the solar disc would seem just as large as the lens itself viewed from its own focus. The greatest burning-glass which has ever been constructed virtually transports an object at its focus to within 250,000 miles of the sun's surface: in other words, to a distance of about 1-400th part of its present amount. In this focus it was found that the most refractory substances, agate, cornelian, platinum, fire-clay, the diamond itself, were melted or even dissipated into vapour. There can be no doubt that if the sun were to come as near to us as the moon, the solid earth itself would melt like wax.

It is difficult to form any numerical statement of the actual temperature of the sun. The intensity of that temperature vastly transcends the greatest artificial heat, and any attempt to clothe such estimates in figures is necessarily very precarious. But assuming the greatest artificial temperature to be about 4,000° Fahr., we shall probably be well within the truth if we state the effective temperature of the sun to be about 18,000° Fahr. This is, indeed, vastly below many of the estimates which have been made. Secchi, for instance, has estimated the sun's temperature to be nearly one thousand times that here given.

The copious outflow of heat from the sun corresponds with its enormous temperature. We can express the amount of heat in various ways, but it must be remembered that considerable uncertainty still attaches to such measurements. The old method of measuring heat by the quantity of ice melted may be used as an illustration. It is computed that a shell of ice 43½ feet thick surrounding the whole sun would in one minute be melted by the sun's heat underneath. A somewhat more elegant illustration was also given by Sir John Herschel, who showed that if a cylindrical glacier forty-five miles in diameter were to be continually flowing into the sun with the velocity of light, the end of that glacier would be melted as quickly as it advanced. From each square foot on the surface of the sun emerges a quantity of heat as great as could be produced by the daily combustion of sixteen tons of coal. This is, indeed, an amount of heat which, properly transformed into work, would keep an engine of many hundreds of horse-power running from one year's end to the other. The heat radiated from a few acres on the sun would be adequate to drive all the steam-engines in the world. When we reflect on the vast intensity of the radiation from each square foot of the sun's surface, and when we combine with this the stupendous dimensions of the sun, imagination fails to realise how vast must be the actual expenditure of heat.

In one way the enormous intensity of the radiation from each unit of sun surface is a consequence of its bulk. Imagine for a moment two suns, one of which had a diameter double the other. If these two suns were of analogous constitution, their stores of heat may be taken as proportional to their volumes. The larger sun would thus have a store of heat eight times as great as the smaller one. But the ratio of the surfaces of the two suns is only four to one. Hence it follows that by the time both suns cooled down, twice as much heat per unit of area must have passed through the surface of the large sun as through the surface of the small one. To emphasize the contrast still more, suppose our present sun compared with a fictitious sun not larger than our earth. If the two suns started under equal circumstances, then before they had both cooled down to the same temperature, nearly one hundred times as much heat must be transmitted through each square foot of the sun's surface as through each square foot of the earth's surface.

In presence of the beneficent, if prodigal, expenditure of the sun's heat, we are tempted to ask a question which has the most vital interest for the earth and its inhabitants. We live from hour to hour by the sun's splendid generosity; and, therefore, it is important for us to know what security we possess for the continuance of his favours. When we witness the terrific expenditure of the sun's heat each hour, we are compelled to ask whether the sun may not be exhausting its resources; and if so, what are the prospects of the future? This question we can partly answer. The whole subject is indeed of surpassing interest, and redolent with the spirit of modern scientific thought.

Our first attempt to examine this question must lie in an appeal to the facts which are attainable. We want to know whether the sun is showing any symptoms of decay. Are the days as warm and as bright now as they were last year, ten years ago, one hundred years ago? We can find no evidence of any change since the beginning of authentic records. If the sun's heat had perceptibly changed within the last two thousand years, we should expect to find corresponding changes in the distribution of plants and of animals; but no such changes have been detected. There is no reason to think that the climate of ancient Greece or of ancient Rome was appreciably different from the climates of the Greece and the Rome that we know at this day. The vine and the olive grow now where they grew two thousand years ago.

We must not, however, lay too much stress on this argument; for the effects of slight changes in the sun's heat may have been neutralised by corresponding adaptations in the pliable organisms of cultivated plants. All we can certainly conclude is that no marked change has taken place in the heat of the sun during historical time. But when we come to look back into vastly earlier ages, we find the most copious evidence that the earth has undergone great changes in climate. Geological records can on this question hardly be misinterpreted. Yet it is curious to note that these changes are hardly such as could arise from the gradual exhaustion of the sun's radiation. No doubt, in very early times we have evidence that the earth's climate must have been much warmer than at present. We had the great carboniferous epoch, when the temperature must almost have been tropical in Arctic latitudes. Yet it is hardly possible to cite this as evidence that the sun was then much more powerful; for we are immediately reminded of the glacial epoch, when our temperate zones were encased in sheets of solid ice, as Northern Greenland is at present. If we suppose the sun to have been hotter than it is at present to account for the vegetation which produced coal, then we ought to assume the sun to be colder than it is now to account for the glacial epoch. It is not reasonable to attribute such phenomena to such oscillations in the radiation from the sun. The glacial epochs prove that we cannot appeal to geology in aid of the doctrine that a secular cooling of the sun is now in progress. The geological variations of climate may have been caused by changes in the earth itself, by changes in the position of its axis, by changes in its actual orbit; but however they have been caused, they hardly tell us much with regard to the past history of our sun.

The heat of the sun has lasted for countless ages; yet we cannot credit the sun with the power of actually creating heat. We must apply even to the majestic mass of the sun the same laws which we have found by our experiments on the earth. We must ask, whence comes the heat sufficient to supply this tremendous outgoing? Let us briefly recount the various suppositions that have been made.

Place two red-hot spheres of iron side by side, a large one and a small one. They have been taken from the same fire; they were both equally hot; they are both cooling, but the small sphere cools more rapidly. It speedily becomes dark, while the large sphere is still glowing, and would continue to do so for some minutes. The larger the sphere, the longer it will take to cool; and hence it has been supposed that a mighty sphere of the prodigious dimensions of our sun would, if once heated, cool gradually, but the duration of the cooling would be so long, that for thousands and for millions of years it could continue to be a source of light and heat to the revolving system of planets. This suggestion will not bear the test of arithmetic. If the sun had no source of heat beyond that indicated by its high temperature, we can show that radiation would cool the sun a few degrees every year. Two thousand years would then witness a very great decrease in the sun's heat.

We are certain that no such decrease can have taken place. The source of the sun's radiation cannot be found in the mere cooling of an incandescent mass.

Can the fires in the sun be maintained by combustion, analogous to that which goes on in our furnaces? Here we would seem to have a source of gigantic heat; but arithmetic also disposes of this supposition. We know that if the sun were made of even solid coal itself, and if that coal were burning in pure oxygen, the heat that could be produced would only suffice for 6,000 years. If the sun which shone upon the builders of the great Pyramid had been solid coal from surface to centre, it must by this time have been in great part burned away in the attempt to maintain its present rate of expenditure. We are thus forced to look to other sources for the supply of the sun's heat, since neither the heat of incandescence nor the heat of combustion will suffice.

There is probably—indeed, we may say certainly—one external source from which the heat of the sun is recruited. It will be necessary for us to consider this source with some care, though I think we shall find it to be merely an auxiliary of comparatively trifling moment. According to this view, the solar heat receives occasional accessions from the fall upon the sun's surface of masses of meteoric matter. There can be hardly a doubt that such masses do fall upon the sun; there is certainly no doubt that if they do, the sun must gain some heat thereby. We have experience on the earth of a very interesting kind, which illustrates the development of heat by meteoric matter. There lies a world of philosophy in a shooting star. Some of these myriad objects rush into our atmosphere and are lost; others, no doubt, rush into the sun, with the same result. We also admit that the descent of a shooting star into the atmosphere of the sun must be attended with a flash of light and of heat. The heat acquired by the earth from the flashing of shooting stars through our air is quite insensible. It has been supposed, however, that the heat accruing to the sun from the same cause may be quite sensible—nay, it has been even supposed that the sun may be re-invigorated from this source.

Here, again, we must apply the cold principles of weights and measures to estimate the plausibility of this suggestion. We first calculate the actual weight of meteoric indraught to the sun which would be adequate to sustain the fires of the sun at their present vigour. The mass of matter that would be required is so enormous that we cannot usefully express it by imperial weights; we must deal with masses of imposing magnitude. It fortunately happens that the weight of our moon is a convenient unit. Conceive that our moon—a huge globe, 2,000 miles in diameter—were crushed into a myriad of fragments, and that these fragments were allowed to rain in on the sun; there can be no doubt that this tremendous meteoric shower would contribute to the sun rather more heat than would be required to supply his radiation for a whole year. If we take our earth itself, conceive it comminuted into dust, and allow that dust to fall on the sun as a mighty shower, each fragment would instantly give out a quantity of heat, and the whole would add to the sun a supply of heat adequate to sustain the present rate of radiation for nearly one hundred years. The mighty mass of Jupiter treated in the same way would generate a meteoric display greater in the ratio in which the mass of Jupiter exceeds the mass of the earth. Were Jupiter to fall into the sun, enough heat would be thereby produced to scorch the whole solar system; while all the planets together would be capable of producing heat which, if properly economised, would supply the radiation of the sun for 45,000 years.

Here, then, is certainly a plausible source for the supply of the sun's heat; but it must be remembered that though the moon could supply one year's heat, and Jupiter 30,000 years' heat, yet the practical question is not whether the solar system could supply the sun's heat, but whether it does. Is it likely that meteors equal in mass to the moon fall into the sun every year? This is the real question, and I think we are bound to reply to it in the negative. It can be shown that the quantity of meteors which could be caught by the sun in any one year can be only an excessively minute fraction of the total amount. If, therefore, a moon-weight of meteors were caught every year, there must be an incredible mass of meteoric matter roaming at large through the system. There must be so many meteors that the earth would be incessantly pelted with them, and heated to such a degree as to be rendered uninhabitable. There are also other reasons which preclude the supposition that a stupendous quantity of meteoric matter exists in the vicinity of the sun. Such matter would produce an appreciable effect on the movement of the planet Mercury. There are, no doubt, some irregularities in the movements of Mercury not yet fully explained, but these irregularities are very much less than would be the case if meteoric matter existed in quantity adequate to the sustentation of the sun. Astronomers, then, believe that though meteors may be a rate in aid of the sun's current expenditure, yet that the greater portion of that expenditure must be defrayed from other resources.

It is one of the achievements of modern science to have effected the solution of the problem—to have shown how it is that, notwithstanding the stupendous radiation, the sun still maintains its temperature. The question is not free from difficulty in its exposition, but the matter is one of such very great importance that we are compelled to make the attempt.

Let us imagine a vast globe of heated gas in space. This is not an entirely gratuitous supposition, inasmuch as there are globes apparently of this character; they have been already alluded to as planetary nebulæ. This globe will radiate heat, and we shall suppose that it emits more heat than it receives from the radiation of other bodies. The globe will accordingly lose heat, or what is equivalent thereto, but it will be incorrect to assume that the globe will necessarily fall in temperature. That the contrary is, indeed, the case is a result almost paradoxical at the first glance; but yet it can be readily shown to be a necessary consequence of the laws of heat and of gases.

Let us fix our attention on a portion of the gas lying on the surface of the globe. This is, of course, attracted by all the rest of the globe, and thus tends in towards the centre of the globe. If equilibrium subsists, this tendency must be neutralised by the pressure of the gas beneath; so that the greater the gravitation, the greater is the pressure. When the globe of gas loses heat by radiation, let us suppose that it grows colder —that its temperature accordingly falls; then, since the pressure of a gas decreases when the temperature falls, the pressure beneath the superficial layer of the gas will decrease, while the gravitation is unaltered. The consequence will inevitably be that the gravitation will now conquer the pressure, and the globe of gas will accordingly contract. There is, however, another way in which we can look at the matter. We know that heat is equivalent to energy, so that when the globe radiates forth heat, it must expend energy. A part of the energy of the globe will be that due to its temperature; but another, and in some respects a more important, part is that due to the separation of its particles. If we allow the particles to come closer together, we shall diminish the energy due to separation, and the energy thus set free can take the form of heat. But this drawing in of the particles necessarily involves a shrinking of the globe; and thus we find that whatever way the matter be viewed, the radiation of heat from the globe must be attended with contraction.

And now for the remarkable consequence, which seems to have a very important application in Astronomy. As the globe contracts, a part of its energy of separation is changed into heat; that heat is partly radiated away, but it is not radiated away as rapidly as it is produced by the contraction. The consequence is, that although the globe is really losing heat and really contracting, yet that its temperature is actually rising. A simple case will suffice to demonstrate this result, paradoxical as it may at first seem. Let us suppose that by contraction of the sphere it had diminished to one-half its diameter; and let us fix our attention on a cubic inch of the gaseous matter in any point of the mass. After the contraction had taken place each edge of the cube would be reduced to half an inch, and the volume would therefore be reduced to one-eighth part of its original amount. The law of gases tells us that if the temperature be unaltered the pressure varies inversely as the volume, and consequently the internal pressure in the cube will be increased eight-fold. As, however, the distance between every two particles is reduced to one-half, it will follow that the gravitation between every two particles is increased four-fold, and as the area is also reduced to one-fourth it will follow that the pressure inside the reduced cube is increased sixteen-fold; but we have already seen that with a constant temperature it only increases eight-fold, and hence the temperature cannot be constant but must rise with the contraction.

We thus have the somewhat astonishing result that a gaseous globe in space radiating heat, and thereby growing smaller, is all the time actually increasing in temperature. But, it may be said, surely this cannot go on for ever. Are we to suppose that the gaseous mass will go on contracting and contracting with a temperature ever fiercer and fiercer, and actually radiating out more and more heat the more it loses? where lies the limit to such a prospect? As the body contracts, its density must increase, until it either becomes a liquid, or a solid, or, at any rate, until it ceases to obey the laws of a purely gaseous body which we have supposed. Once these laws cease to be observed, the argument disappears; the loss of heat may then really be attended with a loss of temperature, until, in the course of time, the body has sunk to the temperature of space itself.

It is not pretended that this reasoning can be applied in all its completeness to the present state of the sun. The sun's density is now so great that the laws of gases cannot be there strictly followed. There is, however, good reason to believe that the sun was once more gaseous than at present; possibly at one time he may have been quite gaseous enough to admit of this reasoning in all its fulness. At present the sun appears to be in some intermediate stage of its progress from the gaseous condition to the solid condition. We cannot, therefore, say that the temperature of the sun is now increasing in correspondence with the process of contraction. This may be true or it may not be true; we have no means of deciding the point. We may, however, feel certain that the sun is still sufficiently gaseous to experience in some degree the rise of temperature associated with the contraction. That rise in temperature may be partly or wholly obscured by the fall in temperature, which would be the more obvious consequence of the radiation of heat from the partially solid body. It will, however, be manifest that the cooling of the sun may be enormously protracted if the fall of temperature from the one cause be nearly compensated by the rise of temperature from the other. It can hardly be doubted that in this we find the real explanation of the fact, that we have no historical evidence of any appreciable alteration in the radiation of heat from the sun.

This question is one of such interest that it may be worth while to look at it from a slightly different point of view. The sun contains a certain store of energy, part of which is continually disappearing in the form of radiant heat. The energy remaining in the sun is partly transformed in character; some of it is transformed into heat, which goes wholly or partly to supply the loss by radiation. The total energy of the sun must, however, be decreasing; and hence it would seem the sun must at some time or other have its energy exhausted, and cease to be a source of light and of heat. It is true that the rate at which the sun contracts is very slow. We are, indeed, not able to measure with certainty the decrease in the sun's bulk. It is a quantity so minute, that the contraction since the birth of accurate astronomy is not large enough to be perceptible in our telescopes. It is, however, possible to compute what the contraction of the sun's bulk must be, on the supposition that the energy lost by that contraction just suffices to supply the daily radiation of heat. The change is very small when we consider the present size of the sun. At the present time the sun's diameter is about 860,000 miles. If each year this diameter decreases by about 220 feet, sufficient energy will be yielded to account for the entire radiation. This gradual decrease is always in progress.

These considerations are of considerable interest when we apply them retrospectively. If it be true that the sun is at this moment shrinking, then in past times his globe must have been greater than it is at present. Assuming the figures already given, it follows that one hundred years ago the diameter of the sun must have been four miles greater than it is now; one thousand years ago the diameter was forty miles greater; ten thousand years ago the diameter of the sun was four hundred miles greater than it is to-day. When man first trod this earth it would seem that the sun must have been many hundreds, perhaps many thousands, of miles greater than it is at this time.

We must not, however, over-estimate the significance of this statement. The diameter of the sun is so great, that a diminution of 10,000 miles would be but little more than the hundredth part of its diameter. If it were suddenly to shrink to the extent of 10,000 miles, the change would not be appreciable to ordinary observation, though a much smaller change would not elude delicate astronomical measurement. It does not necessarily follow that the climates on our earth in these early times must have been very different from those which we find at this day, for the question of climate depends upon other matters besides sunbeams.

Yet we need not abruptly stop our retrospect at any epoch, however remote. We may go back earlier and earlier, through the long ages which geologists claim for the deposition of the stratified rocks; and back again still further, to those very earliest epochs when life began to dawn on the earth. Still we can find no reason to suppose that the law of the sun's decreasing heat is not maintained; and thus we would seem bound by our present knowledge to suppose, that the sun grows larger and larger the further our retrospect extends. We cannot assume that the rate of that growth is always the same. No such assumption is required; it is sufficient for our purpose that we find the sun growing larger and larger the further we peer back into the remote abyss of time past. If the present order of things in our universe has lasted long enough, then it would seem that there was a time when the sun must have been twice as large as it is at present; it must once have been ten times as large. How long ago that was no one can venture to say. But we cannot stop at the stage when the sun was even ten times as large as it is at present; the arguments will still apply in earlier ages. We see the sun swelling and swelling, with a corresponding decrease in its density, until at length we find, instead of our sun as we know it, a mighty nebula filling a gigantic region of space.

Such is, in fact, the doctrine of the origin of our system which has been advanced in that celebrated speculation known as the nebular theory. Nor can it be ever more than a speculation; it cannot be established by observation, nor can it be proved by calculation. It is merely a conjecture, more or less plausible, but perhaps in some degree necessarily true, if our present laws of heat, as we understand them, admit of the extreme application here required, and if also the present order of things has reigned for sufficient time without the intervention of any influence at present unknown to us. This nebular theory is not confined to the history of our sun. Precisely similar reasoning may be extended to the individual planets: the farther we look back, the hotter and the hotter does the whole system become. It has been thought that if we could look far enough back, we should see the earth too hot for life; back further still, we should find the earth and all the planets red-hot; and back further still, to an incredibly remote epoch, when the planets would be heated just as much as our sun is now. In a still earlier stage the whole solar system is thought to have been one vast mass of glowing gas, from which the present forms of the sun, with the planets and their satellites, have been gradually solved. We cannot be sure that the course of events has been what is here indicated, but there are very sufficient grounds for thinking that this doctrine substantially represents what has actually occurred.

Many of the features in the solar system harmonise with the supposition that the origin of the system has been that suggested by the nebular theory. We have already had occasion in an earlier chapter to allude to the fact that all the planets perform their revolutions around the sun in the same direction. It is also to be observed that the rotation of the planets on their axes, as well as the movements of the satellites around their primaries, all follow the same law, with one slight exception in the case of the Uranian system. A coincidence so remarkable naturally suggests the necessity of some physical explanation. Such an explanation is offered by the nebular theory. Suppose that countless ages ago a mighty nebula was slowly rotating and slowly contracting. In the process of contraction, portions of the condensed matter of the nebula would be left behind. These portions would still revolve around the central mass, and each portion would rotate on its axis in the same direction. As the process of contraction proceeded, it would follow from dynamical principles that the velocity of rotation would increase; and thus at length these portions would consolidate into planets, while the central mass would gradually contract to form the sun. By a similar process on a smaller scale the systems of satellites were evolved from the contracting primary. These satellites would also revolve in the same direction, and thus the characteristic features of the solar system could be accounted for.

The nebular origin of the solar system receives considerable countenance from the study of the sidereal heavens. We have already dwelt upon the resemblance between the sun and the stars. If, then, our sun has passed through such changes as the nebular theory requires, may we not anticipate that similar phenomena should be met with in other stars? If this be so, it is reasonable to suppose that the evolution of some of the stars may not have progressed so far as has that of the sun, and thus we may be able actually to witness stars in the earlier phases of their development. Let us see how far the telescope responds to these anticipations.

The field of view of a large telescope usually discloses a number of stars scattered over a black background of sky; but the blackness of the background is not uniform: the practised eye of the skilled observer will detect in some parts of the heavens a faint luminosity. This will sometimes be visible over the whole extent of the field, or it may even occupy several fields. Years may pass on, and still there is no perceptible change. There can be no illusion, and the conclusion is irresistible that the object is a stupendous mass of faintly luminous glowing gas or vapour. This is the simplest type of nebula; it is characterised by extreme faintness, and seems composed of matter of the utmost tenuity. We are, on the other hand, occasionally presented with the beautiful and striking phenomenon of a definite and brilliant star surrounded by a luminous atmosphere. Between these two extreme types of a faint diffused mass on the one hand, and a bright star with a nebula surrounding it on the other, Herschel thought that a graduated series of various other nebulæ could be arranged. We thus have a series of links passing by imperceptible gradations from the most faintly diffused nebulæ on the one side, into stars on the other. It was the perception of this which led Sir William Herschel to his theory of sidereal aggregation.

The nebulæ seemed to Herschel to be vast masses of phosphorescent vapour. This vapour gradually cools down, and ultimately condenses into a star, or a cluster of stars. When the varied forms of nebulæ were classified, it almost seemed as if the different links in the process could be actually witnessed. In the vast faint nebulæ the process of condensation had just begun; in the smaller and brighter nebulæ the condensation had advanced farther; while in others, the star, or stars, arising from the condensation had already become visible.

But, it may be asked, how did Herschel know this? what is his evidence? Let us answer this question by an illustration. Go into a forest, and look at a noble old oak which has weathered the storm for centuries; have we any doubt that the oak-tree was once a young small plant, and that it grew stage by stage until it reached maturity? Yet no one has ever followed an oak-tree through its various stages, the brief span of life has not been long enough to do so. The reason why we believe the oak-tree to have passed through all these stages is, because we are familiar with oak-trees of every gradation in size, from the seedling up to the noble veteran. Having seen this gradation in a vast multitude of trees, we are convinced that each individual has passed through all these stages.

It was by a similar train of reasoning that Herschel was led to adopt the view of the origin of the stars which we have endeavoured to describe. The astronomer's life is not long enough, the life of the human race might not be long enough, to watch the process by which the nebula condenses down so as to form a solid body. But by looking at one nebula after another, the astronomer thinks he is able to detect the various stages which connect the nebula in its original form with the final form. He is thus led to believe that each of the nebulæ passes, in the course of ages, through these stages. And thus Herschel adopted the opinion that stars—some, many, or all—have each originated from what was once a glowing nebula.

Such a speculation may captivate the imagination, but it must be carefully distinguished from the truths of astronomy, properly so called. Remote posterity may perhaps obtain evidence on the subject which to us is inaccessible: our knowledge of nebulæ is too recent. There has not yet been time enough to have seen any appreciable changes: for the study of nebulæ can only be said to date from Messier's Catalogue in 1771.

Since Herschel's time, no doubt, many careful drawings and observations of the nebulæ have been obtained; but still the interval has been much too short, and the earlier observations are too imperfect, to enable any changes in the nebulæ to be investigated with sufficient accuracy. If the human race lasts for very many centuries, and if our present observations are preserved during that time for comparison, then Herschel's theory may perhaps be satisfactorily tested.

CHAPTER XXVII.

THE TIDES.*

Mathematical Astronomy—Recapitulation of the Facts of the Earlier Researches—Another great Step has been taken—Lagrange's Theories, how far they are really True—The Solar System not Made of Rigid Bodies—Kepler's Laws True to Observation, but not Absolutely True when the Bodies are not Rigid—The Errors of Observation—Growth of certain Small Quantities—Periodical Phenomena—Some Astronomical Phenomena are not Periodic—The Tides—How the Tides were Observed—Discovery of the Connection between the Tides and the Moon—Solar and Lunar Tides—Work done by the Tides—Whence do the Tides obtain the Power to do the Work—Tides are Increasing the Length of the Day—Limit to the Shortness of the Day—Early History of the Earth-Moon System—Unstable Equilibrium—Ratio of the Month to the Day—The Future Course of the System—Equality of the Month and the Day—The Future Critical Epoch—The Constant Face of the Moon accounted for—The other Side of the Moon—The Satellites of Mars—Their Remarkable Motions—Have the Tides Possessed Influence in Moulding the Solar System generally?—Moment of Momentum—Tides have had little or no appreciable Effect on the Orbit of Jupiter—Conclusion.

AT various points of our progress through this volume it has been necessary to detail the *methods* of research which have conducted astronomers to those discoveries by which astronomy has been developed into a branch of science. We have explained how the telescope has been employed to scrutinize the heavenly bodies, to survey their positions, to weigh and to measure, and to enable the artist to draw the features by which they are characterised. We have also indicated how another whole class of discoveries has arisen in a different manner. We have shown how observations afford to the mathematician the basis on which his calculations can be rested; we have shown how he can discover truths of the most wide-spreading generality—truths which seem to have no bounds, either in time or in space. We have even shown how the mathematician

* The theory of Tidal Evolution sketched in this chapter is mainly due to the researches of Professor G. H. Darwin. F.R.S.

at his desk has felt the unknown planet, and guided the telescope of the astronomer to the point of the heavens in which it lay. We have in this chapter to unfold another most remarkable modern development of astronomical research, which has enabled us in some degree to explore the recesses of nature in departments entirely closed to every other method of inquiry. These modern researches are entirely based on mathematics; they have, in fact, quite anticipated observation.

To appreciate the importance of these modern researches, it will be well to recall clearly the standpoint which the problems of gravitational astronomy occupied before the new departure was undertaken. We may illustrate the point most simply by taking an example from our solar system. Let us recapitulate briefly the history of our knowledge of the motions of the planets around the sun. The steps of that history are fourfold. We have first the establishment of the Copernican theory, which demonstrated that the planets revolve around the sun. The next great epoch is marked by Kepler's famous discovery, that the orbit of each planet was an ellipse, with the sun in the focus. The third epoch is that of the Newtonian discovery of gravitation, by which the motion of each planet in an ellipse was shown to be a consequence of the law of universal gravitation. Another step brings us to the labours of the great French mathematicians at the close of the last century. By these researches it seemed that the theory of gravitation was well-nigh completed. It was shown that though the planets mutually disturbed each other, the effect of these disturbances must be always small. The eccentricity of the ellipse which each planet was describing was shown not to be constant; it was found that that eccentricity was for ever changing—now waxing, now waning—but always confined between very narrow limits. It was shown that the orbits of the planets were inclined to each other at angles which were always changing; but these changes are merely oscillations, for the angles constantly fluctuate about a mean value, from which they never depart to any considerable distance. All these movements are periodic, like the motion of the planet itself; and after oscillating through small limits for a majestic period of innumerable years, the system would regain almost its original position, again to start on a series of oscillations which will last for untold ages.

These sublime discoveries were regarded as the crowning triumph of gravitational astronomy. The investigations which led to them originated in consequence of the attention of mathematicians being called to apparent irregularities in the motion of the heavenly bodies which demanded explanation. Observation was thus always in advance of theory; but when, at length, the theory had overtaken the results of observation—when, indeed, in some cases (as in the discovery of Neptune), theory had actually outstripped observation—it was naturally supposed that the theory was perfect. No doubt, here and there a small difficulty remained unexplained: the moon would not come up to the meridian within a second of the right time, a planet would be slightly out of its calculated place, or a satellite not be eclipsed at the expected moment. Men thought—and often truly thought—that these were merely minute imperfections, which closer examination would remove; and thus they concluded that gravitational astronomy was fundamentally perfect, and that the trifling discrepancies remaining would be ultimately cleared away.

Dazzled by these brilliant discoveries, theoretical astronomers had done but little for many years. They dwelt with admiration on the great classical achievements, and endeavoured here and there to patch up the small flaws and still outstanding difficulties of the theory. At length, however, theoretical astronomers have been aroused, and have taken another great stride. It is, indeed, curious to note what has been the origin of these modern researches. The energies of a Laplace were called forth to explain the long irregularities of Jupiter and Saturn; the genius of an Adams or a Le Verrier was evoked by the unexplained perturbations of Uranus. In such cases theoretical labours have arisen, in the hope of reconciling the discrepancies between theory and observation. The motives which have excited this new outburst of discovery have been wholly different. There is now no notorious discrepancy or irregularity, to remain a reproach to astronomy until it is accounted for: the theory to which we wish now to draw attention is entirely, or almost entirely, in advance of astronomical observation.

That the great discoveries of Lagrange on the stability of the planetary system are correct, is in one sense strictly true. No one has ever ventured to impeach the mathematics of Lagrange. Given the planetary system in the form which Lagrange assumed, and the stability of that system is assured for all time. There is, however, one assumption which Lagrange makes, and on which his whole theory was founded: his assumption is that the planets are rigid bodies.

No doubt our earth seems a rigid body. What can be more solid and unyielding than the mass of rocks and metals which form the earth, so far as it is accessible to us? In the wide realms of space the earth is but as a particle; it surely was a natural and a legitimate assumption to suppose that that particle was a rigid body. If the earth were absolutely rigid—if every particle of the earth were absolutely at a fixed distance from every other particle—if under no stress of forces, and under no conceivable circumstance, the earth could experience even the minutest change of form—if the same could be said of the sun and of all the other planets—then Lagrange's prediction of the eternal duration of our system must be fulfilled.

But what are the facts of the case? Is the earth really rigid? We know from experiment that a rigid body in the mathematical sense of the word does not exist. Rocks are not rigid; steel is not rigid; even a diamond is not perfectly rigid. The whole earth is far from being rigid even on the surface, while the interior is still, perhaps, more or less fluid. The earth cannot be called a rigid body; still less can the larger bodies of our system be called rigid. Jupiter and Saturn are perhaps hardly even what could be called solid bodies. The solar system of Lagrange consisted of a rigid sun and a number of minute rigid planets; the actual solar system consists of a sun which is in no sense rigid, and planets which are only partially so.

The question then arises as to whether the discoveries of the great mathematicians of the last century will apply, not only to the ideal solar system which they conceived, but to the actual solar system in which our lot has been cast. There can be no doubt that these discoveries are approximately true: they are, indeed, so near the absolute truth, that observation has not yet satisfactorily shown any departure from them.

But in the present state of science we can no longer overlook the important questions which arise when we deal with bodies not rigid, in the mathematical sense of the word. Let us, for instance, take the simplest of the laws to which we have referred, the great law of Kepler, which asserts that a planet will revolve for ever in an elliptic path of which the sun is one focus. This is seen to be verified by actual observation: indeed, it was established by observation before any theoretical explanation of that movement was propounded. If, however, we state the matter with a little more precision, we shall find that what Newton really demonstrated was, that if two rigid particles attract each other by a law of force which varies with the inverse square of the distance between the particles, then each of the particles will describe an ellipse with the common centre of gravity in the focus. The earth is, to some extent, rigid, and hence it was natural to suppose that the relative behaviour of the earth and the sun would, to a corresponding extent, observe the simple elliptic law of Kepler; as a matter of fact they do observe it with such fidelity, that if we make allowance for other causes of disturbance, we cannot, by our most careful observations, detect the slightest variation in the motion of the earth arising from its want of rigidity.

There is, however, a subtlety in the investigations of mathematics which, in this instance at all events, immeasurably transcends the most delicate observations which our instruments enable us to make. The principles of mathematics tell us that though Kepler's laws may be true for bodies which are absolutely and mathematically rigid, yet that if the sun or the planets be either wholly, or even in their minutest part, devoid of perfect rigidity, then Kepler's laws can be no longer true. Do we not seem here to be in the presence of a contradiction? Observation tells us that Kepler's laws are true in the planetary system;

theory tells us that these laws cannot be true in the planetary system, because the bodies in that system are not perfectly rigid. How is this discrepancy to be removed? or is there really a discrepancy at all? There is not. When we say that Kepler's laws have been proved to be true by observation, we must reflect on the nature of the proofs which are attainable. We observe the places of the planets with the instruments in our observatories; these places are measured by the help of our clocks and of the graduated circles on the instruments. These observations are no doubt wonderfully accurate; but they do not, they cannot, possess absolute accuracy in the mathematical sense of the word. We can, for instance, determine the place of a planet with such precision that it is certainly not one second of arc wrong; and one second is an extremely small quantity. A foot-rule placed at a distance of about forty miles subtends an angle of a second, and it is surely a delicate achievement to measure the place of a planet, and feel confident that no error greater than this can have intruded into our result.

When we compare the results of observation with the calculations conducted on the assumption of the truth of Kepler's laws, and when we pronounce on the agreement of the observations with the calculations, there is always a reference, more or less explicit, to the inevitable errors of the observations. If the calculations and observations agree so closely that the differences between the two are minute enough to have arisen in the errors inseparable from the observations, then we are satisfied with the accordance; for, in fact, no closer agreement is attainable, or even conceivable. Now, the influence which the want of rigidity exercises on the fulfilment of the laws of Kepler can be estimated by calculation; it is found, as might be expected, to be extremely small; so small, in fact, as to be contained within that slender margin of error by which observations are liable to be affected. We are thus not able to discriminate by actual measurement the effects due to the absence of rigidity; they are inextricably hid among the errors of observation.

The great law of universal gravitation has been recently applied to astronomical discovery in the direction we have indicated. By its means we are enabled to peer back into the abyss of past time, and trace to some extent the changes which our system has undergone. Let us attempt to render some account of these memorable researches.

At the spring of the year the advent of the birds which visit us yearly, the bursting of the trees into leaf, and the gradual opening of the flowers, bring before us in a most pleasing manner the principle of periodicity, which is of such prevalence in nature. The astronomical phenomena which are most familiar to us are, indeed, all periodic. We have the rising and setting of the sun, we have the changes of the moon, we have the changes of the seasons, as very simple instances; but there are many other periodical phenomena with which astronomers are acquainted. Take, for instance, that great cycle which is accomplished in the precession of the equinoxes. The pole describes a majestic circle in the heavens in a period of about 26,000 years; but even this long period seems brief when compared with some other periodic phenomena. There is a stupendous period in our system when the planets, having gone through all their mutual perturbations, commence a new cycle. All periodic phenomena have one great feature in common: at the close of a cycle the condition of the system is the same as it was at the commencement. For our present purpose phenomena of this kind are immaterial: they are not the source of the mighty changes now to occupy our attention.

There are some astronomical phenomena in progress which are not periodic: they do not increase, and then decrease, with the rhythm of the more obvious phenomena. It behoves us to look very closely into any phenomena of this character: for they are the real architects of the universe.

There are two elements in determining the effect which such a cause can produce. One of those elements is the efficiency of the cause itself; the other element is the time during which that cause has acted. The latter factor is capable of indefinite increase. Even if the cause be extremely small, yet if the time during which that cause has acted be extremely great, the result attained may possess stupendous dimensions. We propose to illustrate one of these constantly acting influences, which has had some share in moulding our solar system generally, and has specially had a paramount share in the development of our earth and our moon.

The argument on which we are to base our researches is really founded on a very familiar phenomenon. There is no one who has ever visited the sea-side, who is not familiar with that rise and fall of the sea which we call the tide. Twice every twenty-four hours the sea advances on the beach to produce high tide; twice every day the sea again retreats to produce low tide. These tides are not merely confined to the coasts; they penetrate for miles up the courses of rivers; they periodically inundate great estuaries. In a maritime country the tides are of the most profound practical importance: they also possess a significance of a far less obvious character, which it is our object now to investigate.

These daily pulses of the ocean have long ceased to be a mystery. It was in the earliest times perceived that there was a connection between the tides and the moon. Ancient writers, such as Pliny and Aristotle, have referred to the alliance between the times of high water and the age of the moon. I think we sometimes do not give the ancient astronomers as much credit as their shrewdness really entitles them to. We have all read—we have all been taught—that the moon and the tides are connected together; but how many of us are in a position to say that we have actually noticed that connection by direct personal observation? The first man who really studied this matter, with sufficient attention to convince himself and to convince others of its reality, was a great philosopher. We know not his name, we know not his nation, we know not the age in which he lived; but our admiration of his discovery must be increased by the reflection that he had not the theory of gravitation to guide him. A philosopher of the present day who had never seen the sea could still predict the necessity of tides as a consequence of the law of universal gravitation; but the primitive astronomer, who knew not of the invisible bond by which all bodies in the universe are drawn together, made a splendid—indeed, a typical—inductive discovery, when he ascertained the relation between the moon and the tides.

We can surmise that this discovery, in all probability, first arose from the observation of experienced navigators. In all matters of entering port or of leaving port, the state of the tide is of the utmost concern to the sailor. Even in the open sea he has sometimes to shape his course in accordance with the currents produced by the tides; or, in guiding his course by taking soundings, he has always to bear in mind that the depth varies with the tide. All matters relating to the tide would thus come under his daily observation. His daily work, the success of his occupation, the security of his life, depend often on the tides; and hence he would be solicitous to learn from his observation all that would be useful to him in the future. To the coasting sailor the question of the day is the time of high water. That time varies from day to day; it is an hour or more later to-morrow than to-day, and there is no very simple rule which can be enunciated. The sailor would therefore welcome gladly any rule which would guide him in a matter of such importance. We can make a conjecture as to the manner in which such a rule was first discovered. Let us suppose that a sailor at Calais, for example, is making for harbour. He has a beautiful night—the moon is full; it guides him on his way; he gets safely into harbour; and the next morning he finds the tide high between 11 and 12.* He often repeats the same voyage, but he finds sometimes a low and inconvenient tide in the morning. At length, however, it occurs to him that *when he has a moonlight night* he has a high tide at 11. This occurs once or twice: he thinks it is but a chance coincidence. It occurs again and again. At length he finds it always occurs. He tells the rule to other sailors; they try it too. It is invariably found that when the moon is full, the high tide always recurs at the same hour at the same place. The connection between the moon and the tide is thus established, and the intelligent sailor will naturally compare other phases of the moon with the times of high water. He finds, for example, that the moon at the first quarter always gives high water at the same hour of the day; and finally he obtains a practical

* The hour varies with the locality: it would be 11.49 at Calais; at Liverpool, 11.23; at Swansea Bay, 5.56, &c.

rule, by which, from the state of the moon, he can at once tell the time when the tide will be high at the port where his occupation lies. A diligent observer will trace a still further connection between the moon and the tides: he will observe that some high tides rise higher than others, that some low tides fall lower than others. This is a matter of great practical importance. When a dangerous bar has to be crossed, the sailor will feel much additional security in knowing that he is carried over it on the top of a spring tide; or if he has to contend against tidal currents, which in some places have enormous force, he will naturally prefer for his voyage the neap tides, in which the strength of these currents is less than usual. The spring tides and the neap tides will become familiar to him, and he will perceive that the spring tides occur when the moon is full or new—or, at all events, that the spring tides are within a certain constant number of days of the full or new moon. It was, no doubt, by reasoning such as this, that in primitive times the connection between the moon and the tides came to be perceived.

It was not, however, until the great discovery of Newton had disclosed the law of universal gravitation that it became possible to give a physical explanation of the tides. It was then seen how the moon attracts the whole earth and every particle of the earth. It was seen how the fluid particles which form the oceans on the earth, were enabled to obey the attraction in a way that the solid parts could not. When the moon is overhead it draws the water up, as it were, into a heap underneath, and thus gives rise to the high tide. The water on the opposite side of the earth is also affected in a way that might not be at first anticipated. The moon attracts the solid body of the earth with greater intensity than it attracts the water at the other side, which lies more distant from it. The earth is thus drawn away from the water, which accordingly exhibits a high tide as well on the side of the earth away from the moon as on that towards the moon. The low tides occupy the intermediate positions.

The sun also excites tides on the earth, but owing to the great distance of the sun, the difference of its attraction on the sea and on the solid interior of the earth is not so appreciable. The solar tides are thus much less than the lunar tides. When the two conspire, they cause a spring tide; when the solar and lunar tides are opposed, we have the neap tide.

There are, however, a multitude of circumstances to be taken into account when we attempt to apply this general reasoning to the conditions of a particular case. Owing to local peculiarities the tides vary enormously at the different parts of the coast. In a confined area like the Mediterranean Sea, the tides have only a comparatively small range, varying at different places from one foot to a few feet. In mid-ocean also the tidal rise and fall is not large, amounting, for instance, to a range of three feet at St. Helena. Near the great continental masses the tides become very much modified by the coasts. We find at London a tide of eighteen or nineteen feet; but the most remarkable tides in the British Islands are those in the Bristol Channel, where, at Chepstow or Cardiff, there is a rise and fall during spring tides to the height of thirty-seven or thirty-eight feet, and at neap tides to a height of twenty-eight or twenty-nine. These tides are surpassed in magnitude at other parts of the world. The greatest of all tides are those in the Bay of Fundy, at some parts of which the rise and fall at spring tides is not less than fifty feet.

It will, of course, be obvious that the rising and falling of the tide is attended with the formation of currents. Such currents are, indeed, well known, and in some of our great rivers they are of the utmost consequence. These currents of water can, like water-streams of any other kind, be made to do useful work. We can, for instance, impound the rising water in a reservoir, and as the tide falls we can compel the enclosed water to work a water-wheel before it returns to the sea. We have, indeed, here a source of actual power; but it is only under very unusual circumstances that it would be found economical to use the tides for this purpose. The question can be submitted to calculation, and the area of the reservoir can be computed which would retain sufficient water to work a water-wheel of given horse-power. It can be shown that the area of the reservoir necessary to impound water enough to produce 100 horse-power would be 40 acres. The whole question is then reduced to the simple one of expense: would the construction and the maintenance of this reservoir be more or less costly than the erection and the maintenance of a steam-engine of equivalent power? In most cases it would seem that the latter would be by far the cheaper; at all events, we do not practically find tidal engines in use, so that the power of the tides is now running to waste, no less than the power of Niagara itself. The economical aspects of the case may, however, be very profoundly altered at some remote epoch, when our stores of fuel, now so lavishly expended, give appreciable signs of approaching exhaustion.

The tides are, however, *doing work* of one kind or another. A tide in a river estuary will sometimes scour away a bank and carry its materials elsewhere. We have here work done and energy consumed, just as much as if the same task had been accomplished by engineers directing the powerful arms of navvies. We know that work cannot be done, without the consumption of energy in some of its forms; whence, then, comes the energy which supplies the power of the tides? At a first glance the answer to this question seems a very obvious one. Have we not said that the tides are caused by the moon? and must not the energy, therefore, be derived from the moon? This seems obvious enough, but unfortunately, it is not true. It is one of those cases by no means infrequent in Dynamics, where the truth is widely different from that which seems to be obviously the case. An illustration will perhaps make the matter clearer. When a rifle is fired, it is the finger of the rifleman that pulls the trigger; but are we, then, to say that the energy by which the bullet has been driven off has been supplied by the rifleman? Obviously not; the energy is, of course, due to the gunpowder, and all the rifleman did was to provide the means by which the energy stored up in the powder could be liberated. To a certain extent we may compare this with the tidal problem; the tides raised by the moon are the originating cause whereby a certain store of energy is drawn upon and applied to do such work as the tides are competent to perform. This store of energy, strange to say, does not lie in the moon; it is in the earth itself. Indeed, it is extremely remarkable that the moon actually gains energy from the tides, by itself absorbing some of the store which exists in the earth. This is not put forward as an obvious result; indeed, it is very far from being obvious, as it depends upon a refined dynamical theorem.

Let us, however, clearly understand the nature of this mighty store of energy from which the tides draw their power, and on which even the moon herself makes large and incessant drafts. In what sense does the earth possess a store of energy? Do we not know that the earth rotates on its axis once every day? It is this rotation which is the source of the energy. Let us compare the rotation of the earth with the rotation of the fly-wheel belonging to a steam-engine. The rotation of the fly-wheel is really a reservoir, into which the engine pours energy at each stroke of the piston. The various machines in the mill worked by the engine merely draw on the store of energy accumulated in the fly-wheel. The earth may be likened to a gigantic fly-wheel detached from the engine, though still connected with the machines in the mill. From its stupendous dimensions and from its rapid velocity, that great fly-wheel possesses an enormous store of energy, and that energy must be expended before the fly-wheel comes to rest. Hence it is that though the tides are caused by the moon, yet the energy they require is obtained by simply drawing on the vast supply ready to hand in the rotation of the earth.

There is, however, a distinction of a very fundamental character between the earth and the fly-wheel of an engine. As the energy is withdrawn from the fly-wheel and consumed by the various machines in the mill, it is continually replaced by fresh energy, which flows in from the exertions of the steam-engine, and thus the velocity of the fly-wheel is maintained. But the earth is the fly-wheel without the engine. When the tides draw on the store of energy and expend it in doing work, that energy is not replaced. The consequence is irresistible: the energy in the rotation of the earth must be decreasing. This leads to a consequence of the very utmost significance. If the engine be cut off from the fly-wheel, then, as every one knows, the massive fly-wheel may still give a few rotations, but it will speedily come to rest. A similar inference must be made with regard to the earth; but its store of energy is so enormous, in comparison with the demands which are made upon it, that the earth is able to hold out. Ages of countless duration must elapse before the energy of the earth's rotation can be completely exhausted by such drafts as the tides are capable of making. Nevertheless, it is necessarily true that the energy is decreasing; and if it be decreasing, then the speed of the earth's rotation must be surely, if slowly, abating. Now we have arrived at a consequence of the tides which admits of being stated in the simplest language. If the speed of rotation be abating, then the length of the day must be increasing; and hence we are conducted to the following most important statement; that the *tides are increasing the length of the day.*

To-day is longer than yesterday—to-morrow will be longer than to-day. The difference is so small that even in the course of ages it can hardly be said to have been distinctly established by observation. We do not pretend to say how many centuries have elapsed since the day was even one second shorter than it is at present; but centuries are not the units which we employ in tidal evolution. A million years ago it is quite probable that the divergence of the length of the day from its present value may have been very considerable. Let us take a profound glance back into the depths of times past, and see what the tides have to tell us. If the present order of things has lasted, the day must have been shorter and shorter the farther we look back into the dim past. The day is now twenty-four hours; it was once twenty hours, once ten hours; it was once six hours. How much farther can we go? Once the six hours is past, we begin to approach a limit which must at some point bound our retrospect. The shorter the day, the more is the earth bulged at the equator; the more the earth is bulged at the equator, the greater is the strain put upon the materials of the earth by the centrifugal force of its rotation. If the earth were to go too fast, it would be unable to cohere together; it would separate into pieces, just as a grindstone driven too rapidly is rent asunder with violence. Here, therefore, we discern in the remote past a barrier which stops the present argument. There is a certain critical velocity which is the greatest that the earth could have without risk of rupture, but the exact amount of that velocity is a question not very easy to answer. It depends upon the nature of the materials of the earth: it depends upon the temperature: it depends upon the effect of pressure, and on other details not accurately known to us. An estimate of the critical velocity has, however, been made, and it has been shown that the shortest period of rotation which the earth could have, without flying into pieces, is about three or four hours. The doctrine of tidal evolution has thus conducted us to the conclusion that, at some inconceivably remote epoch, the earth was spinning round its axis in a period approximating to three or four hours.

We thus learn that we are indebted to the moon for the gradual elongation of the day from its primitive value up to twenty-four hours. In obedience to one of the most profound laws of nature, the earth has reacted on the moon, and the reaction of the earth has taken a tangible form. It has simply consisted in gradually driving the moon away from the earth. We may observe that this driving away of the moon resembles a piece of retaliation on the part of the earth. "You raise tides on me," says the earth, "and I do not like to be troubled with the tides, so I gradually push you away. If you will stop raising the tides on me, then I will stop pushing you away, and we can revolve round each other uniformly and for ever." The consequence of the retreat of the moon is sufficiently remarkable. The path in which the moon is revolving has at the present time a radius of 240,000 miles. This radius must be constantly growing larger, in consequence of the tides. Provided with this fact, let us now glance back into the past history of the moon. As the moon's distance is increasing when we look forwards, so we find it decreasing when we look backwards. The moon must have been nearer the earth yesterday than it is to-day: the difference is no doubt inappreciable in years, in centuries, or in thousands of years; but when we come to millions of years, the moon must have been significantly closer than it is at present: until at length we find that its distance, instead of being 240,000 miles, has dwindled down to 40,000, to 20,000, to 10,000 miles. Nor need we stop —nor can we stop—until we find the moon actually close to the earth's surface. If the laws of nature have lasted long enough, and if there has been no external interference, then it cannot be doubted that the moon and the earth were once in immediate proximity. It is easy to calculate the time in which the moon must have been revolving round the earth. The nearer the moon is, the quicker it must revolve; and at the wondrous epoch, when the moon was close to the earth, it must have completed each revolution in about three or four hours.

This has led to one of the most daring speculations that has ever been made in astronomy. We cannot refrain from enunciating it; but it must be remembered that it is only a speculation, and to be received with corresponding reserve. The speculation is intended to answer the question, What brought the moon into that position, close to the surface of the earth? We will only say that there is the gravest reason to believe that the moon was, at some very early period, fractured off from the earth when the earth was in a soft or plastic condition.

At the beginning of the history we found the earth and the moon close together. We found that the rate of rotation of the earth was only a few hours instead of twenty-four hours. We found that the moon completed its journey round the primitive earth in exactly the same time as the primitive earth rotated on its axis, so that the two bodies were then constantly face to face. Such a state of things formed what a mathematician would describe as a case of unstable dynamical equilibrium. It could not last. It may be compared to the case of a needle balanced on its point; the needle must fall to one side or the other. In the same way, the moon could not continue to preserve this position. There were two courses open: the moon must either have fallen back on the earth, and been re-absorbed into the mass of the earth, or it must have commenced its outward journey. Which of these courses was the moon to adopt? We have no means, perhaps, of knowing exactly what it was which determined the moon to one course rather than to another, but as to the course which was actually taken there can be no doubt. The fact that the moon exists, shows that it did not return to the earth, but commenced its outward journey. As the moon recedes from the earth, it must, in conformity with Kepler's laws, require a longer time to complete its revolution. It has thus happened that from the original period of only a few hours, the duration has increased until it has reached the present number of 656 hours. The rotation of the earth has, of course, also been modified, in accordance with the retreat of the moon. Once the moon had commenced to recede, the earth was released from the obligation which required it constantly to direct the same face to the moon. When the moon had receded to a certain distance, the earth would complete the rotation in less time than that required by the moon for one revolution. Still the moon gets further and further away, and the duration of the revolution increases to a corresponding extent, until three, four, or more days (or rotations of the earth) are identical with the month (or revolution of the moon). Although the number of days in the month increases, yet we are not to suppose that the rate of the earth's rotation is increasing; indeed, the contrary is the fact. The earth's rotation is getting slower, and so is the revolution of the moon, but the retardation of the moon is much greater than that of the earth. Even though the period of rotation of the earth has greatly increased from its primitive value, yet the period of the moon has increased, so that it is several times as large as that of the rotation of the earth. As ages roll on the moon recedes further and further, its orbit increases, the duration of the revolution augments, until at length a very noticeable epoch is attained, which is, in one sense, a culminating point in the career of the moon. At this epoch the revolutions of the moon, when measured in rotations of the earth, attain their greatest value. It would seem that at this time the month was about twenty-nine days. It is not, of course, meant that the month and the day at that epoch were the month and the day as our clocks now measure time. Both were shorter then than now. But what we mean is, that at this epoch the earth rotated twenty-nine times on its axis while the moon completed one circuit.

This epoch has now been passed. No attempt can be made at present to evaluate the date of that epoch in our ordinary units of measurement. At the same time, however, no doubt can be entertained as to the immeasurable antiquity of the event, in comparison with all historic records; but whether it is to be reckoned in hundreds of thousands of years, in millions of years, or in tens of millions of years, must be left in great degree to conjecture.

This remarkable epoch once passed, we find that the course of events in the earth-moon system begins to shape itself towards that remarkable final stage which has points of resemblance to the initial stage. The moon still continues to revolve in an orbit with a diameter, steadily, though very slowly, growing. The length of the month is accordingly increasing, and the rotation of the earth being still constantly retarded, the length of the day is also continually growing. But the ratio of the length of the month to the length of the day now exhibits a change. That ratio had gradually increased, from unity at the commencement, up to the maximum value of somewhere about twenty-nine at the epoch just referred to. The ratio now begins again to decline, until we find the earth makes only twenty-eight rotations, instead of twenty-nine, in one revolution of the moon. The decrease in the ratio continues until the number twenty-seven expresses the days in the month. Here, again, we have an epoch which it is impossible for us to pass without special comment. In all that has hitherto been said, we have been dealing with events in the distant past; here we have at length arrived at the present state of the earth-moon system. The days at this epoch are our well-known days, the month is the well-known period of the revolution of our moon. At the present time the month is about twenty-seven of our days, and this relation has remained sensibly true for thousands of years past. It will continue to remain sensibly true for thousands of years to come, but it will not remain true indefinitely. It is merely a stage in this grand transformation; it may possess the attributes of permanence to our ephemeral view, just as the wings of a gnat seem at rest when illuminated by the electric spark, but when we contemplate the mighty history with conceptions of time spacious enough for astronomy, we realise how the present condition of the earth-moon system can have no greater permanence than any other stage in the history.

Our narrative must, however, now assume a different form. We have been speaking of the past; we have been conducted to the present; can we say anything of the future? Here, again, the tides come to our assistance. If we have rightly comprehended the truth of dynamics (and who is there now that can doubt them?), we shall be enabled to make a forecast of the further changes of the earth-moon system. If there be no interruption from any external source at present unknown to us, we can predict—in outline, at all events—the subsequent career of the moon. We can see how the moon will still follow its outward course. The path in which it revolves will grow with incredible slowness, but yet it will always grow; the progress will not be reversed, at all events, before the final stage of our history has been attained. We shall not now delay to dwell on the intervening stages; we will rather attempt to sketch the ultimate type to which our system tends. In the dim future—countless millions of years to come—this final stage will be approached. The ratio of the month to the day, whose decline we have already referred to, will continue to decline. The period of revolution of the moon will grow longer and longer, but the length of the day will increase much more rapidly than the increase in the duration of the moon's period. From the month of twenty-seven days we shall pass to a month of twenty-six days, and so on, until we shall reach a month of ten days, and, finally, a month of one day.

Let us clearly understand what we mean by a month of one day. We mean that the time in which the moon revolves around the earth will be equal to the time in which the earth rotates around its axis. The length of this day will, of course, be vastly greater than our day. The only element of uncertainty in these inquiries arises when we attempt to give numerical accuracy to the statements. It seems to be as true as the laws of dynamics, that a state of the earth-moon system in which the day and the month are equal must be ultimately attained; but when we attempt to state the length of that day, we introduce a hazardous element into the inquiry. In giving, then, any estimate of its length, it must be understood that the magnitude is stated with great reserve. It may be erroneous to some extent, though, perhaps, not to any considerable amount. The length of this great day would seem to be about equal to fifty-seven of our days. In other words, at some critical time in the incredibly distant future, the earth will take something like 1,400 hours to perform a rotation, while the moon will complete its journey precisely in the same time.

We thus see how, in some respects, the first stage of the earth-moon system and the last stage resemble each other. In each case we have the day equal to the month. In the first case the day and the month were only a small fraction of our day; in the last stage the day and month are each a large multiple of our day. There is, however, a profound contrast between the first critical epoch and the last. We have already mentioned that the first epoch was one of unstability—it could not last; but this second state is one of dynamical stability. Once that state has been acquired, it must be permanent, and would endure for ever if the earth and the moon could be isolated from all external interference.

There is one special feature which characterises the movement when the month is equal to the day. A little reflection will show that when this is the case, the earth must constantly direct the same face towards the moon. If the day be equal to the month, then the earth and moon must revolve together, as if bound by invisible bands; and whatever hemisphere of the earth be directed to the moon when this state of things commences, will remain there so long as the day remains equal to the month.

At this point it is hardly possible to escape being reminded of that characteristic feature of the moon's motion which has been observed from all antiquity. We refer, of course, to the fact that the moon at the present time constantly turns the same face to the earth.

It is incumbent on astronomers to provide a physical explanation of this remarkable fact. The moon revolves around our earth once in a definite number of seconds. If the moon always turns the same face to the earth, then it is demonstrated that the moon rotates on its axis once in the same number of seconds also. Now, this would be a coincidence wildly improbable unless there were some physical cause to account for it. We have not far to seek for a cause: it is the influence of the tides which has caused the phenomenon. We now find the moon has a rugged surface, which testifies to the existence of intense volcanic activity in former times. Those volcanoes are now silent—the internal fires in the moon seem to have become exhausted; but there was a time when the moon must have been a heated and semi-molten mass. There was a time when the materials of the moon were so hot as to be soft and yielding, and in that soft and yielding mass the attraction of our earth excited great tides. We have no historical record of these tides (they were long anterior to the existence of telescopes; they were probably long anterior to the existence of the human race), but we know that these tides once existed by the work they have accomplished, and that work is seen to-day in the constant face which the moon turns towards the earth. The gentle rise and fall of the oceans which form our tides, present a picture widely different from the tides by which the moon was once agitated. The tides on the moon were vastly greater than those of the earth. They were greater because the weight of the earth is greater than that of the moon, so that the earth was able to produce much more powerful tides in the moon than the moon has ever been able to raise on the earth.

That the moon should bend the same face to the earth depends immediately upon the condition that the moon shall rotate on its axis in precisely the same period as that which it requires to revolve around the earth. The tides are a regulating power of the most unremitting efficiency to ensure that this condition should be observed. If the moon rotated more slowly than it ought, then the great lava tides would drag the moon round faster and faster, until it attained the desired velocity; and then, but not till then, they would give the moon peace. Or if the moon were to rotate faster on its axis than on its orbit, again the tides would come furiously into play; but this time they would be engaged in retarding the moon's rotation, until they had reduced the speed of the moon to one rotation for each revolution.

Can the moon ever escape from the thraldom of the tides? This is not very easy to answer, but it seems perhaps not impossible that the moon may, at some future time, be freed from tidal control. It is, indeed, obvious that the tides, even at present, have not the extremely stringent control over the moon which they once exercised. We now see no ocean on the moon, nor do the volcanoes show any trace of molten lava. There can hardly be tides *on* the moon, but there may be tides *in* the moon. It may be that the interior of the moon is still hot enough to retain an appreciable departure from solidity, and if so, the tidal control would still retain the moon in its grip; but the time will probably come, if it have not come already, when the moon will be cold to the centre—cold as the temperature of space. If the materials of the moon were what a mathematician would call absolutely rigid, there can be no doubt that the tides could no longer exist, and the moon would be emancipated from tidal control. It seems impossible to predicate how far the moon can ever conform to the circumstances of an actual rigid body, but it may be conceivable that at some future time the tidal control shall have practically ceased. There would then be no longer any necessary identity between the period of rotation and that of revolution. A gleam of hope is thus projected over the astronomy of the distant future. We know that the time of revolution of the moon is increasing, and so long as the tidal governor could act, the time of rotation must increase sympathetically. We have now surmised a state of things in which the control is absent. There will then be nothing to prevent the rotation remaining as at present, while the period of revolution is increasing. The privilege of seeing the other side of the moon, which has been withheld from all previous astronomers, may thus in the distant future be granted to their successors.

The tides which the moon raises in the earth act as a brake on the rotation of the earth. They now constantly tend to bring the period of rotation of the earth to coincide with the period of revolution of the moon. As the moon revolves once in twenty-seven days, the earth is at present going too fast, and consequently the tidal control at the present moment endeavours to retard the rotation of the earth. The rotation of the moon long since succumbed to tidal control, but that was because the moon was comparatively small and the tidal power of the earth was enormous. Now the matters are reversed. The earth is large and more massive than the moon, the tides raised by the moon are but small and weak, and the earth has not yet completely succumbed to the tidal action. But the tides are constant, they never for an instant relax the effort to control, and they are gradually tending to render the day and the month coincident, though the process is a very slow one.

The theory of the tides leads us to look forward to an ultimate state of things, in which the moon revolves around the earth in a period equal to the day, so that the two bodies shall constantly bend the same face to each other, provided the tidal control be still able to guide the moon's rotation. So far as the mutual action of the earth and the moon are concerned, such an arrangement possesses all the attributes of permanence. If, however, we venture to project our view to the incredibly remote future, we can discern a certain external cause which must prevent this mutual accommodation between the earth and the moon from being eternal. The tides raised by the moon on the earth are so much greater than those raised by the sun, that we have, in the course of our previous reasoning, held little account of the sun-raised tides. This is obviously only an approximate method of dealing with the question. The influence of the solar tide is appreciable, and its importance relatively to the lunar tide will gradually rise as the earth and moon approach the final critical stage. The solar tides will have the effect of constantly applying a further brake to the rotation of the earth. It will therefore follow that after the day and the month have become equal, a still further retardation awaits the length of the day. We thus see that in the remote future we shall find the moon revolving around the earth in a shorter time than that in which the earth rotates on its axis.

A most instructive corroboration of these views is afforded by the discovery of the satellites of Mars. The planet Mars is one of the smaller members of our system. It has a mass which is only the eighth part of the mass of the earth. A small planet like Mars has much less energy of rotation to be destroyed than a larger one like the earth. It may therefore be expected that the small planet will proceed much more rapidly in its evolution than the large one; we might, therefore, anticipate that Mars and his satellites have attained a more advanced stage of their history than is the case with the earth and her satellite.

When the discovery of the satellites of Mars startled the world a few years ago, there was no feature which created so much amazement as the periodic time of the interior satellite. We have already pointed out in Chapter X. how Phobos revolves around Mars in a period of 7 hours 39 minutes. The period of rotation of Mars himself is 24 hours 37 minutes, and hence we have the fact, unparalleled in the solar system, that the satellite is actually revolving three times as rapidly as the planet is rotating. There can hardly be a doubt that the solar tides on Mars have abated its velocity of rotation in the manner just suggested.

It has always seemed to me that the matter just referred to is one of the most interesting and instructive in the whole history of astronomy. We have, first, a very beautiful telescopic discovery of the minute satellites of Mars, and we have a determination of the very anomalous movement of one of them. We have then found a very satisfactory physical explanation of the cause of this phenomenon, and we have shown it to be a striking instance of tidal evolution. Finally, we have seen that the system of Mars and his satellite is really a forecast of the destiny which, in the lapse of incredible ages, awaits the earth-moon system.

It seems natural to inquire how far the influence of tides can have contributed towards moulding the planetary orbits. The circumstances are here very different to those we have encountered in the earth-moon system. Let us first enunciate the problem in a definite shape. The solar system consists of the sun in the centre, and of the planets revolving around the sun. These planets rotate on their axes; and circulating round some of the planets, we have their system of satellites. For simplicity, we may suppose all the planets and their satellites to revolve in the same plane, and the planets to rotate about axes which are perpendicular to that plane. In the study of the theory of tidal evolution, we must be mainly guided by a profound dynamical principle, known as the conservation of the "moment of momentum." The proof of this great principle is not here attempted; suffice it to say, that it can be strictly deduced from the laws of motion, and is thus only second in certainty to the fundamental truths of ordinary geometry or of algebra. Take, for instance, the giant planet, Jupiter. In one second he moves around the sun through a certain angle. If we multiply the mass of Jupiter by that angle, and if we then multiply the product by the square of the distance from Jupiter to the sun, we obtain a certain definite amount. A mathematician calls this quantity the "orbital" moment of momentum of Jupiter. In the same way, if we take the planet Saturn, multiply the mass of Saturn by the angle through which the planet moves in one second, and into the square of the distance between the planet and the sun, then we have the orbital moment of momentum of Saturn. In a similar manner we ascertain the moment of momentum for each of the other planets due to revolution around the sun. We have also to define the moment of momentum of the planets around their axes. In one second Jupiter rotates through a certain angle; we multiply that angle by the mass of Jupiter, and by the square of a certain line which depends on his internal constitution: the product forms the "rotational" moment of momentum. In a similar manner we find the rotational moment of momentum for each of the other planets. Each satellite revolves through a certain angle around its primary in one second; we obtain the moment of momentum of each satellite by multiplying its mass into the angle described in one second, and then multiplying the product into the square of the distance of the satellite from its primary. Finally, we compute the moment of momentum of the sun due to its rotation. This we obtain by multiplying the angle through which the sun turns in one second by the whole mass of the sun, and then multiplying the product by the square of a certain line of prodigious length, which depends upon the details of the sun's internal structure.

If we have succeeded in explaining what is meant by the moment of momentum, then the statement of the great law is comparatively simple. We are, in the first place, to observe that the moment of momentum of any planet may alter. It would alter if the distance of the planet from the sun changed, or if the velocity with which the planet rotates upon its axis changed; so, too, the moment of momentum of the sun may change, and so may those of the satellites. In the beginning, a certain total quantity of moment of momentum was communicated to our system, and not one particle of that total can the solar system, as a whole, squander or alienate. No matter what be the mutual actions of the various bodies of the system, no matter what perturbations they may undergo—what tides may be produced, or even what mutual collisions may occur—the great law of the conservation of moment of momentum must be obeyed. If some bodies in the solar system be losing moment of momentum, then other bodies in the system must be gaining, so that the total quantity shall remain for ever the same. This consideration is one of supreme importance in connection with the tides. The distribution of moment of momentum in the system is being continually altered by the tides; but however the tides may ebb or flow, the total moment of momentum can never alter so long as influences external to the system are absent.

We must here point out the contrast between the endowment of our system with energy, and with moment of momentum. The mutual actions of our system, in so far as they produce heat, tend to squander the energy, a considerable part of which can be thus dissipated and lost; but the mutual actions have no power of dissipating the moment of momentum.

The total moment of momentum of the solar system being taken to be 100, this is at present distributed as follows:—

Orbital moment of momentum of Jupiter	... ...	60
" " Saturn	... ...	24
" " Uranus	... ...	6
" " Neptune	... ...	8
Rotational moment of momentum of Sun	... ...	2
		100

The contributions of the other items are excessively minute. The orbital moments of momentum of the few interior planets contain but little more than one thousandth part of the total amount. The rotational contributions of all the planets and of their satellites is very much less, being not more than one sixty-thousandth part of the whole. When, therefore, we are studying the general effects of tides on the planetary orbits, these trifling matters may be overlooked. We shall, however, find it desirable to narrow the question still more, and concentrate our attention on one splendid illustration. Let us take the sun and the planet Jupiter, and supposing all other bodies of our system to be absent, let us discuss the influence of tides produced in Jupiter by the sun, and of tides in the sun by Jupiter.

It might be hastily thought that, just as the moon was born of the earth, so the planets were born of the sun and have gradually receded by tides into their present condition. We have the means of inquiry into this question by the figures just given, and we shall show that it seems utterly impossible that Jupiter, or any of the other planets, can ever have been very much closer to the sun than they are at present. In the case of Jupiter and the sun, we have the moment of momentum made up of three items. By far the largest of these items is due to the orbital revolution of Jupiter, the next is due to the sun, the third is due to the rotation of Jupiter on its axis. We may put them in round numbers as follows :—

Orbital moment of momentum of			Jupiter ...	... 600,000
Rotational	,,	,,	Sun ...	... 20,000
,,	,,	,,	Jupiter ...	... 12

The sun produces tides in Jupiter, those tides retard the rotation of Jupiter. They make Jupiter rotate more and more slowly, therefore the moment of momentum of Jupiter is decreasing, therefore its present value of 12 must be decreasing. Even the mighty sun himself may be distracted by tides. Jupiter raises tides in the sun, those tides retard the motion of the sun, and therefore the moment of momentum of the sun is decreasing, and it follows from both causes that the item of 600,000 must be increasing, in other words the orbital motion of Jupiter must be increasing, or Jupiter must be receding from the sun. To this extent therefore, the sun-Jupiter system is analogous to the earth-moon system. As the tides on the earth are driving away the moon, so the tides in Jupiter and the sun are gradually driving the two bodies apart. But there is a profound difference between the two cases. It can be proved that the tides produced in Jupiter by the sun are more effective than those produced in the sun by Jupiter. The contribution of the sun may, therefore, be at present omitted; so that, practically, the augmentations of the orbital moment of momentum of Jupiter are achieved at the expense of that stored up by Jupiter's rotation. But what is 12 compared with 600,000. Even when the whole of Jupiter's rotational moment of momentum and that of his satellites has become absorbed into the orbital motion, there will hardly be an appreciable difference in the latter. In ancient days we may indeed suppose that Jupiter being hotter was larger than at present, and that he, therefore, had more rotational moment of momentum. But it is hardly credible that Jupiter can ever have had one hundred times the moment of momentum that he has at present. Yet even if 1,200 units of rotational momentum had been transferred to the orbital motion it would only correspond with the most trivial difference in the distance of Jupiter from the sun. We are hence assured that the tides have not appreciably altered the dimensions of the orbit of Jupiter, or of the other great planets.

The time will however, come, when the rotation of Jupiter on his axis will be gradually abated by the influence of the tides. It will then be found that the moment of momentum of the sun's rotation will be gradually expended in increasing the orbits of the planets, but as this reserve only holds about two per cent. of the whole amount in our system, it cannot produce any considerable effect.

And now we must draw this chapter to a close, though there are many other subjects that might be included. The theory of tidal evolution is, indeed, one of quite exceptional interest. The earlier mathematicians expended their labour on the determination of the dynamics of a system which consisted of rigid bodies. We are indebted to contemporary mathematicians for opening up celestial mechanics on the more real supposition that the bodies are not rigid; in other words, that they are subject to tides. The mathematical difficulties are enormously enhanced, but the problem is more true to nature, and has already led to some of the most remarkable astronomical discoveries made in modern times.

Our Story of the Heavens has now been told. We commenced this work with some account of the mechanical and optical aids to astronomy; we have ended it with a brief description of an intellectual method of research which reveals some of the celestial phenomena that occurred ages before the human race existed. We have spoken of those objects which are comparatively near to us, and then step by step we have advanced to the distant nebulæ and clusters which seem to lie on the confines of the visible universe. Yet, after all, how little is all we can see even with our greatest telescopes, when compared with the whole extent of infinite space! No matter how vast may be the depth which our instruments have sounded, there is yet a beyond of infinite extent. Imagine a mighty globe described in space, a globe of such stupendous dimensions that it shall include the sun and his system, all the stars and nebulæ, and even all the objects which our finite capacities can imagine. Yet, after all, what must be the relation of even this great globe to the whole extent of infinite space. The globe will bear to that a ratio infinitely less than that which the water in a single drop of dew bears to the water in the whole Atlantic Ocean.

THE END.

APPENDIX.

ASTRONOMICAL QUANTITIES.

THE SUN.

THE sun's mean distance from the earth is 92,700,000 miles; his diameter is 865,000 miles; his density, as compared with water, is 1·4; his ellipticity is insensible; he rotates on his axis in a period between 25 and 26 days.

THE MOON.

The moon's mean distance from the earth is 238,000 miles. The least possible distance is 221,000 miles; the greatest is 260,000 miles. The diameter of the moon is 2,160 miles; and her density, as compared with water, is 3·5. The time of a revolution around the earth is 27·322 days.

THE PLANETS.

	Distance from Sun in Millions of Miles.			Periodic Time in Days.	Diameter in Miles.	Axial Rotation.			Density compared with Water.
	Mean.	Least.	Greatest.			h.	m.	secs.	
Mercury	35·9	28·6	43·3	87·969	2,992	24	5	(?)	6·85
Venus -	67·0	66·6	67·5	224·70	7,660	23	21	(?)	4·81
Earth -	92·7	91·1	94·6	365·26	7,918	23	56	4·09	5·66
Mars -	141	128	155	686·98	4,200	24	37	22·7	4·01
Jupiter -	482	459	505	4,332·6	85,000	9	55	—	1·38
Saturn -	884	834	936	10,759	71,000	10	14	23·8	0·75
Uranus -	1,780	1,700	1,860	30,687	31,700	Unknown.			1·28
Neptune	2,780	2,760	2,810	60,127	34,500	Unknown.			1·15

THE SATELLITES OF MARS.

Name.		Mean Distance from Centre of Mars.		Periodic Time. hrs.	mins.	secs.
Phobos	...	5,800 miles	...	7	39	14
Deimos	...	14,500 ,,	...	30	17	54

THE SATELLITES OF JUPITER.

Name.	Mean Distance from Centre of Jupiter.	Periodic Time. days.	hrs.	mins.	secs.
I.	... 262,000 miles ...	1	18	27	34
II.	... 417,000 „ ...	3	13	13	42
III.	... 664,000 „ ...	7	3	42	33
IV.	... 1,170,000 „ ...	16	16	32	11

THE SATELLITES OF SATURN.

Name.	Mean Distance from Centre of Saturn.	Periodic Time. days.	hrs.	mins.	secs.
Mimas	... 118,000 miles ...	0	22	37	27·9
Enceladus	... 152,000 „ ...	1	8	53	6·7
Tethys	... 188,000 „ ...	1	21	18	25·7
Dione	... 241,000 „ ...	2	17	41	8·9
Rhea	... 337.000 „ ...	4	12	25	10·8
Titan	... 781,000 „ ...	15	22	41	25·2
Hyperion	... 946,000 „ ...	21	7	7	40·8
Japetus	... 2,280,000 „ ...	79	7	54	40·4

THE SATELLITES OF URANUS.

Name.	Mean Distance from Centre of Uranus.	Periodic Time. days.
Ariel	... 119,000 miles ...	2.520383
Umbriel	... 166,000 „ ...	4·144181
Titania	... 272,000 „ ...	8·705897
Oberon	... 363,000 „ ...	13·463269

THE SATELLITE OF NEPTUNE.

Name.	Mean Distance from Centre of Neptune.	Periodic Time. days.
Satellite	... 220,000 miles ...	5·87690

INDEX.

B.

C.

H.

I.

J.

M.

R.

S.

T.

PRINTED BY CASSELL & COMPANY, LIMITED, LA BELLE SAUVAGE, LONDON, E.C.

Printed in the United States
By Bookmasters